PRINCIPLES AND PERFORMANCE IN DIESEL ENGINEERING

ELLIS HORWOOD SERIES IN MECHANICAL ENGINEERING

STRENGTH OF MATERIALS
Vol. 1: Fundamentals – Vol. 2: Applications
J. M. ALEXANDER, University College, Swansea
TECHNOLOGY OF ENGINEERING MANUFACTURE
J. M. ALEXANDER, University College, Swansea, G. W. ROWE, Birmingham University and R. C. BREWER
VIBRATION ANALYSIS AND CONTROL SYSTEM DYNAMICS
C. BEARDS, Imperial College of Science and Technology
STRUCTURAL VIBRATION ANALYSIS
C. BEARDS, Imperial College of Science and Technology
COMPUTER AIDED DESIGN AND MANUFACTURE 2nd Edition
C. B. BESANT, Imperial College of Science and Technology
BASIC LUBRICATION THEORY 3rd Edition
A. CAMERON, Imperial College of Science and Technology
SOUND AND SOURCES OF SOUND
A. P. DOWLING and J. E. FFOWCS-WILLIAMS, University of Cambridge
MECHANICAL FOUNDATIONS OF ENGINEERING SCIENCE
H. G. EDMUNDS, University of Exeter
ADVANCED MECHANICS OF MATERIALS 2nd Edition
Sir HUGH FORD, F.R.S., Imperial College of Science and Technology, and J. M. ALEXANDER, University College of Swansea.
MECHANICAL FOUNDATIONS OF ENGINEERING SCIENCE
H. G. EDMUNDS, Professor of Engineering Science, University of Exeter
ELASTICITY AND PLASTICITY IN ENGINEERING
Sir HUGH FORD, F.R.S. and R. T. FENNER, Imperial College of Science and Technology
DIESEL ENGINEERING PRINCIPLES AND PERFORMANCE
S. D. HADDAD, Associate Professor, Western Michigan University, USA, and Director of HTCS Co., UK, and N. WATSON, Reader in Mechanical Engineering, Imperial College of Science and Technology, University of London
DIESEL ENGINEERING DESIGN AND APPLICATIONS
S. D. HADDAD, Associate Professor, Western Michigan University, USA, and Director of HTCS Co., UK. and N. WATSON, Reader in Mechanical Engineering, Imperial College of Science and Technology, University of London
TECHNIQUES OF FINITE ELEMENTS
BRUCE M. IRONS, University of Calgary, and S. AHMAD, Bangladesh University, Dacca
FINITE ELEMENT PRIMER
BRUCE IRONS and N. SHRIVE, University of Calgary
CONTROL OF FLUID POWER: ANALYSIS AND DESIGN 2nd (Revised) Edition
D. McCLOY, Ulster Polytechnic, N. Ireland and H. R. MARTIN, University of Waterloo, Ontario, Canada
UNSTEADY FLUID FLOW
R. PARKER, University College, Swansea
DYNAMICS OF MECHANICAL SYSTEMS 2nd Edition
J. M. PRENTIS, University of Cambridge
ENERGY METHODS IN VIBRATION ANALYSIS
T. H. RICHARDS, University of Aston Birmingham
ENERGY METHODS IN STRESS ANALYSIS: With Intro. to Finite Element Techniques
T. H. RICHARDS, University of Aston in Birmingham
COMPUTATIONAL METHODS IN STRUCTURAL AND CONTINUUM MECHANICS
C. T. F. ROSS, Portsmouth Polytechnic
FINITE ELEMENT PROGRAMS FOR AXISYMMETRIC PROBLEMS IN ENGINEERING
C. T. F. ROSS, Portsmouth Polytechnic
ENGINEERING DESIGN FOR PERFORMANCE
K. SHERWIN, Liverpool University
ROBOTS AND TELECHIRS
M. W. THRING, Queen Mary College, University of London

PRINCIPLES AND PERFORMANCE IN DIESEL ENGINEERING

Editor-in-Chief:
S. D. HADDAD, B.Sc.(Eng.), Ph.D.
Associate Professor
Western Michigan University, USA
and Director HTCS Co., UK

Associate Editor:
N. WATSON, B.Sc.(Eng.), Ph.D.
Reader in Mechanical Engineering
Imperial College of Science and Technology
University of London

ELLIS HORWOOD LIMITED
Publishers · Chichester

Halsted Press: a division of
JOHN WILEY & SONS
New York · Chichester · Brisbane · Toronto

First published in 1984 by

ELLIS HORWOOD LIMITED
Market Cross House, Cooper Street, Chichester, West Sussex, PO19 1EB, England

The publisher's colophon is reproduced from James Gillison's drawing of the ancient Market Cross, Chichester.

Distributors:

Australia, New Zealand, South-east Asia:
Jacaranda-Wiley Ltd., Jacaranda Press,
JOHN WILEY & SONS INC.,
G.P.O. Box 859, Brisbane, Queensland 40001, Australia

Canada:
JOHN WILEY & SONS CANADA LIMITED
22 Worcester Road, Rexdale, Ontario, Canada.

Europe, Africa:
JOHN WILEY & SONS LIMITED
Baffins Lane, Chichester, West Sussex, England.

North and South America and the rest of the world:
Halsted Press: a division of
JOHN WILEY & SONS
605 Third Avenue, New York, N.Y. 10016, U.S.A.

British Library Cataloguing in Publication Data
Principles and performance in diesel engineering.
(Ellis Horwood Series in mechanical engineering)
1. Diesel motor
I. Haddad, S.D. II. Watson, N.
621.43'6 TJ795

Library of Congress Card No. 84-4575

ISBN 0-85312-732-8 (Ellis Horwood Limited)
ISBN 0-470-20075-8 (Halsted Press)

Typeset by Ellis Horwood Limited.
Printed in Great Britain by The Camelot Press, Southampton.

Table of Contents

Preface

The diesel engine is recognized especially for its efficiency and reliability which explains its well established position in commercial vehicles and industrial applications. In recent years more diesels are also being used in passenger cars. Therefore diesel engineering is becoming a fast growing applied engineering discipline to meet exacting demands for designing optimum diesel engines in relation to greater efficiency and power, low noise, acceptable emissions, fuel economy and greater maintainbility.

In October 1982, Dr. Haddad organized an intensive one-week course on Diesel engineering aimed at practising engineers in the diesel and related industries. This course provided concentrated study in the various inter-related aspects of diesel engine technology: combustion and heat release; supercharging and turbocharging; test equipment; computer control and data acquisition; diesel engine design; and diesel cars and diagnostics with emphasis on automotive applications. Most of the lecturers on this course produced comprehensive lecture notes. Subsequent interest in the course content encouraged us to edit these notes into two companion books. The chapters in these books represent the accumulation of years of expertise, since each author is an expert in his field.

Diesel Engineering – Principles and Performance, covers the basic aspects of fuels, air–fuel mixing, combustion plus governing, turbocharging and exhaust emissions. *Diesel Engineering – Design and Applications,* concentrates more on mechanical design, stress, noise and vibration and application primarily to passenger cars since this is of most interest today and space precludes coverage of the variety of diesel applications.

It is important to realize that the reader is assumed to have a basic understanding of the principles of engineering thermodynamics and of internal combustion engines, though some chapters' authors have made some effort to introduce some basics as well. On the other hand, it will be found that the reader would require very little experience with I.C. engine theory but some experience with diesel engine practice in order to obtain maximum benefit from this book.

In preparing the original and revised material for this book it has been our aim to provide a useful handbook on diesel engineering, both for reference and for up to date knowledge to enable diesel engine engineers to acquire the necessary tools for further development of this most fascinating prime mover. No attempt has been made to provide a totally balanced approach to diesel engineering. Instead, the editors have tried to incorporate most detail in areas of most interest to the majority of readers.

S. D. Haddad and N. Watson

1

Combustion and heat release in diesel engines

S. D. HADDAD, Dept. of Engineering Technology, Western Michigan University

1. INTRODUCTION – IDEAL GAS CYCLES

Diesel engine combustion involves a large number of complex processes. In the past, and even at present, experience and intuition have been employed to determine the emphasis on each process. However, considerable theoretical and experimental work has been published during the last three decades, giving a better understanding of these processes and enabling most aspects of diesel engine combustion and performance to be calculated.

Before looking at diesel combustion in some detail it is appropriate to revise the basic thermodynamic principles of ideal gas cycles, in particular as to the effects of pressure, volume and temperature on thermal efficiency as follows:

(a) *Carnot cycle*

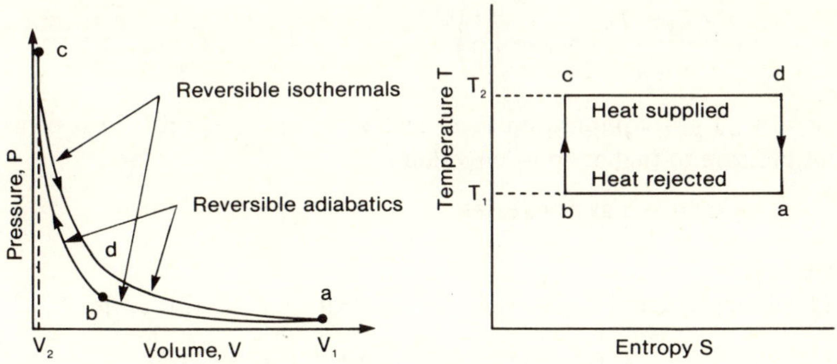

Fig. 1.

The thermal efficiency, $\eta_T = (T_2 - T_1)/T_2$.

The maximum thermal efficiency with any cycle occurs when the heat is supplied at the highest possible temperature and is rejected at the lowest possible temperature. It will be seen, therefore, that the Carnot cycle gives the highest attainable thermal efficiency for a heat engine operating between temperatures T_1 and T_2.

The cycle has two main disadvantages:

(i) *High peak pressure.* This results in increased mechanical friction and also a strengthened (and hence heavier) engine structure is needed.

(ii) *Low specific power output.* It is therefore necessary to employ a large swept volume (large engine bulk and weight) to obtain a usable amount of power.

(b) *OTTO (constant volume) cycle*

This cycle is quite closely approached in practice by the low-speed petrol engine.

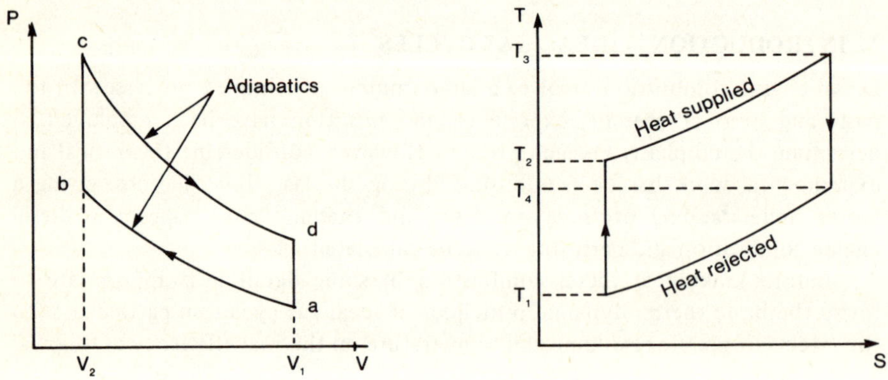

Fig. 2.

$$\eta_T = \frac{T_3 - T_4}{T_3} = 1 - \left(\frac{1}{r}\right)^{\gamma - 1}$$

where $r = V_1/V_2$ = compression ratio and γ = ratio of specific heat of gas at constant pressure to that at constant volume.

η_T increases as r increases

(c) *Diesel cycle*

The low-speed diesel engine operates approximately on this cycle.

$$\eta_T = 1 - \frac{T_1(r_e^\gamma - 1)}{T_2(r_e - 1)^\gamma} = 1 - \left(\frac{1}{r}\right)^{\gamma - 1} \left\{\frac{(r_e^\gamma - 1)}{\gamma(r_e - 1)}\right\}$$

where $r_e = V_3/V_2$.

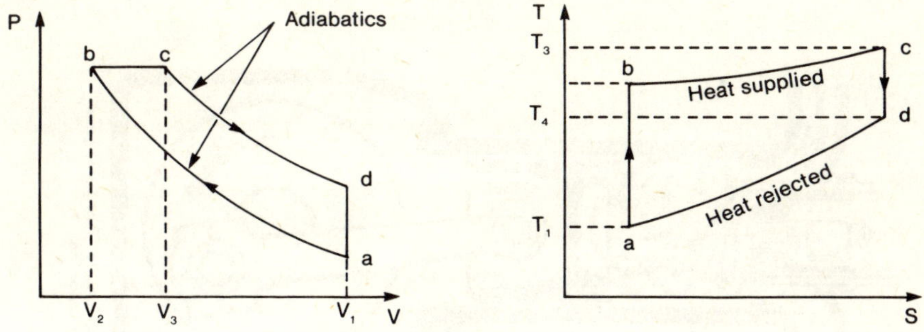

Fig. 3.

The factor $(r_e^\gamma - 1)/\gamma(r_e - 1)$ is usually greater than unity and increases as r_e increases. Therefore,

η_T *decreases as period of heat reception is increased*
η_T *increases as r increases*

(d) *Dual cycle*

Both high-speed petrol and diesel conform approximately to this cycle.

$$\eta_T = 1 - \left(\frac{1}{r}\right)^{\gamma - 1} \frac{(p \, . \, r_e{}^{\gamma - 1})}{(p - 1) + p\,\gamma(r_e - 1)}$$

where, $p = P_3/P_2 = T_3/T_2$.

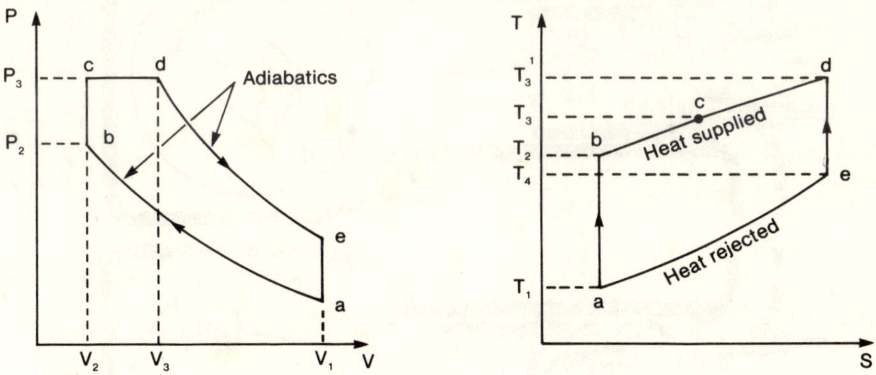

Fig. 4.

For the same compression ratio, the thermal efficiencies of the last three cycles are in the following descending order:

1. Constant volume (Otto) cycle
2. Dual cycle
3. Diesel cycle

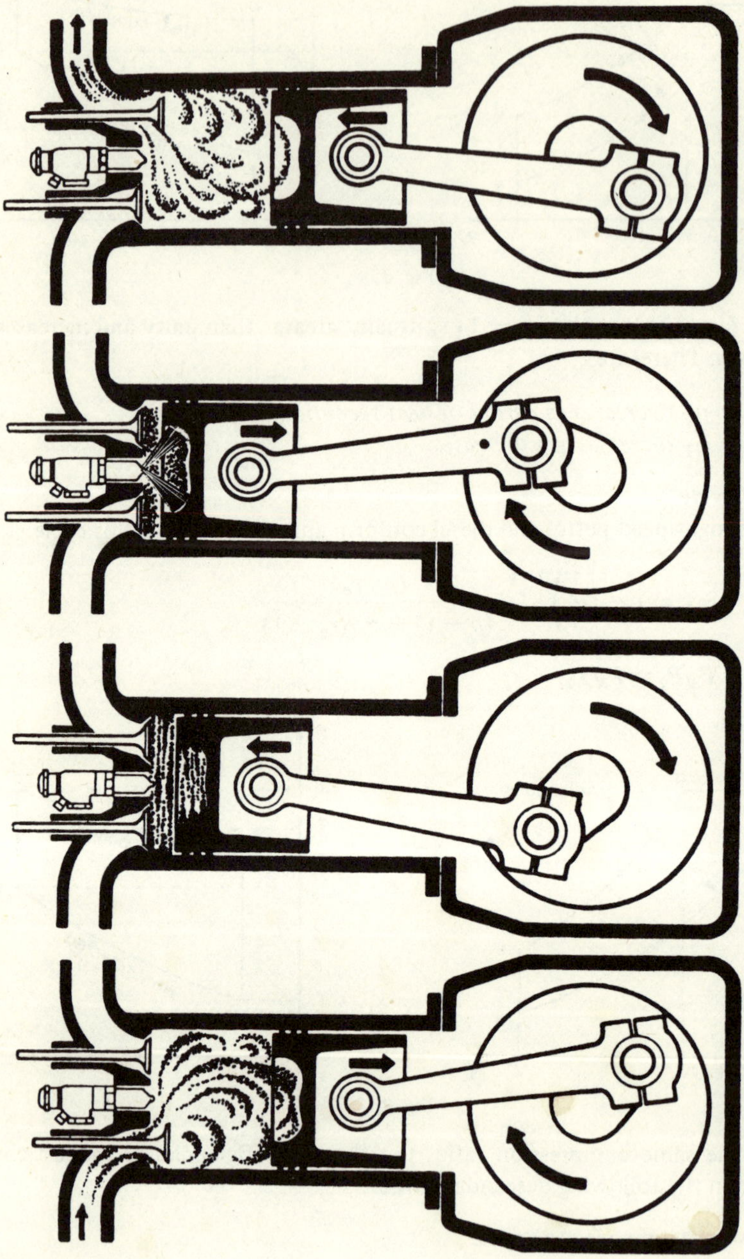

Fig. 5 – The sequence of operations in a four-stroke diesel engine.

In practice, the petrol engine has a limited compression ratio because of the tendency for the fuel to 'pink' or 'detonate'. The diesel engine has only minor limitations, enabling much higher compression ratios to be used. Thus diesel engines are usually more efficient than petrol engines.

2. BASIC TYPES OF DIESEL ENGINE

It is perhaps appropriate to revise the operation of the basic types of diesel engine in common use: four and two-cycle, single and double-acting, with one or more cylinders are used. In the automotive field, engines using four or six cylinders in-line and working on the four-cycle principle are most common.

2.1 Four-stroke Cycle Diesel Engine (Fig. 5)

The inlet valve opens; air is drawn into the cylinder as the piston increases the cylinder volume.

With the inlet and the exhaust valves closed, the piston compresses the air. The air temperature rises. When the piston almost reaches the minimum volume position, Top Dead Centre (T.D.C.), fuel is injected in a finely divided form into the hot swirling air in the combustion space. After a short delay ignition occurs, the gas pressure rises rapidly and a pressure wave is set up.

Work is done by the gas pressure on the piston as the piston moves to maximum cylinder volume, Bottom Dead Centre (B.D.C.). During this expansion or power stroke, the combustion gas temperature and pressure fall.

As B.D.C. is approached by the piston, the exhaust valve opens and the products of combustion are rejected from the cylinder through the exhaust valve by the piston moving to T.D.C. Near T.D.C., the inlet valve opens again and the cycle is repeated.

The four-stroke cycle is performed once every two revolutions of the engine crankshaft.

2.2 Two-stroke Cycle Diesel Engine (Fig. 6)

The piston uncovers the inlet and exhaust ports in the wall of the cylinder. Air from the blower passes through the inlet port into the cylinder whilst the exhaust gas leaves the cylinder via the exhaust port.

The piston moves to reduce the cylinder volume and closes both inlet and exhaust ports. Air is compressed; fuel is injected and spontaneously ignited.

Work is done by the burning gas as the piston moves to increase the cylinder volume. On approaching maximum cylinder volume, the piston again uncovers the exhaust and then the inlet port, and the cycle starts again.

The two strokes of the cycle are completed once during each revolution of the crankshaft.

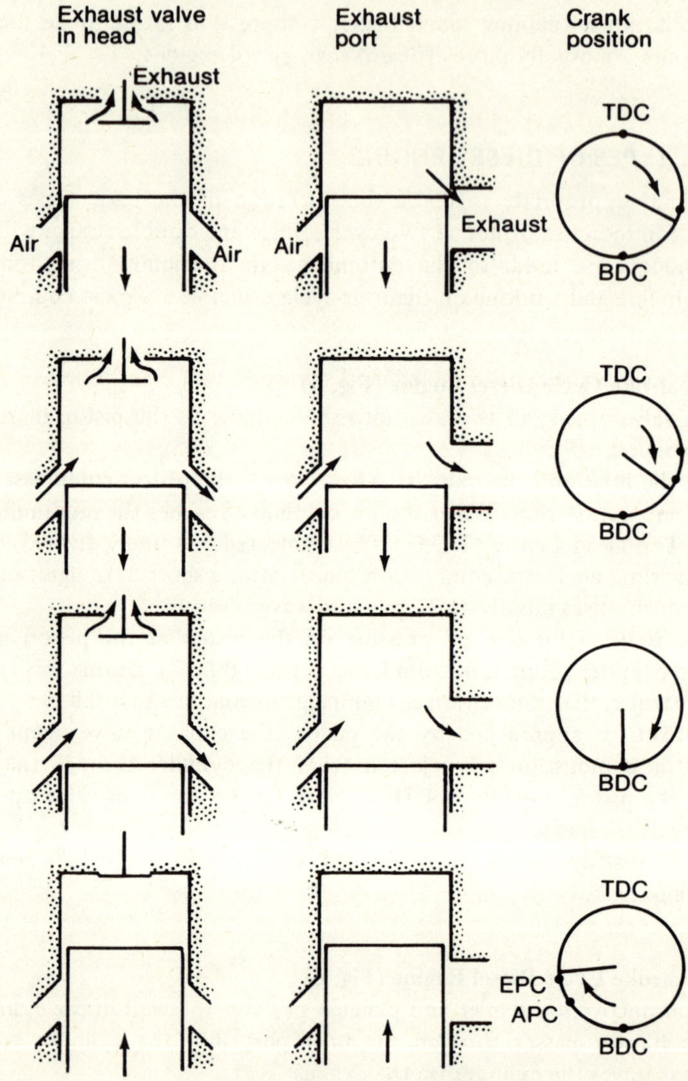

Fig. 6 – Cycle of events in two-stroke cycle gas exchange process.

2.3 Multi-fuel Diesel Engine

This has been designed primarily for military applications, but is of interest to civilian commercial vehicle users in areas where the supply of conventional diesel fuel may be scarce.

Usually, this engine is of the blown, opposed-piston two-stroke type, designed to run on a variety of readily available liquid fuels, such as diesel oil, gasoline and kerosene.

The combustion chambers of multi-fuel engines must be capable of using fuels with widely differing characteristics and this involves careful consideration of air movement and insulation against heat loss during compression.

At this point it is also worth while to summarize the most favourable and unfavourable points to diesel engines compared with the S.I. counterparts.

2.4 Points Favourable to Diesel Engines Compared with Petrol Engines

(a) Lower specific fuel consumption due to higher compression ratio; no fuel is required when coasting; only a small quantity of fuel is used when idling.
(b) Equal power output per cylinder because equal quantities of fuel are supplied to each cylinder. Hence fewer cylinder to cylinder pressure variations.
(c) High torque at low engine speeds, as breathing is less restricted with no carburettor.
(d) Lower exhaust temperature due to increased expansion of combustion gases: mean petrol engine exhaust temperature is about 900°C; mean diesel engine exhaust temperature is about 550°C.
(e) Low carbon monoxide content in exhaust gas: 3% in petrol against 0.2% in diesel.
(f) Reduced fire risk, as fuel oil is less flammable than petrol.

2.5 Points Unfavourable to Diesel Engines

(a) Lower air utilization because of the need for excess air to ensure smoke-free combustion and therefore lower BMEP/unit swept volume. Hence for a given output, the diesel engine normally requires larger cylinders.
(b) Higher maximum cylinder pressure and wide fluctuations of torque; therefore stronger and heavier parts are necessary, thus reducing the power to weight ratio.
(c) Noise level higher; however, this can be attenuated by redesign.
(d) A neglected diesel engine usually shows its condition by smoking.

3. DIESEL ENGINE COMBUSTION

The transient combustion process which continually takes place in an operating diesel engine is basically represented by the pressure–crank angle diagram shown in Fig. 7.

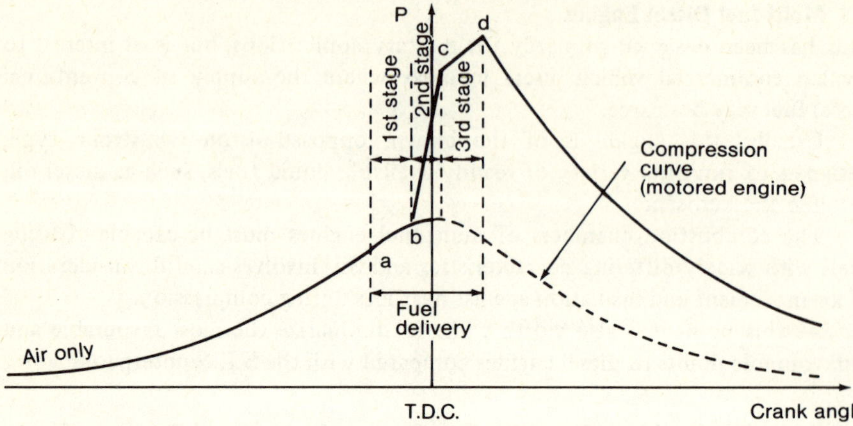

Fig. 7.

At A, the injector starts to inject fuel into the combustion chamber. A finite time (AB) elapses before the fuel droplets reach ignition temperature, although when this happens most of the fuel injected during the first stage ignites spontaneously, causing an abrupt pressure rise at B (second stage). The rest of the fuel injected during stage 3 is burnt as soon as it enters the combustion chamber.

With reference to the ideal cycle analyses, it will be seen that maximum thermal efficiency would result if all the fuel were burnt during stage 2 (at constant volume) although in practice such combustion would cause higher engine vibration and noise.

Power output is controlled by metering the amount of injected fuel. Thus the effective compression ratio remains almost constant, resulting in very good part-load efficiency.

The classical diesel compression and combustion pressure diagram shown in Fig. 7 has been proved to be an acceptable model and can be roughly divided into three broad stages:

The first stage: During this stage, the fuel is sprayed into the hot air in the cylinder. The fuel droplets are heated up and then vaporize. Mixing of the vaporized fuel and air will take place and hence an ignitable prepared mixture is ready to burn after the ignition lag has elapsed. This means that during the above stage, physical and chemical transformations occur to the fuel and air mixture. In fact this stage is the most important, as it is the seat of the so-called diesel 'knock'. It is the well-known *delay period.*

The second stage: The fuel injected and prepared for combustion in the first stage now begins to burn and this gives rise to the typical diesel combustion knock which is caused by the very *rapid pressure rise* resulting from the sudden burning of the prepared fuel accumulated during the delay period.

It follows that the second stage is actually determined by the conditions of the first stage.

The third stage: The burning of the prepared first stage mixture is completed and the burning of the fuel injected in the second and third stages occurs. The conditions are, therefore, controlled by the mixing of the rest of the fuel injected during this stage with air.

In general the process involved in the above stages, though not completely understood, can be divided into two portions. During the physical portion of the above processes, fuel injection, fuel jet distintegration, air penetration by liquid fuel, fuel evaporation and mixing of the evaporated fuel with air take place. On the other hand, during the chemical portion, pre-ignition reactions take place which are characterized in principle by pre-flame oxidation and thermal disintegration of hydrocarbons or pyrolysis.

More important, for the formation of the prepared mixture, the injection procedure such as the time of injection, the range of the jet, the atomization of the fuel spray, density of cylinder air, its relative velocity and temperature, play a significant role.

In the main we have been able to understand diesel combustion processes from combustion photography (using transparent windows) and analysis of measured cylinder pressure diagrams (using piezoelectric transducers). From these studies it has been possible to develop simplified combustion models to compute the pressure–time changes in the cylinder. The most well-known models for diesel combustion are well surveyed by R. S. Benson as follows:

1. *Single-zone Combustion Models*
Using: Lyn's method and Whitehouse–Way's method

2. *Two-zone Combustion Models*
Using: the Conical/Annular burning zone model and the Swirling-wall jet-burning zone model

3. *Multi-zone Combustion Models*

4. *Turbulent Flow Prediction Models*

In all these models the most important starting point is an appropriate model for the fuel injection, evaporation and subsequent ignition to release the fuel energy. Basically the injected liquid fuel is treated as droplets or sprays, or a combination of both. In this context many researchers have published very useful droplet and spray combustion theories, notably Lyn, Parks, Dent, Myers, Faeth, Whitehouse and Benson, and others.

Most of these combustion models include a certain amount of semi-empirical relationships based on experimental results when observing engine combustion processes using transparent windows and employing special Schlieren illumination techniques coupled with high-speed photography [11], as illustrated in Fig. 8.

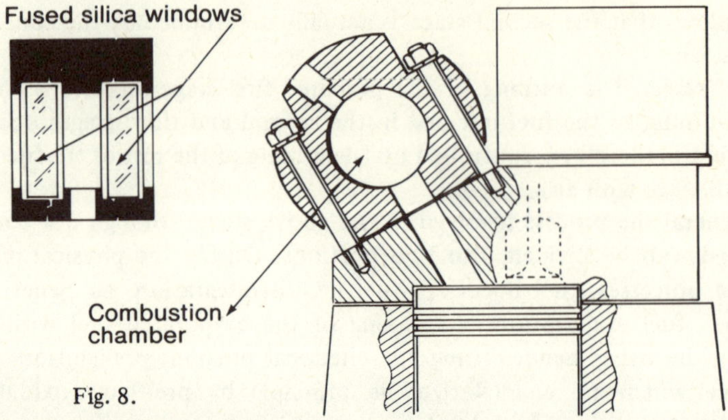

Fig. 8.

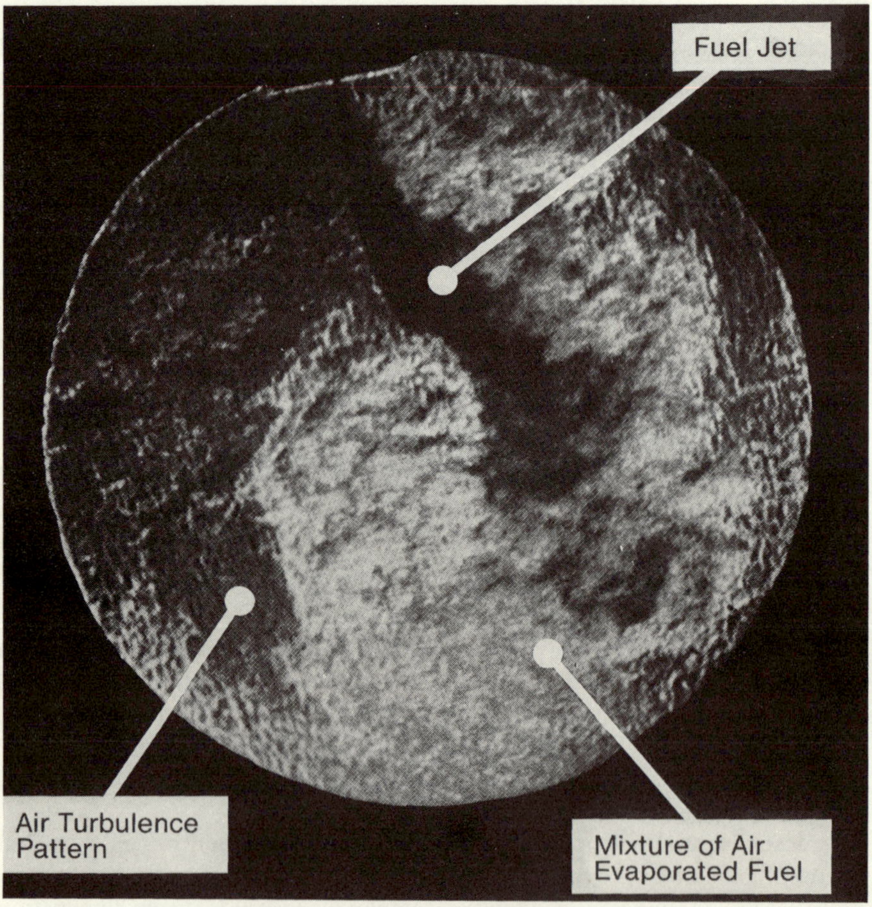

Fig. 9 – A single exposure from a high-speed Schlieren colour film of an evaporating fuel jet in a combustion chamber, showing the pattern of air turbulence, fuel–air mixture and jet of fine drops. (Picture taken at T.D.C.; exposure time about 1 micro-second; combustion suppressed.)

Such studies provide considerable information to substantiate the theoretically developed combustion models. Fig. 9 illustrates a single exposure from a high-speed Schlieren film of an evaporating fuel jet in a combustion chamber, showing fuel–air mixture and jet of fine drops. Careful study of a huge number of these single exposures has assisted combustion researchers in establishing realistic combustion computer models for subsequent development of new, more efficient combustion systems.

4. COMBUSTION CHAMBER DESIGN IN DIESEL ENGINES

The combustion process in the diesel engine must be controlled to avoid both excessive maximum cylinder pressure and an excessive rate of pressure rise. At the same time, the process should be so rapid that substantially all the fuel is burned early in the expansion stroke.

In discussing methods used for attaining these objectives, diesel engine combustion chambers can be divided into two types:

(i) open chamber *or* direct injection engines (D.I.)
(ii) divided chamber *or* prechamber *or* indirect injection engines (I.D.I.)

Examples of D.I. designs are shown in Fig. 10.
Typical I.D.I. chamber designs are shown in Fig. 11.

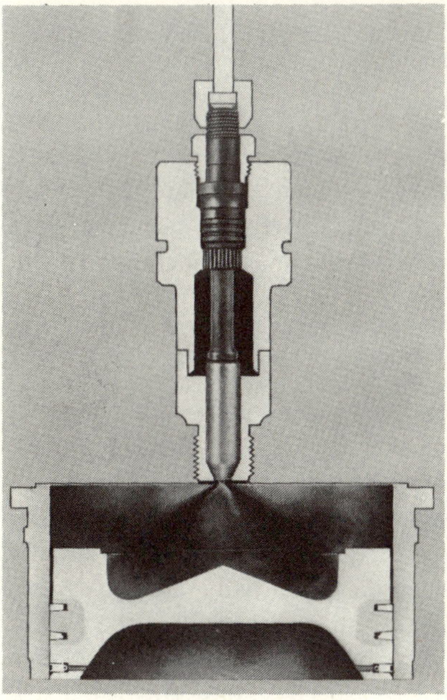

a

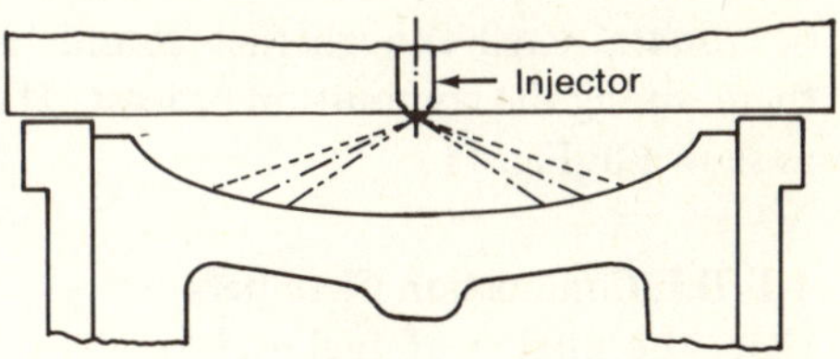

b

Fig. 10(a) – Direct injection system; (b) Quiescent combustion chamber.

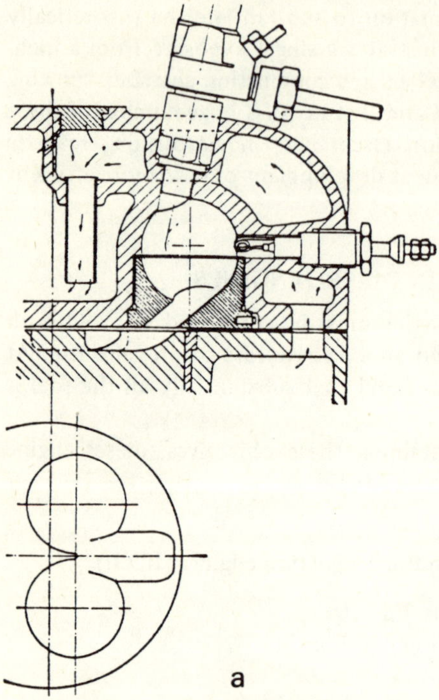

a

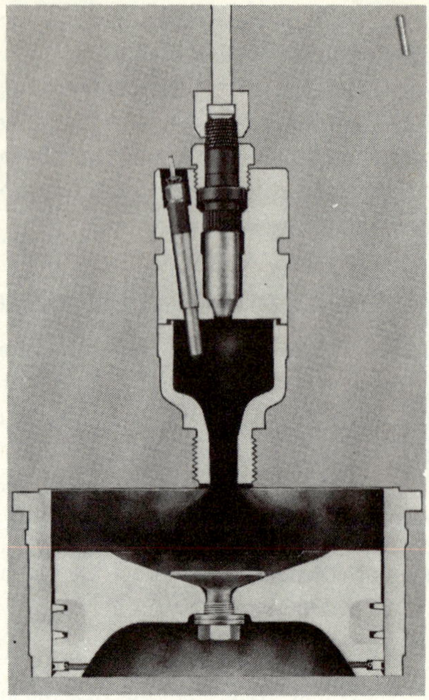

b

Fig. 11(a) – Ricardo Comet Mark III combustion chamber; (b) Precombustion chamber system.

In contrast to the open chamber design, the prechamber diesel engine is divided into two or more compartments between which there are restrictions or *throats,* small enough that considerable pressure differences occur between them during the combustion process. The burning starts in a small compartment, as shown in Fig. 11.

4.1 D.I. Combustion Chambers

Here, the mixing of fuel and air depends entirely on the spray characteristics and on air motion, and it is not vitally affected by the combustion process itself. Engines of this type are very sensitive to the spray characteristics, which must be carefully arranged to secure rapid mixing. Sub-division of the spray and the use of high injection pressures are usually required. The mixing is usually assisted by a *swirl* (induced by directing the inlet air tangentially) or by a *squish,* which is the air motion caused by a small clearance space over part of the piston.

High-speed diesel engines of the open chamber type require fuel of high *ignition quality,* as delay angles increase at high speeds unless this quality is improved.

Figure 12 shows the 'M' M.A.N. wall burning combustion system, optimized to give greater swirl and mixing for complete combustion.

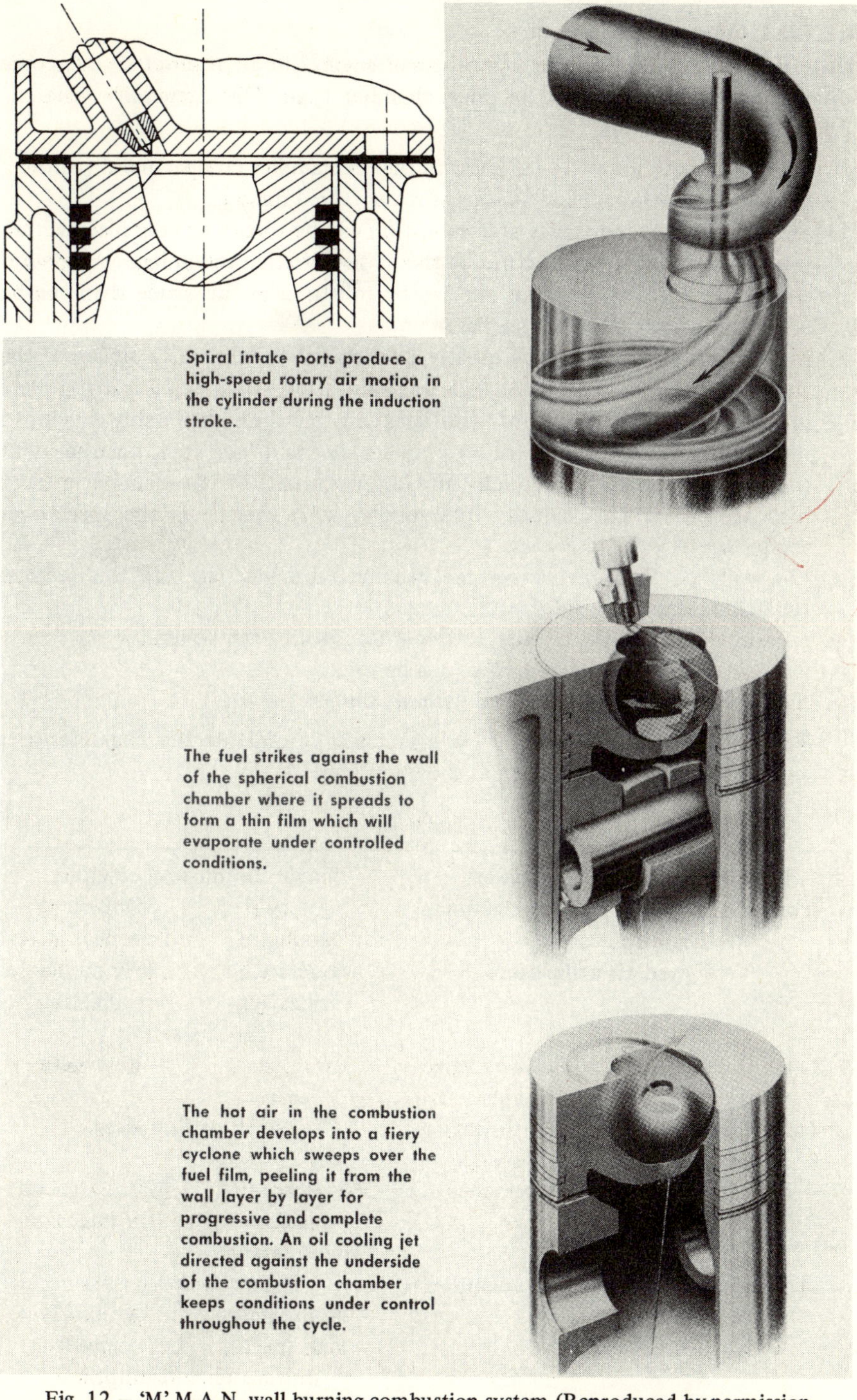

Fig. 12 – 'M' M.A.N. wall burning combustion system (Reproduced by permission of M.A.N. Maschinenfabrik Augsbury-Nurnberg AG, Augsburg).

4.2 I.D.I. Combustion Chambers

These are used chiefly on high-speed, small engines in an attempt to overcome some of the limitations of the open chamber type. They generally have the following characteristics:

(i) extremely high air swirl velocities in the prechamber compartment (through the use of the throat) and this ensures very rapid mixing;

(ii) the first and second stages of combustion can be forced to take place within the prechamber. The structure of this is so stiff that much higher pressures and higher rates of pressure rise can be tolerated in this space than can be allowed in the space over the piston;

(iii) the mixing process may be greatly accelerated by the early stages of the combustion process itself. At high fuel–air ratios, combustion is incomplete in the prechamber because of insufficient air, and the high pressure developed by the early part of combustion projects the unburned fuel, together with the early combustion products, into the main part of the chamber at very high velocities. This causes rapid mixing with the air in the space over the piston;

(iv) the walls of the prechamber are usually extremely hot and this reduces the delay periods considerably;

(v) because of the above, fuels of poor ignition quality can be used;

(vi) also, single hole injection nozzles can be used;

(vii) high fuel–air ratios can be used *without smoke*;

Benson and Whitehouse [28] conveniently summarize the characteristics of the D.I. and I.D.I. combustion, as shown in Table 1.

Table 1.

Divided combustion chamber		Single combustion chamber	
Precombustion	Swirl chambers	Air swirl	Quiescent
High heat losses		Moderate	Low heat losses
Very good air utilization		Moderate air utilization	Fairly moderate utilization
Poorer starting		Good starting	
Some restrictions on valves	Limits on valve area	Good valve area	Best valve area
Throttling losses in passages	Some throttling losses	No throttling loses	
Easy fuel injection requirements		Moderate fuel injection requirements	More difficult for injection
Low pressure rises in main chamber		Higher pressure rises	
High speeds		Moderately high speeds	Medium/low speeds
		Better fuel consumption	

5. EFFECT OF OPERATING CONDITIONS ON DIESEL ENGINE COMBUSTION

The combustion rate in diesel engines, as well as the rate of pressure rise, depends greatly on the design of the combustion chamber and injection system. In practice, the design of these elements varies widely and it is therefore more difficult to generalize trends in this type of engine than the S.I. engine, where there is less variety in design. Therefore the effects noted below are for general guidance.

5.1 Injection Timing

Proper timing of the combustion process is essential to diesel engine operation. In practice, to attain the maximum IMEP (indicated mean effective pressure) and maximum efficiency, the injection timing should be adjusted to give peak pressure at about 15° after T.D.C.

Shortest delay periods occur when the delay period includes T.D.C. – earlier or later injection gives longer delays.

Early injection gives a very high rate of pressure rise – here a long delay occurs together with inward piston motion during the subsequent rapid combustion. The latter occurs prior to, during, or only slightly after T.D.C.

Late injection also gives a very high rate of pressure rise – owing again to a long delay period. This now occurs even though the rapid combustion period is now well after T.D.C. with the piston in rapid descent.

5.2 Engine Speed

Delay time is independent of engine speed. Thus, the delay angle will vary with speed and so engine speed may have an appreciable effect on pressure–crank angle diagrams. However, other factors also change with engine speed, e.g. volumetric efficiency, spray characteristics, etc. Thus, the unique effect of engine speed is masked by such effects.

In most cases, however, increasing speed has caused higher rates of pressure rise and higher maximum pressures and this has caused difficulties in high-speed diesel engine development. To prevent the delay angle becoming too great, fuels of high ignition quality (reduced ignition delay) are specified for high-speed engines.

5.3 Fuel–Air Ratio

There is very little reduction in the maximum rate of pressure rise except at the very low fuel–air ratios. On the other hand, maximum pressure falls steadily with decreasing fuel–air ratio.

No low limit on the overall fuel–air ratio seems to be imposed by combustion characteristics – because burning occurs in small elements at local fuel–air ratios within the usual combustible range (7:1 to 21:1 air–fuel ratio). A practical difficulty at very low overall fuel–air ratios is that of maintaining good spray characteristics with the small amounts of fuel involved.

A practical limit on the maximum fuel–air ratio, however, is set by incomplete combustion accompanied by a smoky exhaust – called the *smoke limit*. This arises from an appreciable fraction of the fuel failing to find the necessary O_2 for combustion and passing through the cylinder unburned – this is due to poor mixing. One of the best measures of the quality of a diesel combustion system is the fuel–air ratio at which smoke becomes pronounced. The higher this ratio, the better is the mixing process and the higher the mean effective pressure.

5.4 Compression Ratio (C.R.)

Figure 13 shows the typical effect of compression ratio on the pressure–crank angle diagram when injection timing, speed and fuel quality are held constant. The delay period is reduced with increasing compression ratio.

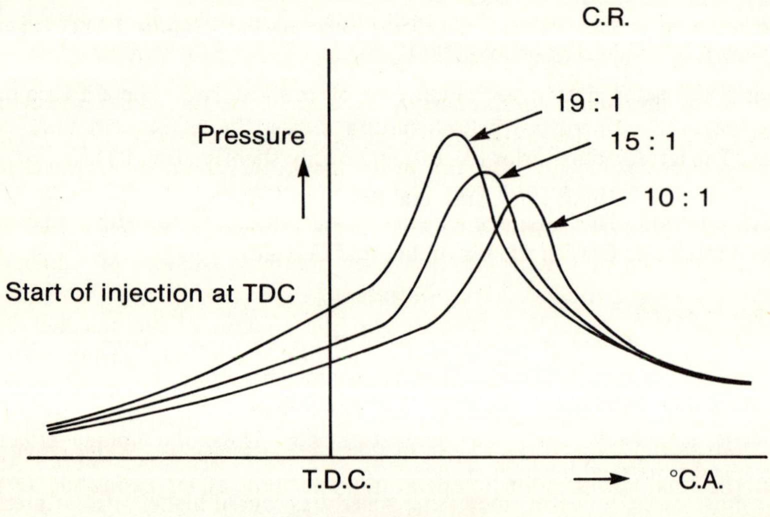

Fig. 13.

In practice, the higher the compression ratio, the higher will be the engine friction, leakage and torque required for starting. Thus, diesel engine designers use the lowest compression ratio consistent with satisfactory starting and operation with the available fuel.

5.5 The Delay period, its Calculation and Control

In most cases, attempts have been directed towards reducing the ignition delay period since it is an important factor when considering the rate of pressure rise and hence combustion noise. If the rate of vaporization is kept constant, the rate of pressure rise is linearly related to the delay period (i.e. the rate of pressure rise increases as the delay period becomes longer).

Let I.d. = Ignition delay period = f (cetane number CN, temperature T in exponential manner, pressure p, speed N, and mixing rate and other fuel properties).

Shipinski [13] presented the following correlation:

$$\text{I.d.} = \frac{6N}{1000}\,\frac{0.0271}{p0.386}\left(\frac{40}{CN}\right)^{0.69}\exp\left(\frac{8360}{T}\right)$$

Another estimate has been given by Wolfer [1,28] as

$$\text{I.d. (in ms)} = \frac{0.44e^{\frac{4650}{T}}}{p^{1.19}} \tag{20}$$

where p is in atm. and T in °K at the point of ignition. Both are approximate correlations based on limited experimental data. The following techniques have been shown to reduce the delay period:

(i) By using an *exhaust turbo charger,* more air can be forced into the cylinder and this reduces the ignition delay (and increases the power output) and the engine runs smoothly. On the other hand, the thermal and mechanical stresses may be increased considerably.

(ii) Another method consists of using an air compressor to inject (or spray) some paraffin or gas oil into the inlet pipe with the idea of reducing the ignition delay period – called *fumigation.*

(iii) The rate of pressure rise is related to the amount of fuel injected during the delay period. A small amount of fuel is desirable during the delay period and this may be achieved using a well-timed *pilot injection.*

(iv) By special arrangement of the *inlet manifold over exhaust manifold*: this could produce for every 1° rise in inlet temperature, a corresponding 2.7° in the final compression temperature. The limit of increasing the air inlet temperature is set by the volumetric efficiency.

6. HEAT RELEASE CALCULATIONS IN DIESEL ENGINES

In order to assess the performance of a diesel engine such as power, efficiency, starting behaviour, smoke and noise, the measured cylinder pressure development is not sufficient. In addition, it is very important to understand the development of heat release. Heat releast (heat output) is basically a thermodynamic problem and can be obtained by analysis of the measured cylinder pressure diagram. Moreover, if the heat release is known, determination of the pressure and temperature variation in the engine can be deduced.

Considerable progress has been made to evaluate the heat release development but present methods still rely on certain assumptions and approximations.

This is because of:

(a) the complex interaction of the physical and chemical nature of the diesel combustion;
(b) the fact that the processes involved cannot easily be ascertained by experiment, as they occur at extremely short time intervals and in the presence of high cylinder pressures and temperatures.

For most studies it is not really necessary to produce an elaborate calculation of heat release. This chapter therefore presents an approximate method for the calculation of heat release from a measured cylinder pressure diagram together with analysis of the various stages of combustion. The assumptions and errors normally associated with these calculations are also discussed.

6.1 Basic Heat Release Theory

In the cylinder of an internal combustion engine, part of the energy carried in with the fuel is changed into mechanical work, part of it leaves the cylinder with the exhaust gas and part of it is given off to the walls by heat transfer. If the pressure and temperature variation and the laws of heat rransfer between gas and walls are known, the heat release pattern can be calculated.

The first law of thermodynamics in differential form related to the crank angle may be written as:

$$\frac{\mathrm{d}Q_{\mathrm{H}}}{\mathrm{d}\theta} = \frac{\mathrm{d}U}{\mathrm{d}\theta} + \frac{\mathrm{d}L}{\mathrm{d}\theta} \tag{1}$$

where Q_{H} is the *heat release (heat output)* which causes a change in the internal energy U of the working gas and produces an output work, L. In other words, Q_{H} represents the heat absorbed by the working gas and it is given as the difference between the input fuel heat and the heat given off to the walls; that is:

$$\frac{\mathrm{d}Q_{\mathrm{H}}}{\mathrm{d}\theta} = \frac{\mathrm{d}Q_{\mathrm{b}}}{\mathrm{d}\theta} - \frac{\mathrm{d}Q_{\mathrm{w}}}{\mathrm{d}\theta} \tag{2}$$

where Q_{b} is the heat liberated by the combustion of fuel during each crank angle increment (i.e. heat input of the fuel) and Q_{w} is the heat loss from cylinder gases to the cylinder head, piston crown and exposed portion of liner during the increment.

For a perfect gas $\mathrm{d}L = P\,\mathrm{d}V$. Therefore,

$$\frac{\mathrm{d}L}{\mathrm{d}\theta} = P\frac{\mathrm{d}V}{\mathrm{d}\theta} \tag{3}$$

Therefore the change in the internal energy of the gas (from equations (1)–(3)) becomes

$$\frac{dU}{d\theta} = \frac{dQ_b}{d\theta} - \frac{dQ_w}{d\theta} - P\frac{dV}{d\theta} \tag{4}$$

Now the equation of state of a perfect gas is $PV = mRT$. Therefore, $P = mRT/V$. Therefore,

$$P\frac{dV}{d\theta} = \frac{mRT}{V} \cdot \frac{dV}{d\theta} \tag{5}$$

Neglecting the effect of local concentration on internal energy and assuming it is only a function of cylinder gas temperature, then

$$\frac{dU}{d\theta} = m\,C_v(T)\,\frac{dT}{d\theta} \tag{6}$$

Substituting equations (6) and (5) for $dU/d\theta$ and $P\,dV/d\theta$, respectively in equation (4), then equation (4) becomes:

$$mC_v(T)\,\frac{dT}{d\theta} = \frac{dQ_b}{d\theta} - \frac{dQ_w}{d\theta} - \frac{mRT}{V}\,\frac{dV}{d\theta} \tag{7}$$

that is $(mRT/V)(dV/d\theta)$ is the rate of work done on the piston.

To simplify the calculation of equation (7) a step by step method is used. Normally a crank angle increment of one degree is employed for which the following relationships may be derived:

V_{step} = average volume between two steps

$$= V_n + \frac{V_{n+1} - V_n}{2} = V_n + \frac{dV}{2}$$

$\frac{dT}{d\theta}$ = temperature difference over increment

$$= T_{n+1} - T_n$$

Average temperature over the increment is $T_{av.} = \frac{1}{2}\,(T_{n+1} + T_n)$.
Substituting in equation (7)

$$\frac{dQ_H}{d\theta} = mC_v(T)\,[T_{n+1} - T_n] + \frac{mR\,dV}{V_n + dV/2} \cdot T_{av.} \tag{8}$$

Putting

$$K = \frac{1}{R}\left(V_n + \frac{\mathrm{d}V}{2}\right)\cdot\frac{1}{\mathrm{d}V}$$

we obtain

$$\frac{\mathrm{d}Q_\mathrm{H}}{\mathrm{d}\theta} = mC_\mathrm{v}(T)\,[T_{n+1} - T_n] + m\,\frac{T_\mathrm{av}}{K} \tag{9}$$

where

m = mass of charge (gas + fuel)
R = the gas constant
$C_\mathrm{v}(T)$ = specific heat at constant volume as a function of temperature and air–fuel ratio
V = volume
T = temperature

and (n) refers to a particular station in the crank angle field and av. = average.

Q_H is the difference between the fuel heat input Q_b and the heat given off to the cylinder walls. Q_w, as depicted in equation (2). Therefore for the actual (absolute) rate of heat release $Q_\mathrm{HA,}$ heat loss to the walls must be added to $\mathrm{d}Q_\mathrm{H}/\mathrm{d}\theta$, i.e.

$$\frac{\mathrm{d}Q_\mathrm{HA}}{\mathrm{d}\theta} = \frac{\mathrm{d}Q_\mathrm{H}}{\mathrm{d}\theta} + \frac{\mathrm{d}Q_\mathrm{w}}{\mathrm{d}\theta} \tag{10}$$

where, from heat transfer relationships,

$$\frac{\mathrm{d}Q_\mathrm{w}}{\mathrm{d}\theta} = hA(T_\mathrm{g} - T_\mathrm{w}) \tag{11}$$

where

T_g = gas temperature
T_w = cylinder wall temperature
N = engine speed
A = sum of exposed areas in combustion chamber
h = coefficient of heat transfer (using Eichelberg's or other relationships)

For example,

$$h = 1.015 \times 10^{-4}\; 3\sqrt{\overline{V}_\mathrm{p}} \cdot \sqrt{(P_\mathrm{g} T_\mathrm{g})} \tag{12}$$

where

$\overline{V}_\mathrm{p}$ = piston mean velocity
S = stroke
P_g = gas pressure

The volume at each step can be derived from the geometry of the piston–crank mechanism as

$$V = V_{CL} + A_p \left\{ l + \frac{S}{2}(1 - \cos\theta) - \sqrt{l^2 + \frac{S^2}{4}(\cos^2\theta - 1)} \right\}$$

where

V_{CL} = clearance volume
A_p = piston crown area
l = connecting rod length ℄ to ℄

An approximate correlation for the specific heat at constant volume C_v is given by

$$C_v(T) = 0.6113 + 0.00001096 \sim \text{kJ/kg °K} \tag{13}$$

where T is in °K.

Also normally it is assumed that

$$PV^{1.35} = \text{constant and } TV^{0.35} = \text{constant} \tag{14}$$

and these relations are used in the calculations.

A number of heat release computer programs have been written by various researchers. But based on the foregoing theory, two programs have been developed with varying complexity. The simpler version requires measurement of the cylinder pressure diagram and engine parameters. The more accurate version requires other measurements, outlined in the following section.

These computer programs start with known inlet pressures and temperatures and valve timings. The digitized measurements are then fed into the heat release computer program together with the engine dimensions to yield heat release output and other engine performance parameters.

A typical heat release diagram is shown in Fig. 14 with explanation of the various parts in relation to the combustion process.

During the combustion period (that is, the period in which the rate of heat release is positive), the burning can be said to proceed in three identified phases. In the first place, the rate of heat release is generally very high and lasts for only a few degrees crank angle (about 6°). This corresponds to the period of rapid cylinder pressure rise. Owing to the low temperature, the burning in this phase is under the control of chemical kinetics. The second phase corresponds to a period of gradually decreasing rate of heat release and lasts for about 40° crank angle. This is the main heat release period. The height of the second peak is largely controlled by turbulent mixing (i.e. rate of turbulent mixing of fuel and air). Generally, about 80% of the total heat is released in the first two phases. Finally, as the concentrations fall, the rate of heat release is once again controlled by the chemical kinetics. This rate of heat release persists throughout the expansion stroke, and the heat release during this period amounts to about 20% of the total.

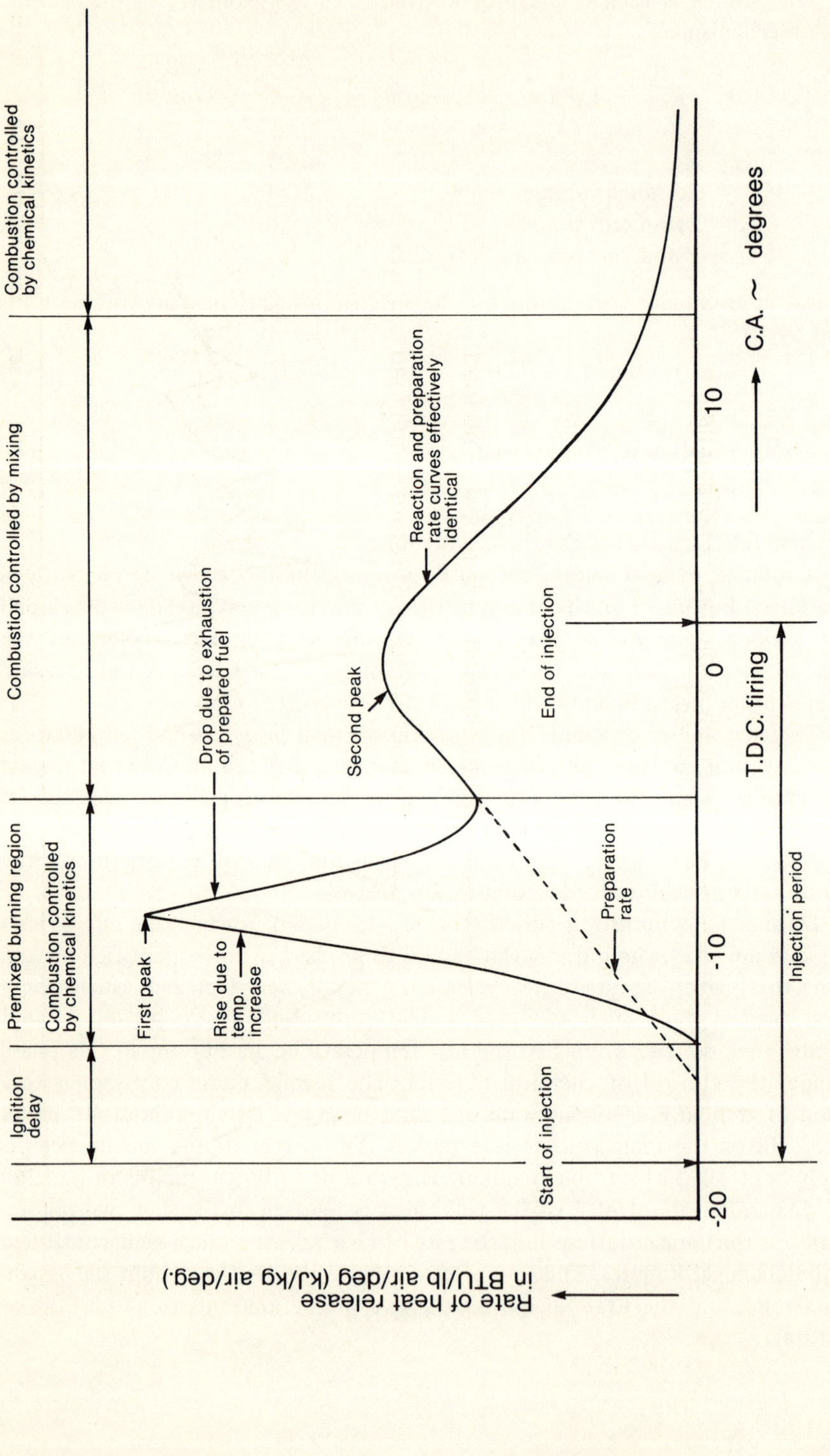

Fig. 14 – A typical heat release diagram.

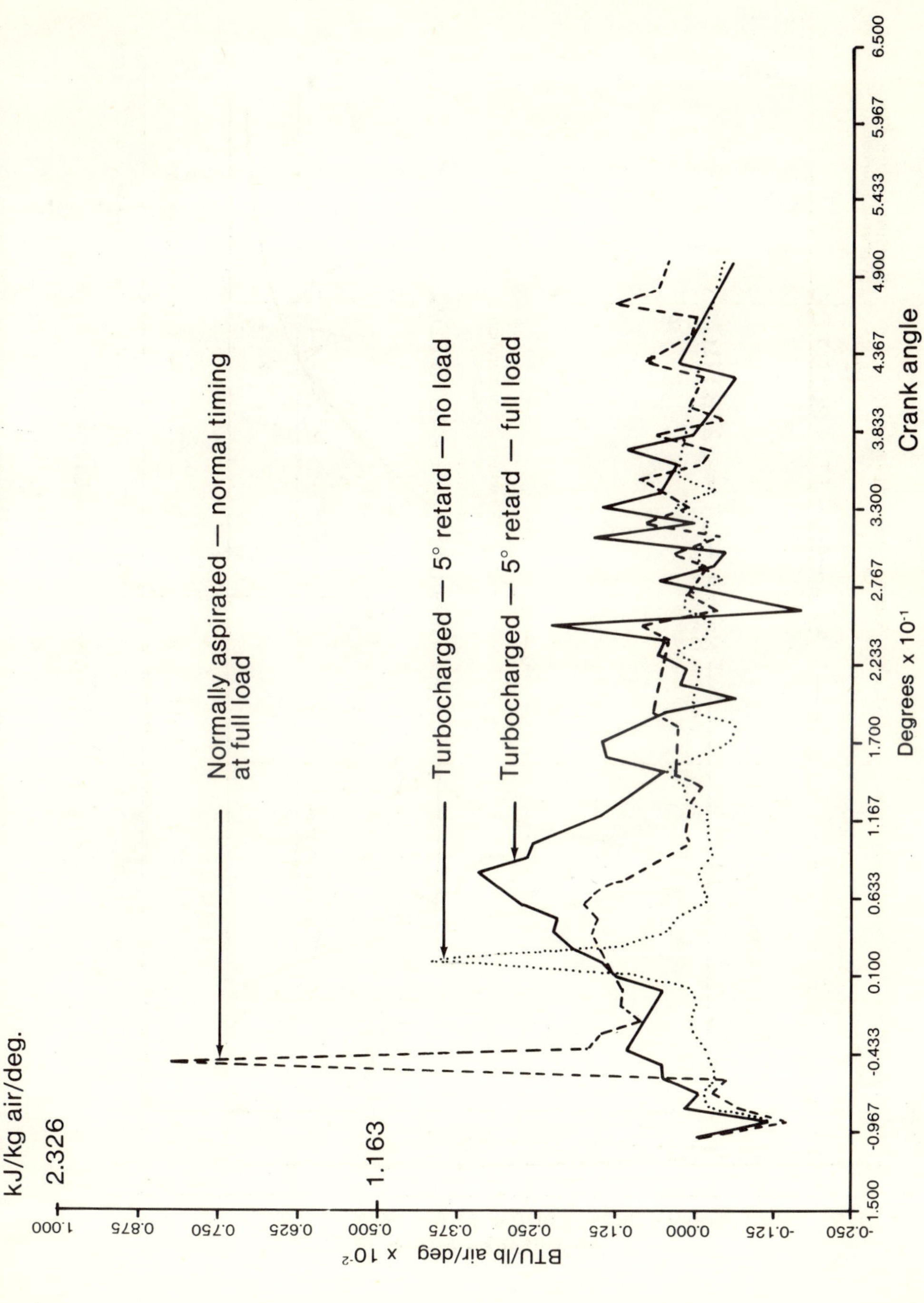

Fig. 15 – Calculated heat release for engine 'B' at 1000 rpm.

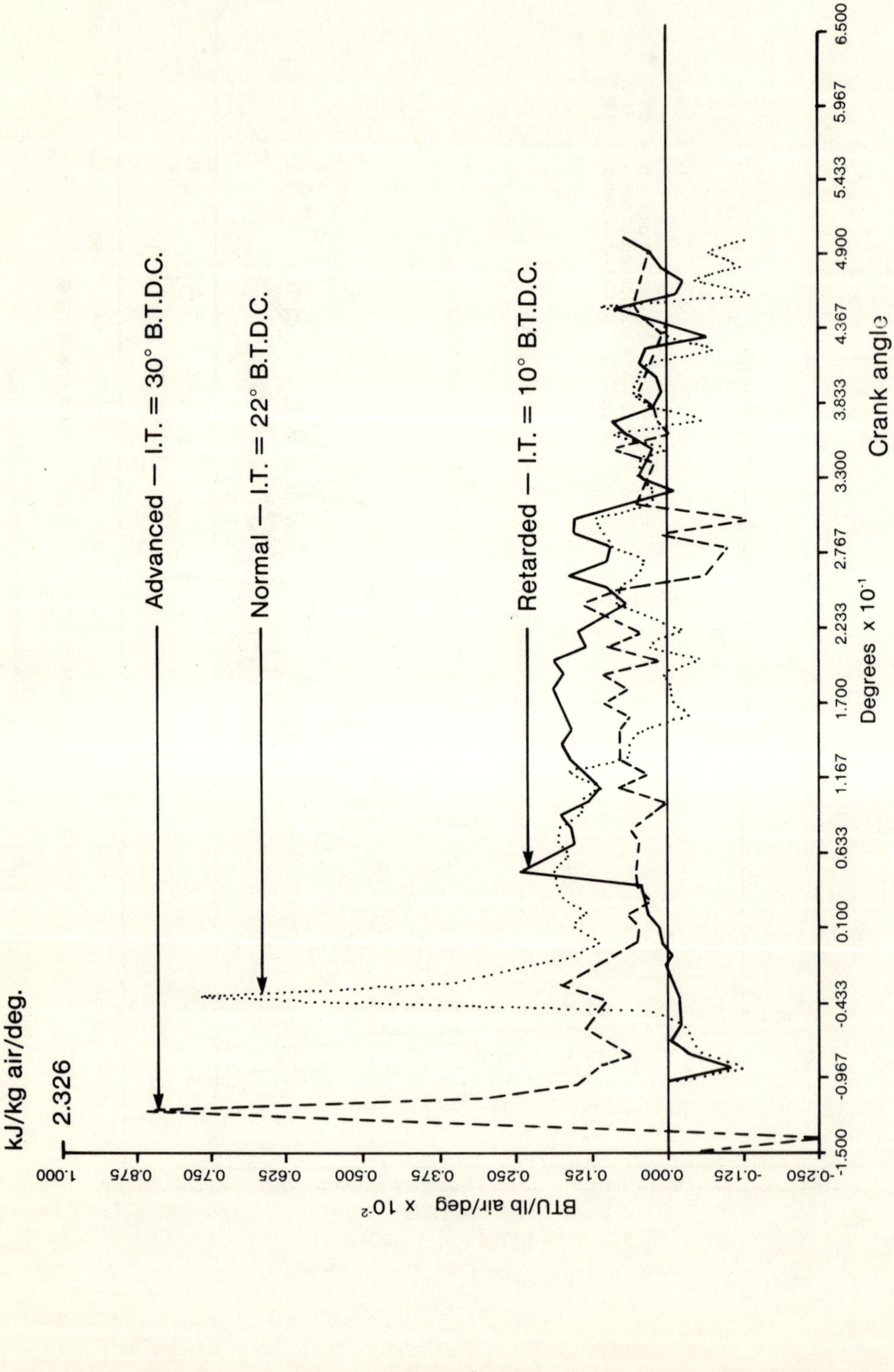

Fig. 16 – Caclulated heat release for engine 'A' at 1000 rpm full load naturally-aspirated'.

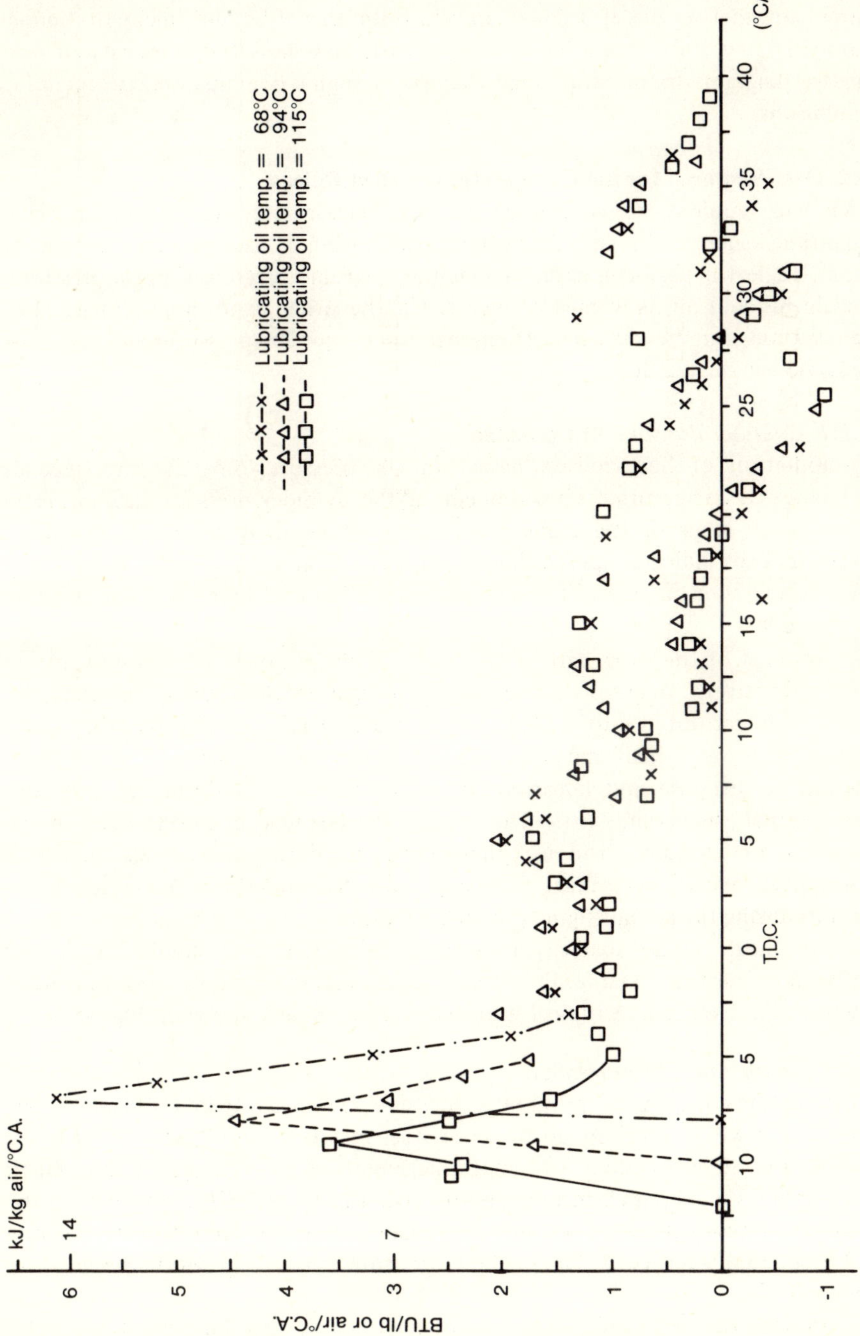

Fig. 17 – Effect of lubricating oil temperature on heat release at 3000 rpm fl (engine C).

Figures 15, 16 and 17 present typical computed heat release diagrams for three automative diesel engines to illustrate the effect of injection timing, turbocharging and lubricating oil temperature, and show the sensitivity of heat release diagrams to monitor small changes in engine combustion and operating conditions.

6.2 Data Required for the Computation of Heat Release

The data required for computing the heat release rate, apart from the fixed quantities such as engine dimensions, calibration constants, etc., consist of crank angle (in degrees), cylinder pressure, fuel line pressure (nozzle pressure), needle lift and air flow measurements. On the other hand, approximate H.R. calculations can be obtained from measured cylinder pressure and fuel flow rate and engine details.

6.2.1 Cylinder Pressure Measurement

Computation of heat release based on the pressure–time diagram depends primarily on an accurate measurement of the cylinder pressure diagram. Even when small steps of the crank angle are considered, quite small changes of pressure can indicate appreciable changes in heat release rate. Therefore a high accuracy of the order of (say) 4 parts in a thousand, based on peak pressure, is desirable.

In most of the work carried out in this field, piezo-electric transducers are used extensively. A piezo-electric transducer is a device whereby the change in pressure is converted into a change in electrical charge. Unlike the other transducers, it is very small and can easily fit into a small space and above all its frequency response and linearity are good. Its main disadvantages, however, are thermal shock and drift. The former disadvantage manifests itself in the form of an excessive drop in cylinder pressure during the expansion stroke. The latter, caused by leakage of charge from the transducer and lead, is less serious during the actual engine running condition.

For heat release analysis, a dynamic calibration is desirable but is very difficult to perform. As a result of this, a static calibration is always recommended by the manufacturers. A typical static calibration curve is shown in Fig. 18.

6.2.2 Crank Angle Measurement

Two methods are commonly used to measure the crank angle. One involves the use of a magnetic pick-up up to sense steel inserts at (say) every 5° near the T.D.C. and at every other 10° in the periphery of a non-magnetic disc. Much care should be exercised in locating the position of the T.D.C., especially if the magnetic pulses generated are not very sharp and short. However, the phasing between crank angle (C.A.) and cylinder pressure can be kept within the accuracy of ± 1/4°C.A.

In the other method, pulses are obtained at the same intervals as above by

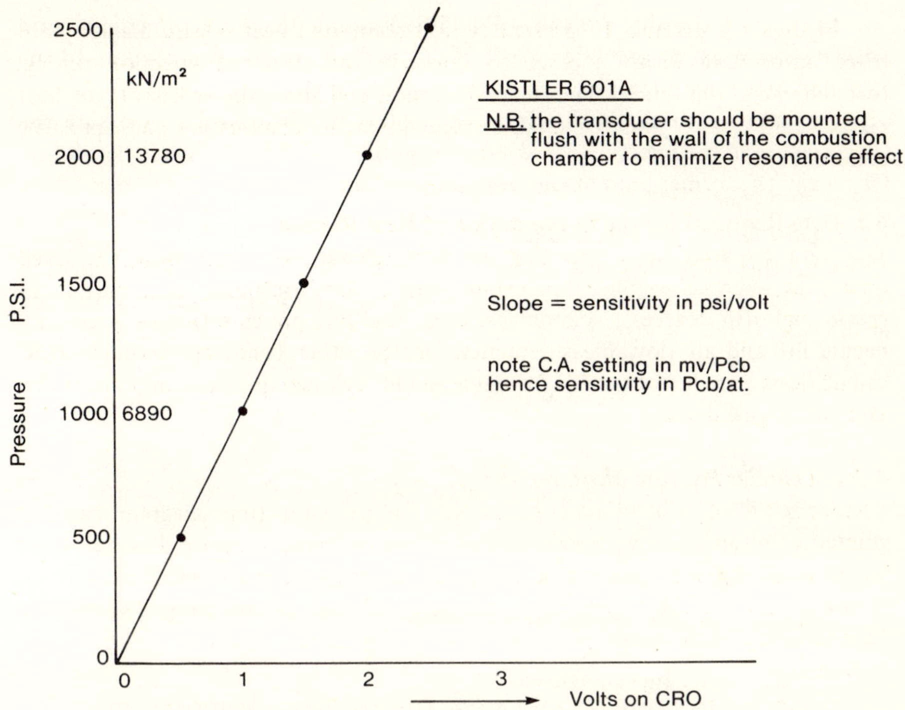

Fig. 18 – A typical calibration curve of C.P. transducer.

shining light from a quartz bulb through slits onto a photo cell. These pulses are finally processed electronically.

6.2.3 Measurement of Needle Lift

Estimation of the rate of injection and ignition delay period requires the measurement of the needle lift and fuel line pressure (nozzle pressure). Needle movement can be measured by means of a variable inductance transducer. Its basic features include a solenoid attached to the injector spindle which moves inside a coil. Any movement of the injector spindle in the coil changes the inductance. The output of the gauge may then be fed via a frequency modulation system with a nominal carrier frequency of 2 MHz to the frequency discriminator F.M. unit.

Like the piezo-electric transducer, the needle lift gauge is generally calibrated statically.

Ultrasonic-based needle lift gauges are also available now whereby very little modification is made to the instrumented injector.

6.2.4 Resistance Strain Gauge for Fuel Pressure Measurement

This type of hydraulic transducer is used extensively for measuring the pipe-line fuel pressure. Its basic principle is the same as that of the stretching of a wire whereby its electrical resistance is increased.

Normally a Cussons P. (4561) Fuel-line Pressure Transducer in conjunction with its strain gauge amplifier P (4551) is used and operated according to the manufacturer's instructions. AVL systems are also commonly used.

For more accurate heat release computation it is useful to measure the fuel pressure in the injector sack volume in addition to the pipe-line fuel pressure. Obviously the former presents installation problems.

6.2.5 Calculation of the Total Fuel Injected

Knowing the foregoing measurement and using the basic hydraulic equation for the fuel injection system, the rate of injection $\mathrm{d}Q/\mathrm{d}t$ (fuel) can be estimated as follows:

$$\frac{\mathrm{d}Q}{\mathrm{d}t}\,(\text{fuel}) = A_e C_d \sqrt{[2\,(P_2 - P_1)]}$$

or

$$\frac{\mathrm{d}Q}{\mathrm{d}t}\,(\text{fuel}) = \pi\, d L_n C_d \sqrt{[2g(P_2 - P_1)\rho]} \qquad (15)$$

where

$A_e = \pi.d.L_n$ = effective area of the discharge orifice
d = diameter of the orifice (nozzle)
L_n = needle lift
C_d = discharge coefficient
P_2 = fuel pressure
P_1 = cylinder pressure at corresponding degree of crank angle
$\mathrm{d}t$ = degree crank angle increment
ρ = density of the fuel

Neglecting the effect of temperature on ρ, and assuming $\rho = 0.825$ gm/cm^3 and $g = 980$ mm/s^2,

$$\frac{\mathrm{d}Q}{\mathrm{d}t}\,(\text{fuel}) = \pi d\; 4.03\; L_n C_d\, (P_2 - P_1)^{0.5}\ \text{mm}^3/\text{ms} \qquad (16)$$

where

$\mathrm{d}t$ is in ms
L_n is in mm
P_2 and P_1 are in p.s.i.
d is in mm

Given the engine speed, the total fuel injected per cycle can be found from equation (16).

6.2.6 Measurement of Inlet Air Flow Rate

Air flow rate can be measured by means of a metering system designed in accordance with the relevant British Standards. The air flow rate through a sharp-edged orifice with a flange tapping may be calculated from the modified form of the Bernoulli and Continuity equations. After simplification, the air flow rate is given by:

$$Q_{air} = \text{Constant} \times \sqrt{h} \sim \text{kg/hr} \tag{17}$$

where h is the pressure difference across the orifice in mm H_2O. According to equation (17), in order to calculate Q_{air}, one measures only h.

Viscous flow air meters are also used but are less accurate than the orifice-based air flow meters.

If needle lift, fuel line pressure and air flow measurements are not available it is possible to use the following equations to calculate the mass of air (m_{air}) [21] and fuel (m_{fuel}) [21], respectively:

$$m_{air} = \eta_v(r_c - 1)V_{CL} \cdot \frac{M_a}{359} \times 0.454 \sim \text{kg} \tag{18}$$

where

η_v is the volumetric efficiency ($\hat{=}$ 80%)
r_c is the nominal compression ratio
M_a is the average molecular weight of air
V_{CL} is the clearance volume

and

$$m_{fuel} = \frac{\eta_v}{x}(r_c - 1)V_{CL}\frac{M_a}{359} \times 0.454 \sim \text{kg} \tag{19}$$

where x is the measured air–fuel ratio by weight.

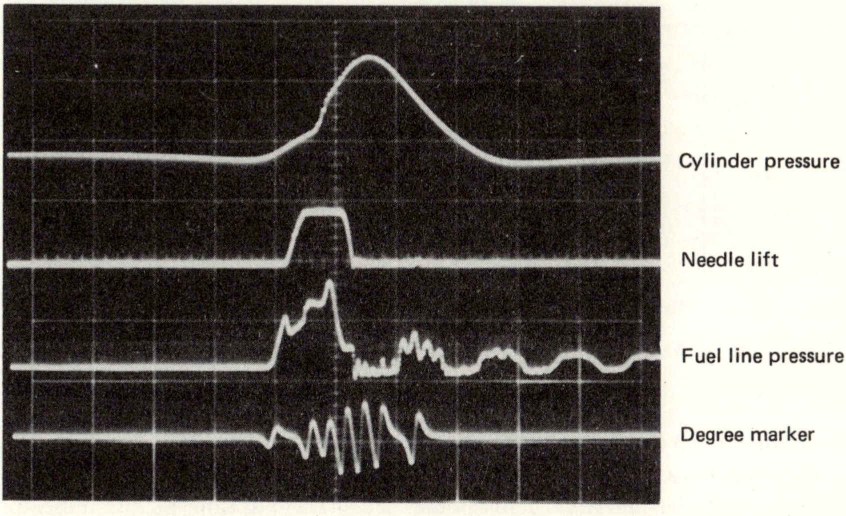

Fig. 19 – A typical record for heat release calculations.

Figure 19 shows a typical actual record of cylinder pressure, needle lift, fuel line pressure and degree marks around T.D.C. for an automotive diesel engine. Figure 20 shows idealized, but typical, measured indicator diagrams of combustion related to the corresponding calculated rate of the heat release diagram.

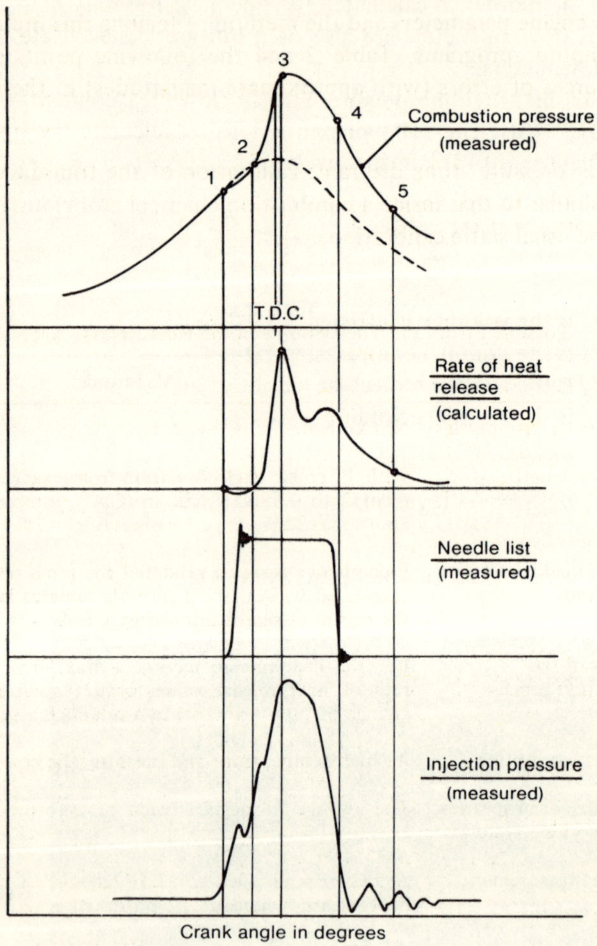

Fig. 20 – Typical indicator diagrams of combustion.

6.3 Comments on the Accuracy of the Computed Rate of Heat Release in Diesel Engines

It is important that extreme care should be taken in scaling all the indicator diagrams measured for the engine under investigation. One way of doing this is to enlarge these diagrams. The enlarged diagrams are then digitized by hand (or by computerized technique) for each crank angle step. A step size of 1° crank angle is normally used, especially throughout the major heat release period, in order to increase the accuracy of the computed heat release rates. Also it must be remembered that the accuracy of these computed heat release diagrams is to a great extent dependent on obtaining accurate measurements of the various engine parameters and the method of feeding this input into the heat release computer programs. Table 2 and the following points summarize the possible sources of errors (with approximate magnitudes) in the calculations of heat release rates:

(i) Cylinder pressure–time diagram: calibration of the transducer in an atmosphere similar to that inside a combustion chamber is obviously more accurate than the usual static calibration.

Table 2.
Possible Errors and Their Source in the Heat Release Analysis

	Error	Magnitude
1.	Mass-averaged properties	Not known.
2.	Time-step length	With 1° crank angle deviation from exact calculation can be from 2 to 0.680%. With an (C.A.), interval of 2° C.A. the above max. difference becomes 6 to −2%.
3.	Rates of heat transfer calculations	Eichelberg's formula rendered the gross rate of heat release calculated by Annand's formula underestimated by 10% in the region of significant energy release.
4.	Progress of the combustion process	For the heat release record, a max. error of −10% in the rate of heat release values occurs when an instantaneous change of total reactants to products is assumed.
5.	Ideal gas assumption	At high temperature and pressure, the error is about 2–5%.
6.	Variability of properties of the working system	The average properties based on ambient temperature can lead to 60% error.
7.	Pressure measurement	A 2% error entailed a +2.5% change in the gross rate of heat release, whereas −2% produced a −3.5% error.
8.	Recording of crank angle rotation	A shift of ± 2° in the T.D.C. position can result in an error of roughly ± 15% in the peak value of gross rate of heat release while the cumulative heat release is affected by ≏ 10%.
9.	Determination of the mass of the charge	An error of ± 2% in the determination of the mass of the charge can result in an error of 4% and −5.5% in the gross heat release rate, respectively.
10.	Design paramters	An error of ± 2% in the compression ratio can result in about 40% in the gross heat release rate.

(ii) The connecting passage of the transducer should be as short as possible to minimize the resonance effect on the pressure trace.
(iii) Small changes of pressure can indicate large changes in heat release rate, therefore high accuracy is desirable.
(iv) Compression ratio quoted by the manufacturers may not be exact ($\pm \frac{1}{2}$ ratio tolerances) and this must be checked, otherwise an error of about 30% in the heat release diagram is possible.
(v) Incorrect determination of T.D.C. could also lead to a large error.
(vi) Accurate measurement of air and fuel flow rate is essential for absolute heat release rates.
(vii) A thermally loaded transducer should be guarded against, which could give a false cylinder pressure trace (a good position for the transducer in the cylinder head should be selected).

6.4 The use of the Heat Release Rate Diagram

The heat release pattern determines the shape and magnitude of the cylinder pressure development, which in turn determines the combustion noise, indicated efficiency, etc.

If the heat release can be predicted with sufficient accuracy, it should be possible to evaluate and optimize all the various aspects of diesel engine performance such as the rate of pressure rise (hence combustion noise), the smoke level at the design stage, etc.

For example, it has been established that the magnitude and incidence of the first peak of the rate of heat release diagram bear a direct relationship to the severity of combustion-induced noise which in turn is related to the ignition delay period.

Finally, as a result of extensive experimental and theoretical research work into combustion and heat release, it is worth while to summarize the following broad deductions to clarify the combustion processes in diesel engines.

(a) It has been demonstrated experimentally that completely varporized fuels do not burn instantaneously, even under conditions of high temperature and pressure. From this finding and other data it can be concluded that both physical and chemical processes may be required in interpreting the ignition delay period.
(b) The lifetime of a liquid droplet depends on the nature of the fuel, on relative velocity, initial size, temperature and density of the charge.
(c) Because of non-homogeneity of the fuel–air mixture, before ignition, the local mixture concentration is high in the area where vapour emanates from liquid fuels. Initial ignition has the advantage of high oxygen concentration around molecular groups of fuel. Also of importance, the diffusion flame occurs between bulk vaporized fuel and oxygen.

(d) The position of the nozzle in a combustion chamber is very important, so too is the interaction of the spray and air movement and the subsequent formation of the mixture which determines the ignition behaviour.
(e) Combustion begins long before vaporization is complete.
(f) Air motion inside the cylinder is approximately proportional to the mean piston speed or engine speed. Thus an increased rate of mixing occurs at high engine speed, although this can be offset by less time being available.
(g) The first rapid burning after ignition is associated with a low-luminosity premixed flame. This blue flame has a short life of approximately 6° crank angle with a peak temperature of over 2000°C. After the premixed flame, a bright white carbon diffusion flame predominates with a maximum temperature of about 2500°C, lasts some 40° crank angle and fades to red as it cools by expansion in a D.I. four-stroke engine.
(h) The rate of injection is higher at low engine speeds and the relative duration of injection is longer (in terms of °CA) at high engine speeds.
(i) Provided the spray is not intentionally directed towards the cylinder walls (as in the MAN system) the longest delay has the highest initial heat release and heat release continues later in the cycle with later injection.
(j) The initial slope and height of the rate of the heat release diagram are determined by the amount of fuel injected and vaporized during the ignition delay period. If this amount is more than 6–10% of stoichiometric fuel, the resulting rapid pressure rise is sufficient for 'knock' to emerge from the general noise level.

REFERENCES

[1] Wolfer, H. H., 'Ignition Lag in the Diesel Engine', VDI Forshung 392 (1938).
[2] Spalding, D. B., *Some Fundamentals of Combustion,* Butterwoths Scientific Publications, 1955.
[3] Meurer, J. S., 'Evaluation of Reaction Kinetics Eliminates Diesel Knock', *SAE Transactions* (1955).
[4] Wakil, M. M., *et al.,* 'Fuel Vaporisation and Ignition Lag in Diesel Combustion', *SAE Transactions,* **64** (1956).
[5] Lyn, W. T., 'Calculations of the Effect of Rate of Heat Release on the Shape of Cylinder-pressure Diagram and Cycle Efficiency', *Proc. Auto. Division,* Inst. Mech. Engrs. (1960–61).
[6] Lyn, W. T., 'The Study of Burning Rate and Nature of Combustion in Diesel Engines', *Ninth International Symposium on Combustion,* New York (1962).
[7] Lyn, W. T. and Valdmanis, E., 'The Application of High Speed Schlieren Photography to Diesel Combustion Research', *Journal of Photographic Science,* **10** (1962).
[8] Austen, A. E. W. and Lyn, W. T., 'The Application of Heat Release Analysis to Engine Combustion Study', 1067, *CIMAC* (1962).

[9] Alcok, J. F. and Scott, W. M., 'Some More Light on Diesel Combustion', *Proc. Inst. Mech. Auto. Group,* March (1963).

[10] Faeth, G. M., 'The Kinetics of Droplet Ignition in a Quiescent Air Environment', *Ph.D. Thesis,* The Pennsylvania State University, (1964).

[11] Scott, W. M., 'Looking in on Diesel Combustion', *SAE Paper SP345,* December (1968).

[12] Faeth, G. M. and Olson, D. R. 'The Ignition of Hydrocarbon Fuel Droplets in Air', *SAE Paper No. 680465* (1968).

[13] Shipinski, J., Myers, P. S. and Uyehara, O. A., 'A Spray Droplet Model for Diesel Combustion', *Proc. Inst. Mech. Engrs.,* **184,** Part 3J, 28 (1969–70).

[14] Adler, D. and Lyn, W. T., 'The Evaporation and Mixing of a Liquid Fuel Spray in a Diesel Air Swirl', *Proc. Inst. Mech. Engrs.,* **184,** Part 3J, 171 (1969–70).

[15] Grigg, J. C. and Syed, M. H., 'The Problem of Predicting Rate of Heat Release in Diesel Engines', *Proc. Inst. Mech. Engrs.,* **184,** Part 3J, 192 (1969–70).

[16] Whitehouse, N. D. and Way, R. J. B., 'Rate of Heat Release in Diesel Engines and its Correlation with Fuel Injection Data', *Proc. Inst. Mech. Engrs.,* **184,** Part 3J, 17 (1969–70).

[17] Whitehouse, N. D. and Way, R. J. B., 'A Simple Method for the Calculation of Heat Release Rates in Diesel Engines Based on the Fuel Injection Rate', *SAE Paper 710134* (1971).

[18] Henein, N. A., 'MIRA Abstract – Combustion and Emission Formation in Fuel Sprays Injected in Swirling Air', *SAE Paper 710220,* January (1971).

[19] Jellings, W., 'An Elementary Approach to the Computation of I.C. Engine Exhaust Gas Emissions', *UMIST Internal Report* (1971).

[20] Launder, B. E. and Spalding, D. B., *'Mathematical Models of Turbulence',* Academic Press, London, 1972.

[21] Ogegbo, A. M., 'Automative Diesel Engine Combustion and Noise', *Ph.D. Thesis,* I.S.V.R., University of Southampton, 1973.

[22] Haddad, S. D., Pullen, H. L. and Priede, T., 'Relation Between Combustion and Mechanically-induced Noise in Automative Diesel Engines', *15th Congress FISITA Paper A-2-4,* Paris, May (1974).

[23] Woschni, G. and Anisits, F., 'Experimental Investigation and Mathematical Representation of the Rate of Heat Release in Diesel Engines Dependent on Engine Operating Conditions', *SAE Paper 740086* (1974).

[24] Whitehouse, N. D. and Sareen, B. K., 'Prediction of Heat Release in a Quiescent Chamber Diesel Allowing for Fuel Air Mixing', *SAE Paper 740084* (1974).

[25] Miffek, K., 'Formation of Mixture and Combustion in Oil Engines', *Lecture Notes 2,* UNIDO In-plant Group Training Programme for Engineers in the field of Oil Engines – Institute of Technical and Economic Services, Praha, Mikulandska 7, Czechoslovakia (tes) 1974.

[26] Shahed, S. M., Chiu, W. S. and Lyn, W. T., 'A Mathematical Model of Diesel Combustion', in *Combustion in Engines,* p. 119, Inst. Mech. Engrs, 1975.

[27] Watson, N. and Meguerdichian, 'Prediction of Mixture Formation and Heat Release in Diesel Engines', *SAE Paper 780225* (1978).

[28] Benson, S. and Whitehouse, N. D., *'Internal Combustion Engines', Volumes 1 and 2,* Pergamon International Library, 1979.

2

Fuels for diesel engines

W. TIPLER, Senior Consultant

1. INTRODUCTION

In the early years of this centruy, when the diesel engine was in its infancy, the question of suitable fuels for the engine was a very live subject among engineers. In 1912, Diesel wrote [1]: 'As early as 1899, I utilized in my experimental engine the by-products of coal distillation and coke plants, such as tar and creosote oils . . . it is possible also to burn vegetable and animal oils without difficulty; the first trials were made in 1900 with earth-nut oil . . . it is certain that motor power can still be produced from the heat of the sun, which is always available for agricultural purposes, even when all our natural stores of solid and liquid fuels are exhausted.' Far-sighted comments made 70 years ago!

Up until the end of World War II, the multi-fuel ability of the diesel engine continued to recieve considerable attention, but in the next 25 years, owing to the world-wide availability of relatively cheap, technically suitable petroleum-based fuels, the diesel engine came to be regarded as a petroleum dependent prime mover. Indeed, until the past seven years, papers concerning fuels for diesel engines were rarely seen; now a publication such as the present book which did not include a discussion of fuels would be considered gravely incomplete. Mechanical engineers have rediscovered fuel as a variable to be given serious consideration during the design and development of engines; fuel experts abound and papers proliferate, making it difficult to contribute any truly original information on the subject. Therefore, this chapter attempts nothing more than a balanced review of the fuel prospects of the high-speed diesel engine.

It is not intended to add to the excessive number of papers discussing trends in the rates of usage of various fossil fuels and estimates of their dates of exhaustion, which have followed the 'energy crisis' of 1973, but to base this chapter upon the generally accepted statements that:

(a) Petroleum-based fuels will become scarcer, dearer and technically less suitable during the 1990s.

(b) Liquid fuels derived from coal will become available in the industrialized areas before the end of the century.
(c) Vegetable-based oils will make a significant contribution to energy supplies in some countries before the end of the century.

These trends can already be seen, and bear out Diesel's statements of 1912 [1].

2. THE TRUE THERMAL EFFICIENCY

When considering the thermal economy of any prime mover burning any fuel, it is no longer adequate merely to study the thermal economy of the engine alone ('Useful energy out divided by fuel energy in' as generations of engineers have been taught). Account must also be taken of the energy used in the production of the fuel fed to the engine; in other words the significant efficiency, under the changing fuel supply situation which is now recognized, is the overall efficiency at which natural energy is converted into useful shaft power [2].

$$\eta_o = \eta_w \cdot \eta_c \cdot \eta_e$$

where

η_o = overall efficiency of conversion.
η_w = efficiency of winning and transporting natural fuel (e.g. crude oil).
η_c = efficiency of converting natural fuel to commercial fuels.
η_e = efficiency of the prime mover.

The oil companies refer to η_o as the 'well to wheel' efficiency; in the case of coal it might be termed 'pit to power' or 'mine to motion'.

In the cases of petroleum and coal, η_w is usually quite high (say 95%) and will not be discussed; it is much lower for vegetable oil production; η_c can be quite low, particularly for the production of liquid fuels from coal, and therefore has a significant bearing upon η_o; η_e is a familiar concept, subject to Carnot's limitation in the case of heat engines, and will not be discussed in this chapter which concentrates mainly upon η_c. Nevertheless it must be remembered that optimization of the overall efficiency of utilization of energy resources (η_o) is unlikely to be achieved by the present general policy of unco-ordinated development of fuels and engines. Very real benefits can accrue from joint fuel/engine studies [3]. This statement will be developed in later sections of this chapter.

3. PETROLEUM-BASED DIESEL FUEL

The past 35 years have seen an almost unbroken upward trend in demand for distillate fuels for both land and air transport. Since the basic oil refining process (distillation) also produces a 'heavy' residual component which is used in industrial fuel oils, this trend could cause severe problems if there were no corresponding increase in demand for such heavy products. In Europe the oil industry was successful in displacing coal from many of its traditional markets (firing of boilers, furnaces and kilns), and was thus able to maintain a balance between the demands for its various products. In the U.S.A. on the other hand, a far more robust coal industry offered more spirited resistance to such incursions and the oil industry therefore adopted a different strategy; production of industrial fuels was reduced by the introduction of additional processes into the refineries – the heavy residual material was 'cracked' to produce additional distillate type products (Fig. 1). (In the 'cracking' process, large molecules are broken into smaller ones, of greater volatility, by the effect of heat in the presence of a catalyst.)

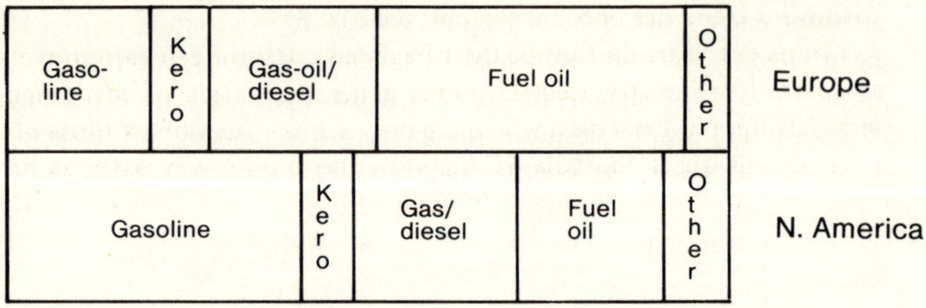

Fig. 1 – Patterns of demand for petroleum products.

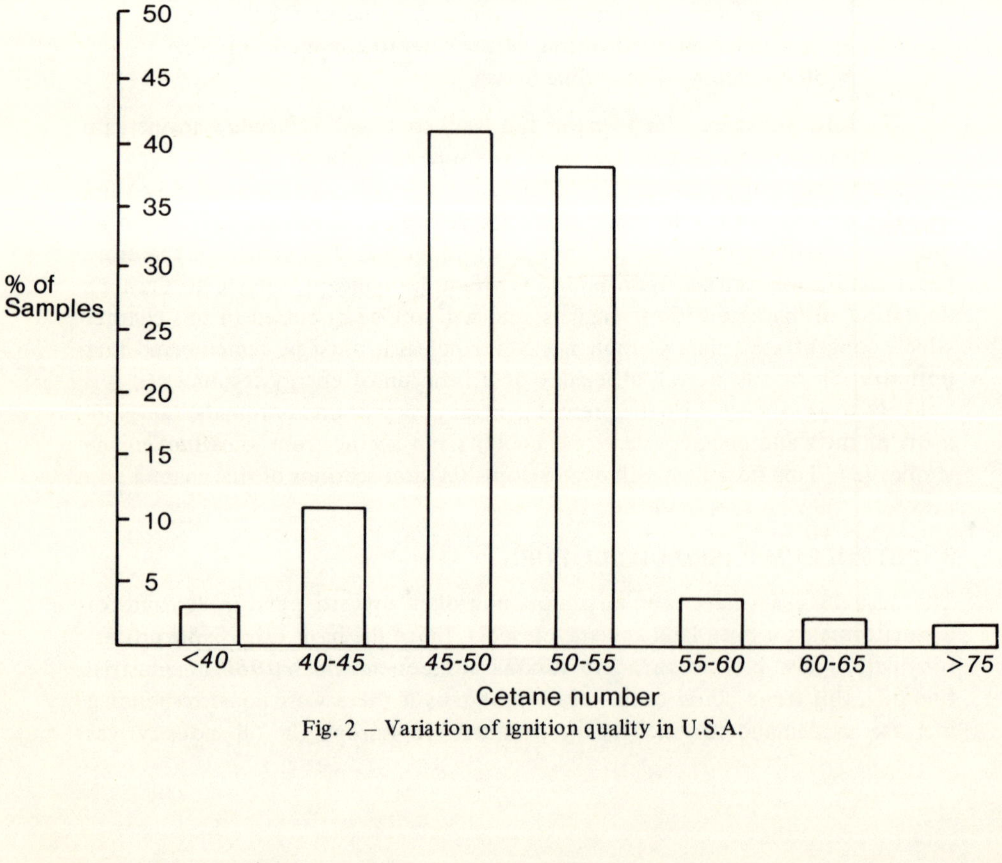

Fig. 2 – Variation of ignition quality in U.S.A.

Whereas 'cracked' products form excellent components for blending into gasoline, they are in general inferior components for use in diesel fuels.

There has therefore been a marked contrast between the quality of distillate diesel fuel in Europe and North America. Here production has been mainly based upon distillation, and the resultant 'straight run' diesel fuel has been of a very good, and steady, cetane number (ignition quality) usually 50–52. In North America, the use of varying proportions of 'cracked' components in diesel fuel has resulted in a far more variable product; see Fig. 2. The average level of cetane number has been maintained at an acceptable level by selection of crude supplies, but as this has become more difficult owing to the international situation, a downward trend has become evident; see Fig. 3.

In recent years, in Europe there has been a definite effort to convert industrial processes (particularly electric power generation) back to coal firing to reduce demands upon oil reserves, which are only a small fraction of those of coal. Such a trend will upset the balance between the production patterns of European

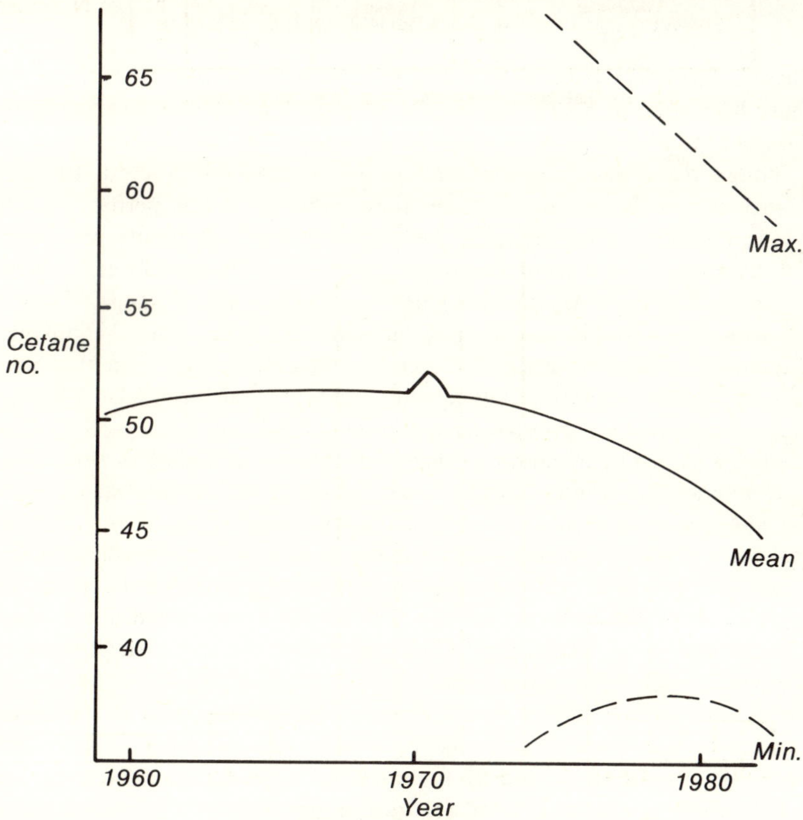

Fig. 3 – Trends in ignition quality in U.S.A.

refineries and the demand for their products. To correct this unbalance, Europe must move towards the American refining system by the incorporation of cracking plants to reduce the proportionate production of heavy residual components. This in turn implies that the general quality of European diesel fuel will fall, and oil companies are already anxious to broaden specification limits to give them greater flexibility in production.

Diesel fuel quality will not only be affected by the proportion of cracked products blended into the fuel, but also by the steadily increasing demand for aviation kerosine, which is forcing diesel fuel further down the barrel. Generally speaking it is expected that the quality of European diesel fuel will show the following trends over the next 10 years:

Density will increase
Viscosity will increase
Cetane number will decrease
Volatility will decrease
Carbon residue will increase
Low temperature flow characteristics will deteriorate

These subjects are covered in detail in U.S. technical literature since such fuels have been used in North America for 30 years, but are now receiving more attention in Europe (e.g. [4]).

Although the trends mentioned above are of academic interest, the true significance lies in the extent to which they influence engine performance in its various aspects, such as specific fuel consumption, availability and maintenance, compliance with legislative requirements, etc. This picture is confused by the undoubted fact that diesel engines (even within a close range of power and speed) do not respond equally to changes in fuel quality. The oil companies are frequently at pains to stress this fact to deflect any suggestion that they should take action to impose close quality limits in their products. Although these arguments can be supported by technical and economic analyses, they are nevertheless a source of irritation to engine manufacturers and users, since these latter are required to produce and maintain engines to various legal standards of exhaust and noise emissions. Such standards tend to be regarded as somewhat artificial when it is known that variations in the quality of commercially available fuels can influence whether the engine on the road is complying with the standards.

Discussions of the influence of the fuel quality trends listed above upon engine behaviour are available (e.g. [5]), and therefore only a brief resumé of this involved subject will be given here.

An increase in fuel density would appear to be advantageous to the engine operator since most diesel fuel is sold by volume. Thus an increase in density represents an increased purchase of fuel mass for a given financial outlay. Unfortunately there is a general trend for fuel energy content on a mass basis to fall as density increases. In addition, denser fuel tends to be slower burning thus modifying the engine heat release diagram. The combined effect of these two factors is that the improvement in volumetric fuel consumption expected with the use of a denser fuel is only about 50% of that which might be expected.

Increase of fuel density implies an increased energy input to the engine cylinder, but slower combustion as already stated. These two factors can result in an increase in exhust smoke, and in emissions of CO and hydrocarbons.

By definition, a reduction of fuel cetane number should result in an increase in ignition delay, and this in general results in increased maximum cylinder pressure and increased emissions of NO_x, HC and CO. However, the magnitudes of these trends vary markedly between engines, and some are almost insensitive over a wide range. This tends to throw doubt upon the validity of the cetane number test as at present defined; this point is discussed in section 6 of this chapter.

Decrease of fuel volatility and increase of carbon residue will both tend to increase exhaust smoke levels and may necessitate more conservative engine rating in future. European engine manufacturers who have tended to scoff at the conservative ratings of U.S. Automative diesel engines may soon be obliged to adopt similar levels when they are required to use fuels more akin to those used in North America for the last 30 years. The carbon residue criterion has only a slight relevance to the combustion conditions of a diesel engine cylinder, as discussed in section 6.

The above are just a few comments on an involved subject, but should suffice to warn the reader that the quality of future diesel engine fuels is a subject to be treated with caution by both engine manufacturers and users.

4. COAL AS A DIESEL FUEL

Since the world's reserves of coal are an order of magnitude greater than those of petroleum, it is logical to consider the possibilities of using them to replace petroleum products. The trend back to coal for industrial applications and its effects upon petroleum-based diesel fuels have already been mentioned, but coal can make more direct contributions to the fuel requirements of diesel engines. The production of liquid diesel fuels from coal is a proven technology; the use of pulverized coal in diesel engines has been explored and is worthy of further study.

4.1 Coal-Based Liquid Fuels

The liquid by-products of the coal carbonization process (creosote and coar tar) were among the earliest diesel fuels, as already mentioned [1], but they are not suitable for use with high-speed engines owing to their low cetane numbers and slow burning characteristics [6].

These simple coal liquids were followed by hydrogenated products (e.g. Bergius and Fischer–Tropsch processes) which were capable of producing fuels of excellent quality and which made significant contributions during World War II. Subsequently this work was continued in South Africa by SASOL, and excellent diesel fuel is produced.

The past 10 years have seen a great upsurge of interest in the production of liquid fuels from coal, or of 'syncrudes' which can be used as feedstocks for existing petroleum refineries (e.g. [7]). The wide range of processes under study are capable of yielding a wide range of fuels and their economies approach those of petroleum fuels; see Table 1. Unfortunately all these processes have a common factor (in addition to the staggering capital outlays required) which is the low value of their thermal efficiencies (η_c). Although some of the quoted figures are of the order of 70%, it must be remembered that the diesel fuel produced usually requires additional processing to raise its ignition quality to an acceptable level for use in high-speed engines; this usually involves additional hydrogenation, and hydrogen production requires appreciable energy usage, thus depressing the value of η_c. This latter factor is not always made clear in publications (e.g. [8]). These low values of η_c are largely the result of the desire to use coal as a source of liquid fuels matching the characteristics of present day petroleum fuels so that they can be used in existing engine types.

Table 1.
Comparative Economics of Coal Conversion Process [7].

Process		Capital cost (£/kW output)	Efficiency η_c (%)	Mean product cost (£/GJ)
Gasification:	SNG (Lurgi)	330	65	3.5
Synthesis:	Sasol 2	c.550	38	5.8
	Methanol	380	56	4.0
	Mobil–M	440	47	4.7
Direct hydrogenation:				
	H-Coal	340	56	3.4
	Exxon Donor Solvent	335	66	3.5
	NCB Liquid Solvent	305	73	3.1
	NCB Supercritical Gas	410	65	3.7

Basis: mid-1980 costs, 5% DCF rate, 20-year plant life, coal price £1.38/GJ (£36/t), 90% load factor.

As was first pointed out by Texaco, even in the case of petroleum fuels, a marked improvement of η_o could be achieved by abandoning present day road transportation patterns (both fuel and engine) and embarking upon a joint development of a wide-cut fuel and a fuel tolerant engine (e.g. [9]). Similar reasoning indicates even greater benefits when applied to the case of liquid fuels from coal [3].

4.2 The Alternative Approach to Coal-Based Road Transport Fuels

As a starting point, first consider current values of η_o applicable to petroleum-based road transport.

Typical values of η_c for gasoline are 83–84% in the U.S.A. and 90% in Europe [3]. The corresponding values for diesel fuel are 93% and 97%, respectively. The reason for the lower figures applicable to the U.S.A. lies in the high energy demands of the 'cracking' processes more widely used in the U.S.A., as has already been mentioned.

For the purposes of this discussion, the average over-the-road efficiency of a gasoline engine will be assumed to be $\eta_e = 15\%$, and for a diesel engine $\eta_e = 20\%$. (A considerable time could be spent in refining these figures, but they are adequate for present purposes.)

Taking $\eta_w = 100\%$, it thus appears that in the U.S.A., for a gasoline engined vehicle $\eta_o = 12\%$ and for a diesel vehicle $\eta_o = 18\%$.

Now in the case of gasoline and diesel fuel produced from coal $\eta_c \simeq 40\%$, so that the corresponding values of η_o become 6% and 8%, respectively. Both these figures are considered to be unacceptable under conditions of growing energy scarcity from fossil sources.

But consider the alternative approach based upon the pioneering work of Texaco. Work begun in the 1940s showed that a spark-assisted diesel engine could burn a very wide range of distillate fuels, had virtually no cetane or octane requirements, and did not have the volatility limitations of the gasoline engine; in other words it could burn almost the complete distillate 'cut' of the crude oil without further processing [9].

If Texaco's reasoning is extended to coal-based fuels, the following figures are relevant:

$\eta_c = 70\%$, because a simple distillate fuel is acceptable without octane, cetane or volatility limits

$\eta_e = 20\%$, since the spark-assisted diesel engine is virtually a diesel engine.

Thus in this case $\eta_o = 14\%$, which is appreciably better than the figures derived earlier for coal-based gasoline and diesel fuel.

The same reasoning could be applied to support the development of the gas turbine or the Stirling engine for road use, and their cases might appear to be stronger since they are continuous combustion machines and therefore inherently fuel tolerant.

It is beyond the scope of this chapter to discuss the major technical and economic problems barring the progress of these alternatives, which may eventually penetrate certain special road transport applications. It is considered that the major part of this field will continue to be held by the internal combustion piston engine, in which case the above simple analysis gives a first indication of the potential savings possible if a joint programme of fuel and engine optimization were embarked upon. Sadly, recent years have seen little sign of such an approach, although limited field trials of Texaco engines are now in hand in the U.S.A.

Not only would this approach yield energy savings but it would also reduce the capital outlay necessary to build the new 'coal refineries'. If a simpler fuel were adequate, the production plant would be correspondingly less complex; in addition if a lower fuel quantity were needed for a given road transport loading, then the amount of coal to be processed would be reduced, making the plant smaller.

There would thus appear to be a strong case for a reassessment of fuel-from-coal policy.

4.3 Pulverized Coal as a Diesel Fuel

The direct use of coal in an internal combustion engine predated Diesel, who quickly abandoned the idea in favour of a wide range of liquid fuels [1]. Sustained efforts to achieve this goal were made in Germany from 1916 until 1945 [2, 10]. Since 1950 there have been intermittent trials in the U.S.A. [2] with a continuous effort during the past 8–10 years (11). Although the problems to be solved are formidable, the great attraction is that η_c can be close to 100% [2].

Most of the work just referred to was carried out on large low-speed engines, mainly below 400 rev/min. on account of the inherently low burning speed of coal, and is therefore not directly applicable to the type of engine now under consideration. However, the possibility of coal-firing of a high-speed engine warrants consideration, in view of the prize on offer – a high value of η_o.

Study of the literature already referred to would suggest that there is a choice between two approaches: either to pick up the German work on low-speed engines to burn untreated coal and edge the operating speed forward by lengthy engine development, or to put considerable efforts into modifying the coal to make it suitable for use in engines closely similar to current production, as is being done in the U.S.A. Ultimately this latter route leads to liquid fuel production. It is suggested that the correct policy is to embark upon a joint programme of fuel and engine development to obtain the optimum (technical and economic) blend of their characteristics.

It is therefore considered necessary to study the following problem areas from both fuel and engine aspects:

(a) Fuel selection and preparation
(b) Fuel injection
(c) Combustion
(d) In-cylinder wear
(e) Turbocharger fouling and erosion

These problems result from the facts that (i) coal is an extremely variable substance both in its hydrocarbon formation and in the quantity and composition of its mineral content, (ii) it is a solid, (iii) it burns more slowly than even the residual component of crude oil, and (iv) the solids remaining after the combustion process are large in quantity and highly abrasive to in-cylinder components.

The problems involved in this field of development are without doubt of major proportions, but their ultimate solution on both technical and economic planes should not be discounted.

5. DIESEL FUELS DERIVED FROM VEGETABLE MATTER

Strictly speaking the petroleum and coal-based fuels already discussed were vegetable in origin, but this section is concerned with fuels derived from vegetable crops by human intervention.

As Diesel foresaw [1] the use of vegetable oils as engine fuels was widely exploited (e.g. [12, 13]), but improving communications made it possible to supply petroleum products to even the most remote areas at reasonable prices; at the same time it became possible to export the vegetable oils at advantageous prices for use in food products, etc. This trend was virtually complete by 1950, but some far-sighted countries saw vegetable oils as an emergency fuel buffering them against external political and economic pressures.

The past 10 years have seen a renewed interest in the use of vegetable oils as fuels for diesel engines, joined by studies of the use of light alcohols derived from vegetable crops. These alcohols have long been a component of gasoline owing to their high octane numbers, but this inevitably implies low cetane numbers so that they were not considered as diesel fuel components in days of plenteous alternatives. In many ways these possibilities parallel those available with coal. Vegetable products can either be processed to yield highly refined, closely defined fuels (alcohols) but at the cost of appreciable capital outlay and a modest value of the efficiency of conversion ($\eta_c \triangleq 50\%$?), or relatively simple methods can be used to extract oils from suitable crops, which are then used after minimal further processing. This latter approach yields a much higher value of η_c but the product is of much more variable quality.

5.1 Alcohols as Diesel Fuels

The light alcohols (methanol, ethanol) which are readily produced from vegetable crops by distillation (and also from solid, liquid or gaseous fuels) are now receiving wide attention as alternative fuels fro gasoline engines, particularly in Brazil where the combination of climatic and economic circumstances is especially favourable; but that story is beyond the scope of this chapter.

In spite of their very low cetane numbers (of the order of zero to five) these same alcohols are being considered as components for diesel fuels, particularly in the energy hungry U.S.A. (e.g. [11, 14]). The possibility of developing the spark-assisted diesel engine to operate on neat alcohols does not appear to be under development, but other well-known avenues of attack upon low cetane fuels are being considered.

The alcohol can be emulsified with the diesel fuel and then injected in the normal manner; alternatively, dual injection can be used, the alcohol being either injected into the cylinder by a second high pressure system, or into the inlet manifold by a low pressure system; in either case a charge of diesel fuel is used to initiate the combustion process. Slight increases in engine thermal efficiency (η_e) have been reported under high load conditions, but the low value of η_c is usually ignored [14, 15]. However, the onset of misfire limits the use of alcohol to about 30% (on an energy basis).

The other possibility is to use ignition-improving additives, but the very low cetane number of the base fuel makes additive concentrations of the order of 10% necessary. The economics of this approach would therefore appear to be doubtful.

Overall, the spark-assisted diesel would appear to be the best prospect, but light alcohols derived from crops are not seen as major fuel sources except in certain favoured areas such as Brazil. On the other hand such fuels derived from natural gas in locations from which it is not economic to export the gas as such, may make a useful contribution in the gasoline field.

5.2 Vegetable oils as Diesel Fuels

In contrast to the light alcohols, the more common vegetable oils (e.g. sun flower oil) have quite reasonable ignition qualities (cetane numbers over 30), but they are slower burning, and on a chemical basis far more complex and variable than the relatively simple alcohols.

Interest in these fuels is again widespread [14, 15] and their significance may be greater than that of vegetable-based alcohols owing to the relatively simple methods of production, which lend themselves more readily to small-scale local activities.

Some industrialized countries see these fuels as possible adjuncts to other energy supplies, but, as with the alcohols, the vast areas required to grow a significant energy crop relative to the needs of an industrialized country must limit their contribution. A more realistic approach is to regard such fuels as an emergency energy source to maintain the operation of farms during periods of fuel shortage; it has been estimated that a typical U.S. farm could make itself energy self-sufficient by putting 10% of its area under an oil crop, but this would not take the farm produce beyond the gate.

At the other end of the scale of affluence are many emergent countries whose economies have been shattered by recent rapid increases in oil prices, and which are in desperate need of energy to build up their standards of living to something approaching comfort. For them, 'home grown' oil fuels could be a considerable boon.

At first consideration, in the light of the experience of 40 years ago [12, 13] such a reversion to vegetable oils might appear to be straightforward, but this is not the case. In the past the engines used in remote areas where robust low-speed units, usually running at less than 1000 rev/min. Consequently combustion time was long, and cylinder volume large for a given power output; thus fuel properties had comparatively little influence on engine performance. Today, engines running at 1500, 2000 or more rev/min. are used, bought because they were lower in capital cost than lower-speed units, and totally adequate for use with petroleum fuels. However, such units are inevitably more sensitive to fuel characteristics, and most tend to show early signs of distress when operated on vegetable oils.

The most common oil, and that upon which most work has been done, is sun flower oil, some properties of which are given in Table 2. (Other common oils are peanut, linseed, rape; this list could easily be extended to 20 or more by including all the oils for which engine experience exists.)

Table 2.
Typical Properties of Sunflower Oil and Diesel Fuel

	Sunflower oil	Diesel fuel
Density (kg/l)	0.92–0.93	0.83–0.85
Kin.visc. at 20°C	65–75	4–6
(mm^2/s) 40°C	30–35	2.5–4
100°C	7–13	1–2
Net calorific value (MJ/kg)	36.8–37	42.7–43
Cetane number	38–47	45–55

Data from several sources.

The markedly higher viscosity of vegetable oils can cause problems in fuel handling, pumping, atomization and fuel jet penetration, but these can be overcome by redevelopment of the engine fuel system, making use of techniques which are already familiar to those having experience of diesel engine operation with high viscosity fuel oils.

The slower burning characteristics of vegetable oils can give rise to exhaust smoke, fuel impingement on cylinder walls and lubricating oil contamination. In direct injection engines this tends to lead to additive depletion and polymerization of the lubricant. Redevelopment of the combustion system to speed up air–fuel mixing should counter these problems. In direct injection engines of at least one type such difficulties have not arisen, an engine having been operated for over 2000 hrs. on neat sun flower oil without problems, whereas typical DI engines are handicapped by nozzle fouling and lubricating oil contamination after 200–400 hrs. operation on a 80% diesel/20% sun flower oil blend. The reasons for this difference no doubt lie in the marked differences between the two types of combustion system. The IDI nozzle with its single relatively large orifice is less sensitive to fouling than the three or four-hole nozzles used in DI engines; the air motion is far more violent than in a DI engine; the mean temperature in the prechamber is lower than in a DI chamber owing to operating over-rich at high load, thus reducing the danger of fuel break-down in the nozzle due to its inferior thermal stability. On the other hand it must not be assumed that all IDI engines are superior to all DI engines with regard to lubricating oil contamination, since one well-known IDI is notorious for its oil contamination when burning petroleum fuels and is used as a standard test engine for this characteristic.

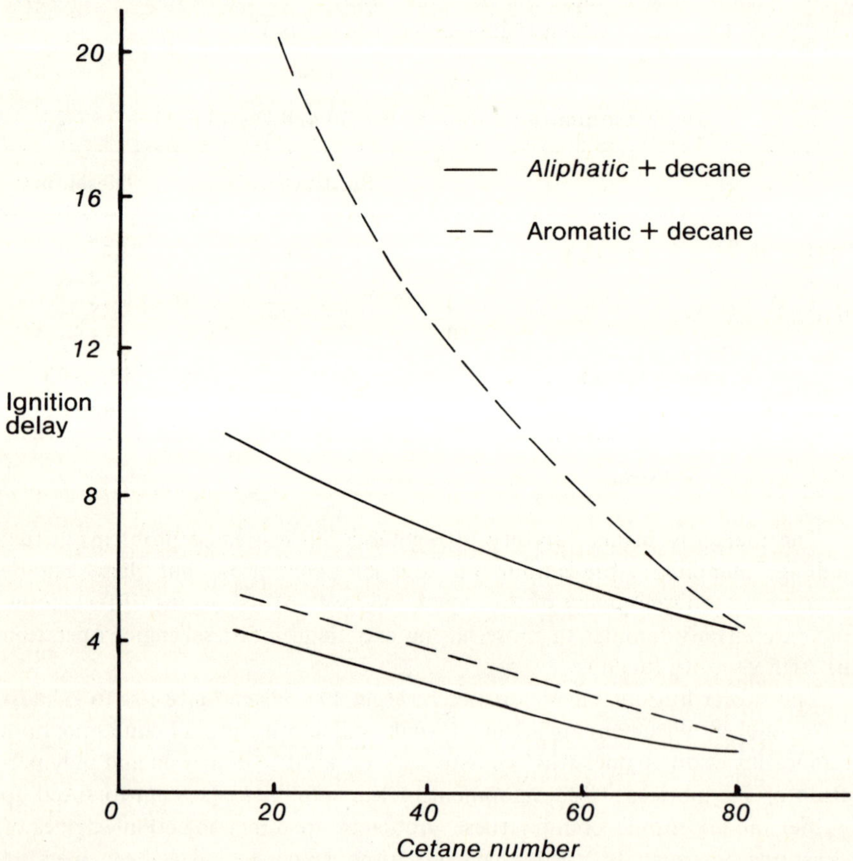

Fig. 4 – Relationship between cetane number, temperature, fuel type and ignition type.

The developed world wishes to maintain its present standards of engine reliability and availability when using vegetable oils. At the same time it does not wish to abandon the superior fuel economy of the DI engine. Considerable efforts are therefore being expended on means of modifying the fuel to overcome the shortcomings referred to (e.g. [15]). For example trans-esterification shortens the molecular chain length from about 57 to about 20, thus reducing viscosity and improving thermal stability. (For comparison, a typical petroleum diesel fuel chain length is about 16.) This process requires careful control, and can only be considered as practical in areas where chemical engineering skills are available; it is receiving careful study in South Africa, Australia and the U.S.A. The value of η_c does not appear to be known.

In the Third World a very different situation exists, requiring a different approach. Unskilled manpower is abundant, and can be used to keep engines running, even if it means more attention to the cleaning of filters and nozzles than would be tolerated in developed areas. Complex chemical processing of small locally produced batches of oil is not practical because of need for capital outlay and skilled manpower. Therefore for these territories it would appear that we should work out technically-simple, manpower-intensive solutions of the problems, involving the minimum capital outlay.

6. FUEL SPECIFICATIONS

The most unexpected outome of the present uncertain fuel situation, so far as most mechanical engineers are concerned, is the need to reconsider the validity of the time-honoured fuel specifications and test method applicable to liquid diesel fuels.

The physical limits, such as viscosity and cold filter plugging point, which determine the mobility of the fuel are adequate, as are the criteria for stability, corrosivity and cleanliness. The flash point is meant to set a criterion for safety in storage and handling, but some years ago questions were raised concerning the validity of the rather artificial test method; however, no action followed.

In contrast with the above generally satisfactory situation, it has been found that the combustion characteristics of fuels are not adequately defined by current specifications. Under the stable fuel supply situation prior to the mid-1970s, engineers became accustomed to the fuel supplied to them, and the chemistry of the combustion process was ignored; surveys of diesel combustion research at that time (e.g. [16]) showed almost complete neglect of chemical factors. Fuels of widely differing chemical compositions can satisfy current specifications, yet yield a range of engine performance data, not all of them satisfactory.

Consider, for example, the cetane number criterion for ignition quality, which has been mentioned frequently in this chapter. The engine determined cetane number is measured by comparing the ignition delay of the test fuel with those of blends of two reference fuels, using a specified test engine under defined operating conditions. If a change of engine or operating conditions is made, the measured cetane number may alter; in other words the cetane number is not a unique defintion of the ignition quality of the fuel. The data shown in Fig. 4 give a clear indication of the variation of correlation between cetane number and ignition delay when either compression temperature or fuel composition is varied [17].

Many fuel test laboratories lack the standard test engine, and quote ignition quality in terms of one of the many statistical correlations based upon properties of the fuel which are readily measured in the laboratory; unfortunately the results obtained are often quoted as 'cetane number' with no indication of the formula used in their calculation. Some of these correlations can be quite reliable (errors of less than 2 numbers) provided that they are used only on fuels lying within the range of chemical compositions from which they were deduced. For example, the Calculated Cetane Index is quoted by both ASTM and BSI, and is given as a convenient nomogram in Fig. 5; the only input data are 50% boiling temperature and specific gravity. The limitations of this reputable formula can readily be seen, if it is used to estimate the ignition quality of gasoline or sun flower oil.

That the Calculated Cetane Index and other empirical formulae should be of limited validity is only to be expected, since the few input data (usually 2 or 3) can hardly reflect the full chemical complexity of a normal commercial fuel.

Today, with the wide availability of various chromatographic techniques for the in-depth analysis of fuels, and of computer facilities of prodigious abilities, it should be possible to evolve an overall correlation between the chemical properties of any fuel and its ignition delay under any physical conditions to be met in an engine. With such a tool to hand, design and development engineers could tackle the fuel problems of the future with greater confidence.

The combustion characteristics of the fuel, once ignited, are specified by an indication of the distillation range of the fuel and by the Conradson carbon residue; the former gives an impression of the likely burning rate and of the likelihood of incomplete combustion whereas the latter is claimed to give an approximate measure of the amount of unburnt carbon to be expected. But the Conradson test, although convenient for laboratory use, bears practically no resemblance to combustion conditions in an engine, and is therefore of very limited value in characterizing a fuel. Therefore, as in the case of ignition delay, the engineer is gravely handicapped in his attempts to predict the performance of novel fuels, and needs in-depth correlations between fuel properties, in-cylinder conditions and combustion rate. Fortunately some such work is in hand, but the total effort allocated to it is woefully small.

7. CONCLUSIONS

The modern high-speed diesel engine is capable of bearing out the predictions

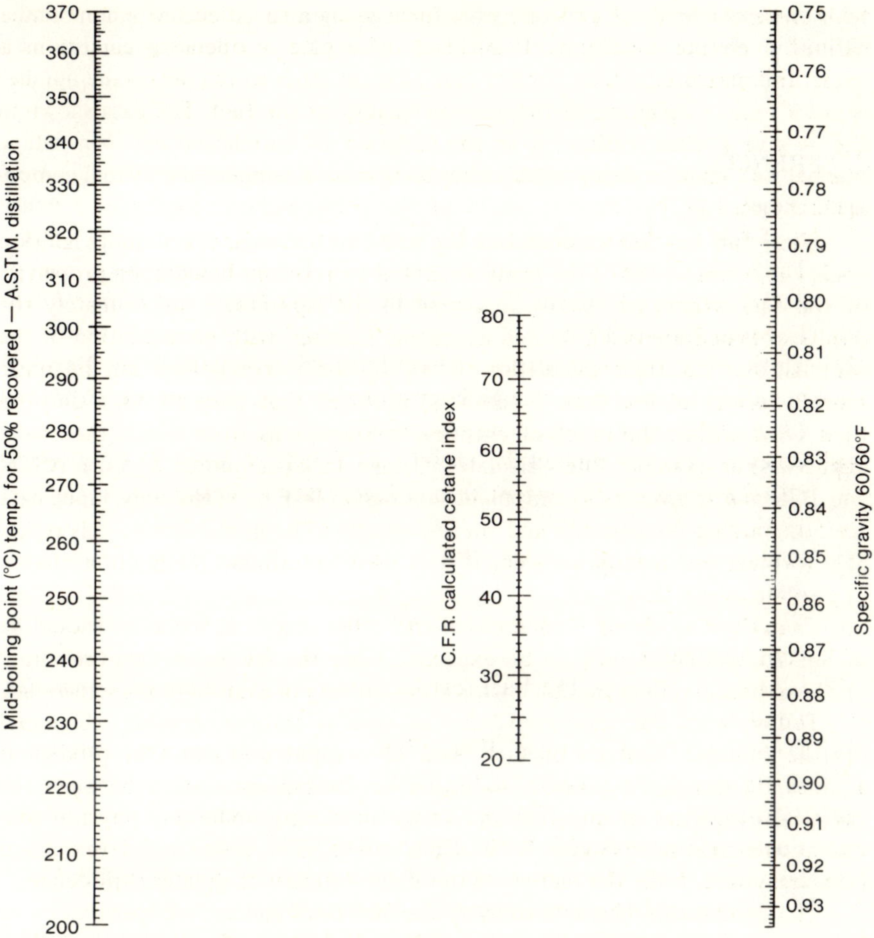

Fig. 5 – Nomogram for deriving CFR Calculated Cetane Index from specific gravity and mid-boiling point. (Derived from original in *Oil and Gas Journal*, 1.6.50, p. 81).

made by Dr. Diesel 70 years ago, in that it is capable of efficient operation using a very wide range of fuels of petroleum, coal or vegetable origins.

Combined development of engine and fuel can increase the overall thermal efficiency of utilization of natural energy resources.

In the 1940s the gas turbine was seen as the omnivorous engine of the future. 'Any fuel, even candle-ends', said one enthusiast, but the next decade showed the gas turbine to have a rather delicate digestion – and it remains so. We now see the diesel engine re-instated as the omnivorous engine, as well as being the motive power unit of highest proven thermal efficiency and in-service reliability. Several problems remain and these offer challenging careers in a range of disciplines.

REFERENCES

[1] Diesel, R., 'The Present Status of the Diesel Engine in Europe', *Lecture to A.S.M.E.,* 30.4.1912.

[2] Tipler, W., 'Prospects for the Operation of Diesel Engines upon coal or its Derivatives', I. Mech. E. Conference: *Power Plants and Future Fuels,* London, January 1975.

[3] Tipler, W., 'The Significance of Fuel/Engine Interactions in the Development of the Stratified Charge Engine', *I. Mech. E. Conference: Stratified Charge Engines,* London, November 1976.

[4] Coley, T. and Drummond, J., 'Diesel Fuel Trends', *I. Mech. E. Conference: Diesel Engines for Passenger Cars and Light Duty Vehicle,* London, October 1982.

[5] Schleyerbach, C. G., *et al.,* 'Performance of Future Type Diesel Fuel', *ibid.*

[6] Tipler, W., 'Energy Economics of Automative Power Generation', *S.A.E. Paper 750761.*

[7] Gibson, J., 'Coal – The Fuel for the Future', *I. P. Forum on Syncrudes,* London, 6.5.81.

[8] Davies, G. O. and Freese, R. G., 'The Preparation and Performance of Coal Derived Diesel Fuel', *CIMAC,* 1981, Paper D74.

[9] Tierney, W. T., *et al.,* 'Energy Conservation Optimisation of the Vehicle–Fuel–Refinery System', *S.A.E. Paper 750673.*

[10] Soehngen, E. E., 'Development of Coal-burning Diesel Engines in Germany', *U.S. Dept. of Commerce Report FE/WAPO/3387-1.*

[11] Ryan, T. W. and Worthen, R. P. (editors), *Alternate Fuels, S.A.E. SP-480,* 1981.

[12] Walton, J., 'The Fuel Possibilities of Vegetable Oil', *Gas and Oil Power,* **33**, July 1938, p. 167.

[13] Chowhury, D. H., 'Indian Vegetable Fuel Oils for Diesel Engines', *Ibid,* **37**, May 1942, p. 80.

[14] Bernstein, L. S. (Editor), 'Alternative Fuels for Diesel Engines', *S.A.E. SP-503,* 1981.

[15] Various: *National Conference on Fuels from Crops,* S.A.E. of Australia, September 1981.

[16] Various: I. Mech. E. Conference; *Combustion in Engines,* Cranfield, July 1975.

[17] Private communication to the author.

3

Heat transfer in diesel engines

NESSIM HAY, Senior Lecturer, Department of Mechanical Engineering, University of Nottingham

---

1. INTRODUCTION

Heat transfer plays a major role in the design and development of diesel engines, and particularly in determining the life of components and therefore reliability and availability of the engine and its life span between major overhauls. Most common failures can be traced back to a heat transfer cause. Premature failures of exhaust valves are due to overheating. Failures of pistons are due to thermal fatigue. Failures of piston rings are due to sticking which is itself due to overheating. Corrosion can be caused by overcooling, and so on.

In this chapter the main areas in diesel engineering in which heat transfer plays an important role are outlined and the main methods of approach for solving the heat transfer problems that arise in these areas are presented.

The chapter has the following structure:

First a quick review of the necessary heat transfer background is given.
Then the types of problems which arise in diesel engines are identified.
A more detailed treatment of each type of problem follows with illustrative examples drawn from the author's work or from other published work.
Finally, broad directives are given on how to tackle heat transfer problems which arise in diesel engines.

2. HEAT TRANSFER BACKGROUND

Let us start from a very simple situation, a plane wall with fluid flowing on either side in steady state as shown in Fig. 1.

With the wall the Fourier heat conduction equation holds:

$$q = kA\,\frac{T_{w_1} - T_{w_2}}{L} \tag{1}$$

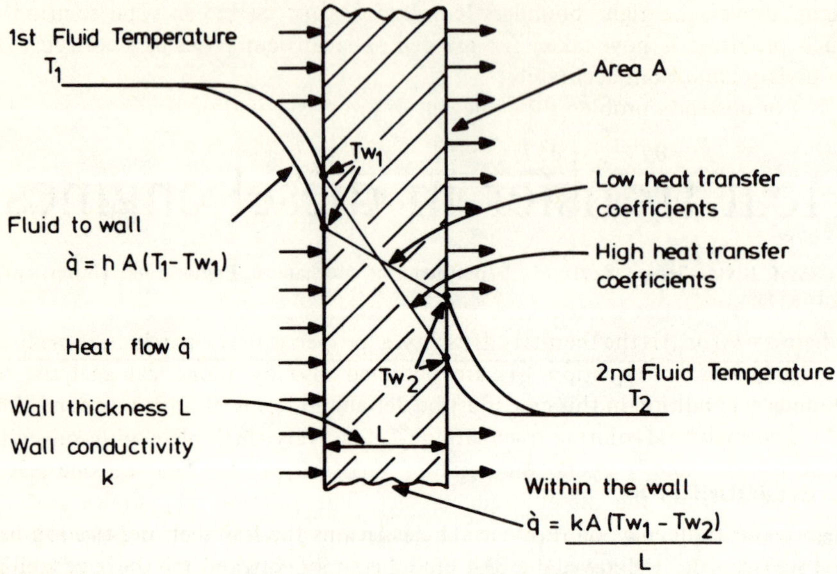

Fig. 1 – Heat transfer through a plane wall.

On either side of the wall Newton's heat convection equation holds:

$$\dot{q} = h_1 A(T_1 - T_{w_1}) = h_2 A(T_{w_2} - T_2) \qquad (2)$$

h_1 and h_2 are the heat transfer coefficients for the particular flow situations at the two walls.

Note that if h_1 is high the temperature difference $(T_1 - T_{w_1})$ will be small and vice versa. Thus the temperature distribution within the wall depends on the values of h_1 and h_2 as illustrated in Fig. 1.

The important conclusion from this is that in conduction problems the boundary conditions have a strong bearing on the solution, and the solution is only right if the boundary conditions are right.

For two or three-dimensional steady conduction problems the situation is much the same. The Fourier equation becomes

$$\frac{\partial^2 T}{\partial x^2} + \frac{\partial^2 T}{\partial y^2} + \frac{\partial^2 T}{\partial z^2} = 0 \qquad (3)$$

and the solution *for a given set of boundary conditions* yields the temperature distribution within the solid (e.g. piston). The boundary conditions are usually in the form of specified temperatures and heat transfer coefficients. The solution is obtained by numerical analysis using either finite difference or finite element or hybrid approaches. Computer codes for this purpose abound. But to get the right answer the right boundary conditions must be fed in. The solution of such problems is now taken for granted. The difficulty lies in identifying the boundary conditions accurately.

For unsteady problems the relevant differential equation is:

$$\rho c_p \frac{\partial T}{\partial t} - k \left[\frac{\partial^2 T}{\partial x^2} + \frac{\partial^2 T}{\partial y^2} + \frac{\partial^2 T}{\partial z^2} \right] = 0$$

i.e.

$$\frac{\partial T}{\partial t} = \alpha \nabla^2 T \tag{4}$$

where $\alpha = k/(\rho c_p)$ is the thermal diffusivity, a 'property' of the conducting medium.

Solutions to equation (4) are obtained also by numerical analysis; the boundary condition in this case must be defined in terms of both space and time.

The analytical solution to equation (4) for a very thick plate with convective heat transfer and a cyclic temperature variation of the fluid on one side is interesting from the diesel engine aspect. The solution [1] shows that the higher the frequency of the thermal oscillations the less their penetration into the surface. Also the lower the Biot modulus ($= hL/k$, where k is the conductivity of the metal) the more the thermal oscillations are damped at the surface. For a typical high-speed engine, the temperature wave over a cycle would be a mere 10°C in amplitude and would only penetrate to a depth of 2 mm in the metal. Thus a steady state approach to the temperature distribution within the metal is a valid one except of course when there is a transient due to a change in loading which will be of a much lower frequency.

3. TYPES OF HEAT TRANSFER PROBLEM WHICH ARISE IN DIESEL ENGINES

These can be broadly classified as

(a) Heat transfer rates for cycle calculations
(b) Thermal loading calculations (pistons, liners, cylinder heads)
(c) Performance of individual components (exhaust valves, fuel injectors, cooling systems)
(d) Improvements to current designs of engines
(e) Problems relating to low heat loss engines.

4. HEAT TRANSFER RATES FOR CYCLE CALCULATIONS

The calculation of the instantaneous work done within the cylinder by applying the 1st law of thermodynamics requires knowledge of the instantaneous heat transfer rate to the walls. This has occupied researchers for many decades. The most favoured correlation for the heat transfer coefficient is that of Woschni [2], which gives the space averaged heat transfer coefficient in the cylinder as a function of temperature and pressure (i.e. crank angle) and constants which relate to the various stages of the cycle and to the geometrical configuration of the engine. The correlation allows the calculation of the load on the cooling system as well.

Although this correlation is very useful it lacks spatial detail. To get over this difficulty results of local and instantaneous heat flux measurements such as those given in [3] can be used to introduce a weighting of the heat transfer coefficient for different locations within the cylinder.

5. THERMAL LOADING CALCULATIONS

Thermal loading calculations are necessary to predict the life of the components and to ascertain that no premature failure occurs. The calculation yields the thermal stresses within the component. The first step in a thermal loading analysis is the determination of the temperature distribution within the component. This could be a steady state or a load transient analysis for which either equation (3) or equation (4) is used. In either case it is necessary at the outset to specify the boundary conditions.

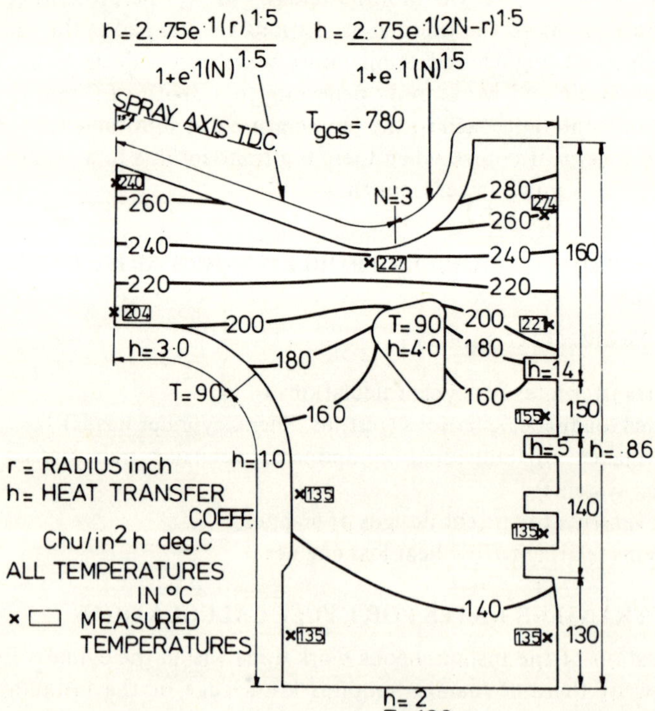

Fig. 2 – Calculated temperature distribution (250 mm dia. piston). (Reproduced from [4], Courtesy of CIMAC.)

If we take, for example, the piston shown in Fig. 2 [4] we require a fluid temperature and a heat transfer coefficient on each and every surface. Since cyclic effects are very small we can assume a mean temperature and heat transfer coefficient. For the top land of the piston we obtain a mean temperature from the cycle calculation and a mean heat transfer coefficient from the Woschni correlation. Spatial variation could be introduced as explained in the last paragraph.

Temperatures and heat transfer coefficients need to be specified for all the other surfaces. This is done by studying the fluid flow next to the surface, searching the literature for a heat transfer correlation that applies to that fluid flow situation, and deriving the heat transfer coefficient from the correlation.

To illustrate how this is done we will take as an example the 'cocktail shaker' cooling gallery (see Fig. 2).

A study of a motion picture of the cocktail shaker action revealed that for some of the time the fluid impinged onto the surface of the gallery and for some of the time it simply flowed along the surface as shown in Figs. 3 and 4.

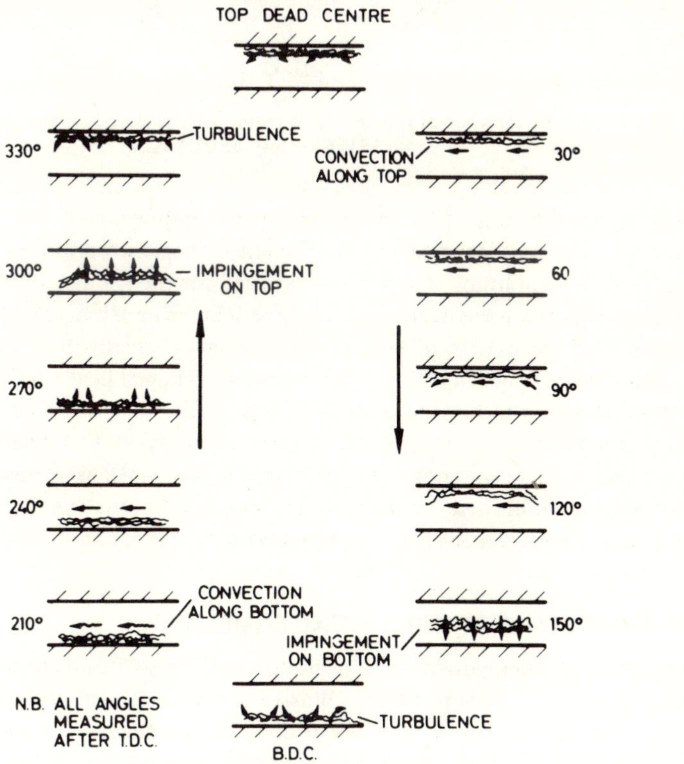

Fig. 3 – Fluid flow regimes for the closed gallery. (Reproduced from [5], Courtesy of CIMAC.)

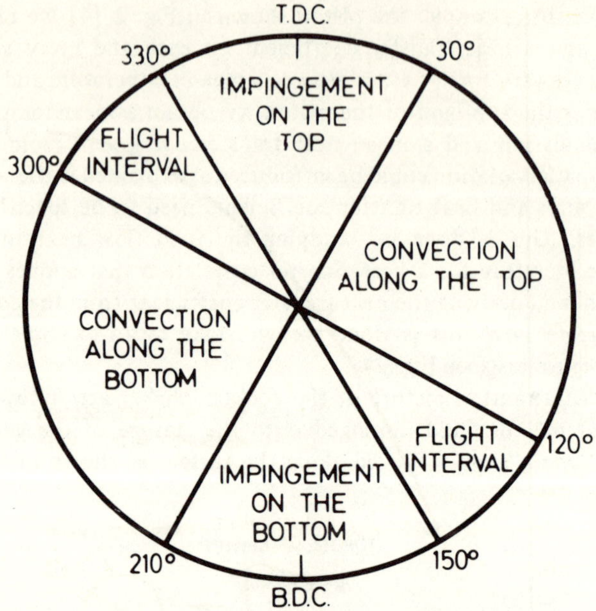

Fig. 4 – Cyclic variation of the fluid flow regimes in the closed gallery. (Reproduced from [5], Courtesy of CIMAC.)

Hence, as described in [5], a correlation for impingement heat transfer was applied to yield the heat transfer coefficient during the period of impingement, and a 'flow in an annulus' correlation was used for the simple convection period. Each heat transfer coefficient was weighted in proportion to the time it was operative and the extent of surface over which it acted, both being obtained from the study of the fluid flow. The resultant averaged heat transfer coefficient is compared with the measured value in Fig. 5. It is obvious that the prediction is very good. Knowing which correlation to apply is the major step in this process and some awareness of the heat transfer field as a prerequisite. The review papers published in the proceedings of the four-yearly International Heat Transfer Conferences form a good starting point for any search.

6. PERFORMANCE OF INDIVIDUAL COMPONENTS

The type of problem under this heading could arise from unduly severe conditions or unusual operation. To illustrate this, take the case of the cooled injector nozzle. This problem arises in dual fuelled engines (unusual operation). The throughput of fuel through the nozzle is small so additional cooling is necessary. This is done using a water jacket as shown in Fig. 6. The nozzle needs to operate within a narrow range of temperature as under-cooling leads to

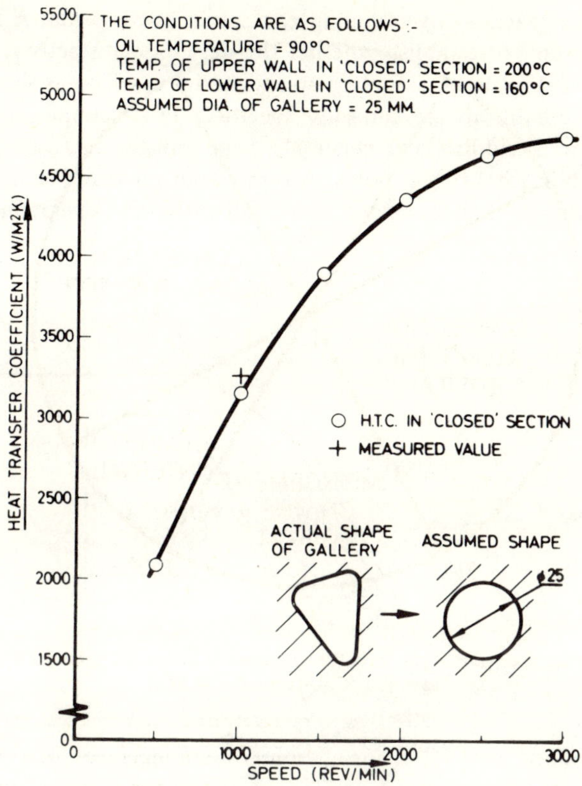

Fig. 5 – Variation of the heat transfer coefficient with rotational speed for 240 mm dia. closed gallery piston. (Reproduced from [5], Courtesy of CIMAC.)

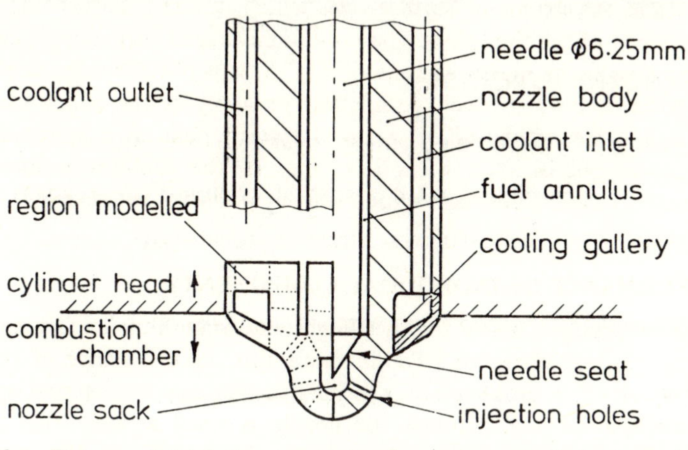

Fig. 6 – Water cooled injector nozzle showing region modelled. (Reproduced from [6], Courtesy of Hemisphere Publishing Corporation.)

softening of the seat and over-cooling can result in corrosion. A heat transfer analysis is necessary to establish whether the design is satisfactory. The analysis is described in [6]. Briefly, a finite element model is set up as shown in Fig. 7. The boundary conditions are carefully specified. For example, in the cooling gallery the flow is swirling and hence the heat transfer coefficient is obtained from a correlation for swirling flows. A typical output is shown in Fig. 8 which incidentally indicates that the temperatures fall within the acceptable limits.

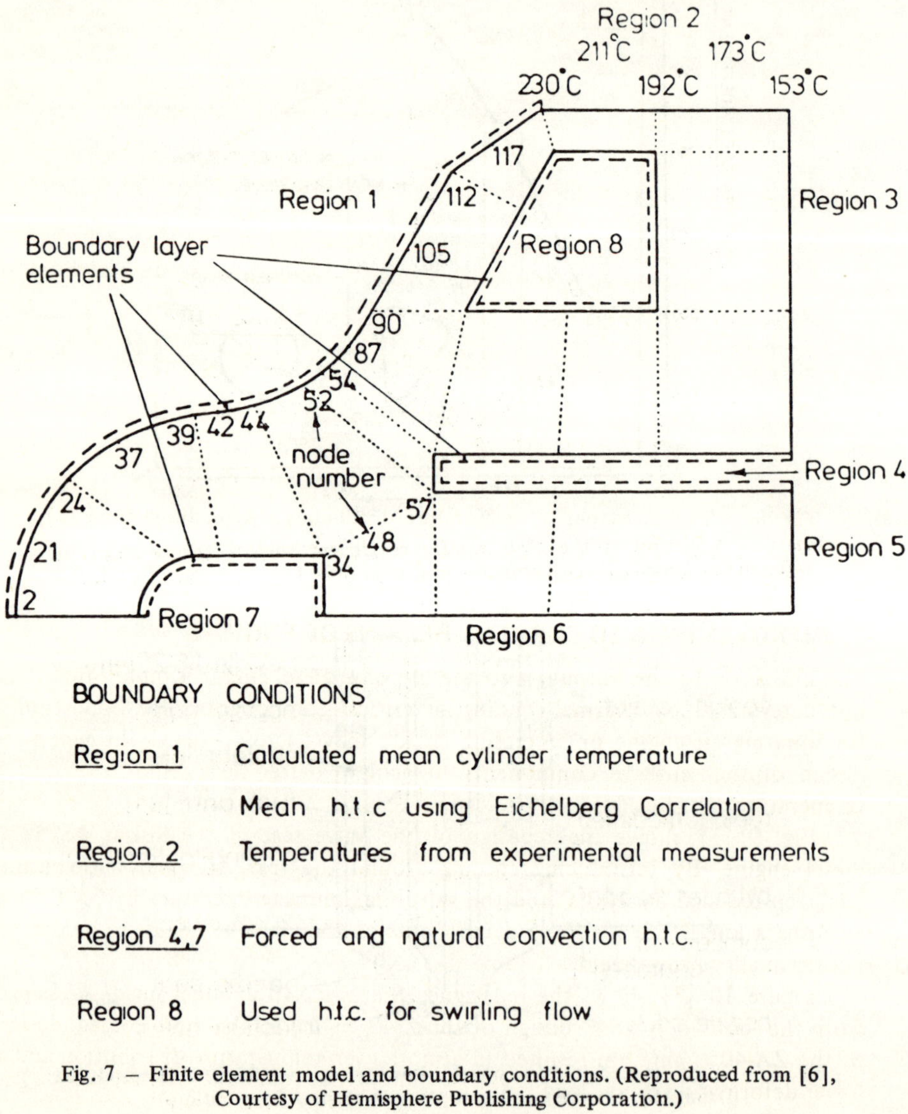

Fig. 7 – Finite element model and boundary conditions. (Reproduced from [6], Courtesy of Hemisphere Publishing Corporation.)

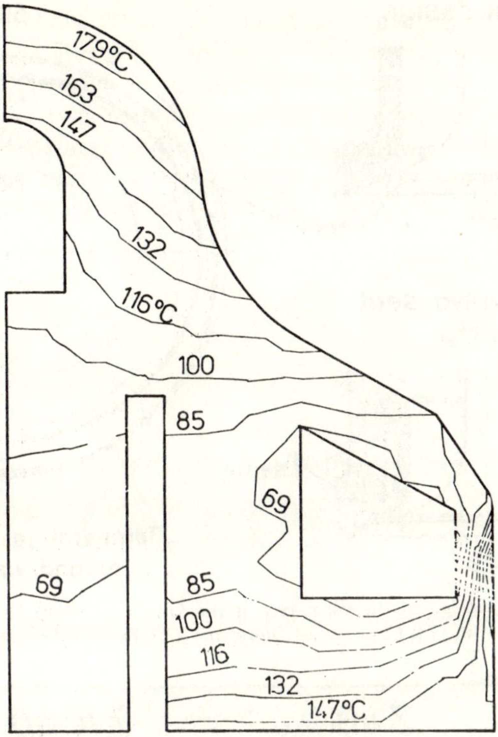

Fig. 8 – Typical output of temperature distribution. (Reproduced from [6], Courtesy of Hemisphere Publishing Corporation.)

7. IMPROVEMENTS TO CURRENT DESIGNS OF ENGINES

The limitation to the output or to the life of a diesel engine is more often than not determined by thermal loading of critical components. The opportunity for uprating an engine or increasing its life may therefore exist through a redesign of the sensitive components to achieve better heat removal or better temperature distribution.

Fig. 9 [7] shows the redesign of the valve seat of the Sulzar AS25/30 diesel engine. By introducing cooling around the seat, the seat temperature has been reduced by 100°C and the valve maximum temperature by 33°C thus ensuring a longer life for these components or allowing the engine to be run on poorer quality heavy fuels.

Figure 10 [8] shows the redesigned Sulzar ZA40 cylinder head. A change from the 'double bottom' design of the Z-40 engine to the 'bore cooled' design of the ZA-40 engine has resulted in a more even temperature distribution and a lower deformation under load.

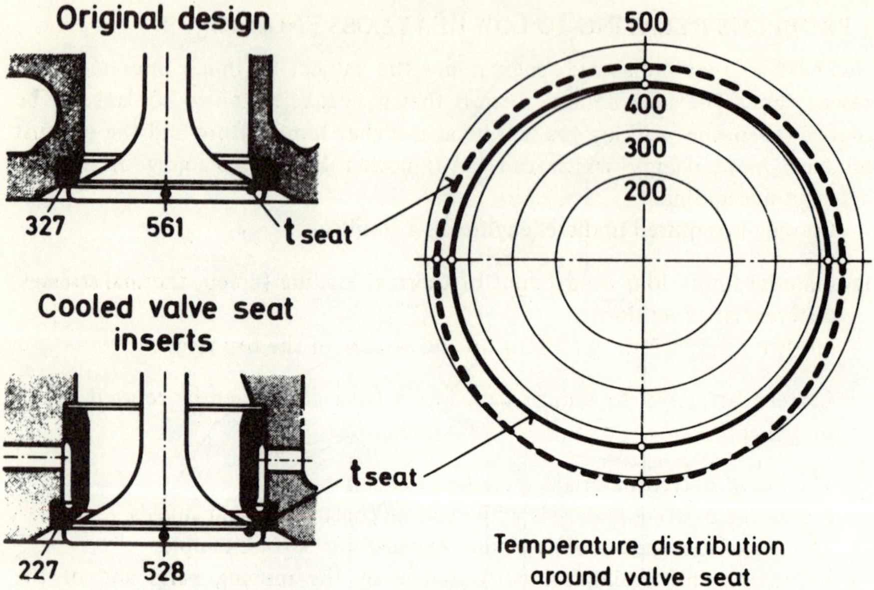

Fig. 9 – Redesign of the valve seat of the Sulzar AS25/30 diesel engine. (Reproduced from [7], Courtesy of *Diesel and Gas Turbine Progress Worldwide.*)

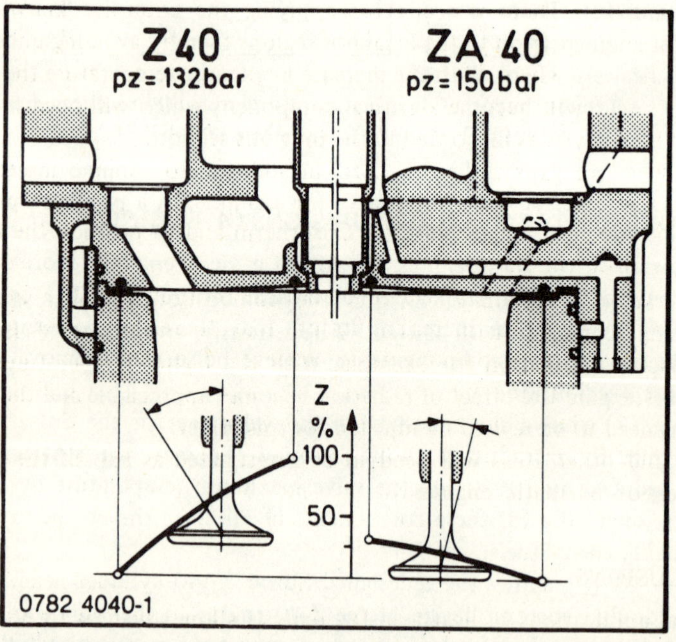

Fig. 10 – Redesign of the cylinder head for the Sulzar ZA 40 diesel engine. (Reproduced from [8], Courtesy of *Diesel and Gas Turbine Progress Worldwide.*)

8. PROBLEMS RELATING TO LOW HEAT LOSS ENGINES

Low heat loss or 'adiabatic' engines are the subject of much interest at the present time. The idea behind them is that if heat rejection to coolant can be reduced, then the working gas will be at a higher temperature and the exhaust will have more energy which can be exploited in turbocharging or through engine compounding.

Cooling is required in diesel engines as a result of

(a) material limitations on account of thermal loading (creep, thermal stresses, differential expansion),
(b) lubricant limitations (gumming up and seizure of the top ring).

Current attempts to reduce heat losses have consequently centred round the following:

(a) The use of better materials, metallic or ceramic.
(b) Protecting existing materials with ceramic coatings or heat shields.
(c) The use of localized cooling, round exhaust ports, for example.
(d) Improving the tribilogical performance of the moving parts and of the lubricant to allow operation at higher temperatures.

Heat transfer aspects figure under all these headings and in particular under (b) and (c). There is scope for applying the extensive knowledge of heat transfer augmentation to the local hot regions thereby avoiding unnecessary rejection of heat elsewhere. With the increase in exhaust temperature the exhaust valve and its seat will become a critical component which will need individual treatment and analysis as illustrated in the previous section.

The use of air gaps as heat barriers in addition to ceramic inserts forms another area of interest. It is necessary, for example, in a piston so protected to determine the position of the 200°C isotherm and to position the top ring below it.

Heat transfer by radiation within the combustion chamber in current engines is small compared with convective heat transfer and is included indirectly in the Woschni correlation for example. With the higher temperatures in the low heat loss engine the effect of radiation becomes appreciable and the current correlations need to be revised or adjusted for this effect.

These and other areas will need to be investigated as part of the development process of 'adiabatic' engines.

9. CONCLUSIONS

The judicial applications of heat transfer analysis allow advances to be made in diesel engine designs and provide the tools for the solution of heat transfer problems that arise in diesel engines. The analysis and solution tools are well developed. The main difficulty lies in identifying the problem and in inputting the right boundary conditions for its solution. The following method of approach is recommended once the problem has been identified:

(a) Analyse the heat transfer situation and establish a realistic model.
(b) Simplify the model if possible.
(c) Identify the boundary conditions to be applied.

Where a heat transfer coefficient is required for a boundary condition:

(a) Analyse the fluid flow situation.
(b) Find a correlation that is relevant to the particular fluid flow situation.
(c) Make sure that the correlation is correctly applied to derive the heat transfer coefficient.
(d) Weight the heat transfer coefficient if necessary in accordance with the time and area over which it is operational.

The solution to the problem can then be obtained using an appropriate computer code.

REFERENCES

[1] Eckert, E. R. G. and Drake Jr., R. M., *'Heat and Mass Transfer'*, McGraw-Hill, 1959, pp. 99–108.

[2] Woschni, G., 'A Universally Applicable Equation for the Instantaneous Heat Transfer Coefficient in the Internal Combustion Engine', *SAE Paper No. 670931*, 1967.

[3] Yoshida, M., Harigaya, Y. and Sato, K., 'Variation of Heat Flux through a Combustion Chamber Wall of Pre-chamber Type Diesel Engine', *Bulletin of JSME*, **25**, No. 201, March 1982, pp. 426–437.

[4] Munro, R. and Griffiths, W. J., 'Diesel Piston Design and Performance and Prediction', *Proceedings of the 11th CIMAC Congress*, Barcelona, 1975.

[5] Evans, G. A. and Hay, N., 'Heat Transfer Model for the Cocktail Shaker Piston', *Proceedings of the 13th CIMAC Congress*, Vienna, 1979.

[6] Hay, N. and Sheldon, K. G., 'Heat Transfer in a Cooled Diesel Injector Nozzle', *Proceedings of the 6th International Heat Transfer Conference*, **2**, pp. 1–5, Toronto, 1978.

[7] 'Economic Diesels for Inland Waterways', *Diesel and Gas Turbine Progress Worldwide*, Nov. 1979, p. 80.

[8] 'Sulzar Z-40 Special Supplement', *Diesel and Gas Turbine Worldwide*, Jan–Feb. 1983.

4

The governing of diesel engines

G. C. GANT, Chief Engineer, Regulateurs Europa Limited

1. INTRODUCTION

For many years the governing of engines was considered something of a black art. It has since been put on a more scientific footing by the application of control theory but in practice there is still a need for the 'sprinkling of holy water' in some of the more difficult applications. The problems arise because an engine is far from being a simple analytic device. It has many delays, lags, non-linearities and limitations in its characteristics which make it difficult to control.

Governing is essentially a special branch of control theory dealing with systems usually referred to as regulators. A regulator is a system or device for maintaining some controlled variable (e.g. engine speed) constant. This is in contrast to a servomechanism which usually adds precision, speed of response and 'muscle' to a variable action. A regulating system will often, in fact, employ a servomechanism within it to give it the necessary response and power to achieve the required regulating performance. Such is the case with most modern engine governors.

Before proceeding any further, let us ask the question: 'Why is a governor necessary at all?'

In a nutshell, it is necessary for two basic reasons:

(a) to provide accurate speed control;
(b) to protect the engine against damaging operating conditions.

Referring to common experience, a small petrol engine, such as in a motor car, does not have, nor need, a governor. It is under the direct control of the driver who sets and controls the operating conditions that he wants and the engine speed is self-regulating through the throttling action of the carburettor. Interestingly, simple governors are to be found on some petrol driven lawn mowers. On these, speed is detected by a vane in the cooling air stream which, balanced against a spring, operates on the carburettor butterfly valve.

The fuel and air supplies in a diesel engine, on the other hand, increase with speed and so it tends very readily to run away and in extreme circumstances will destroy itself. Applications employing large diesels normally require very accurate control of engine speed, load, etc. together with in-built protection against adverse operating conditions which cannot be controlled directly by the user. A properly designed governing system will provide all the necessary functions and the engine design will include an independent overspeed trip to save the engine if the governor fails.

This book deals specifically with diesel engines, therefore all the following discussions deal only with the governing of such engines. In particular this chapter deals in detail with the governing of the larger medium and slow-speed diesels for which the governing requirements are most stringent.

Small automotive types of diesel usually require less precise governing performance such as can be provided by relatively simple mechanisms. Some attention is given to those devices later in the chapter.

Other prime movers, notably gas turbines, steam turbines, water turbines, windmills, dual fuel and spark-ignited gas engines, also employ governors, often much the same units as used on diesels but each also has its own peculiarities and requirements which will not be covered here.

To keep this text within reasonable bounds a general appreciation of the fundamentals of control systems and the relevant branches of mathematics has had to be assumed.

2. THE DIESEL ENGINE AS A CONTROL ELEMENT

2.1 Introduction

Before commencing on a detailed description of governors and governing it is necessary first to discuss the fundamental characteristics of the diesel engine viewed as an element in a control system.

As already touched upon, the diesel engine is a very complex device for which it is difficult to write a comprehensive mathematical model. The problem is doubly difficult in the case of turbocharged engines, as most large diesels are today, and several universities have devoted considerable effort to the problem over the years.

A really detailed mathematical model is fortunately not necessary for a basic understanding of governor action, and system stability can be assessed fairly readily using quite a simple 'small signal' model such as will be described.

The complex models are needed for large transient analysis (i.e. speed changes due to large load changes) in which turbocharger characteristics predominate and many other non-linearities, etc. have a significant effect.

In practice it is not normally necessary for governor manufacturers to do theoretical transient analyses, the problem being dealt with sufficiently by empirical methods based on actual test results.

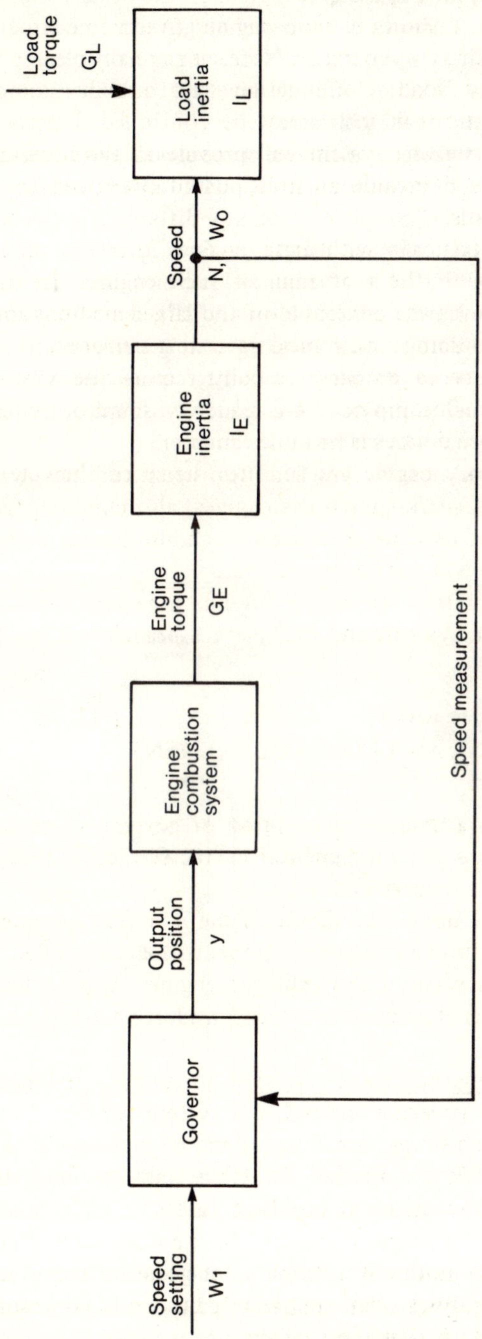

Fig. 1 – Basic engine/governor control loop.

2.2 Mathematical Model for Engine

The block diagram Fig. 1 shows a basic engine/governor control loop with a solidly coupled load such as a generator. Note, engines coupled to loads and to each other in parallel via flexible couplings give rise to further complex stability problems which will not be dealt with here.

In this model the engine comprises a combustion system which produces a torque G_E as a function of governor output position y ($\equiv$ fuel rack position).

Torque G_E provides load torque G_L and any difference between G_E and G_L accelerates (or retards) the combined inertia $I = I_E + I_L$.

It can be seen from Fig. 1 that engine, load and governor are all closely linked to make up a complete control loop and the characteristics of all parts are equally important in determining good governing performance. A first class governor cannot be expected properly to control an engine with slack linkage between governor and fuel pump or one cylinder misfiring or a bad drive to the governor or with any of many other possible faults.

Consider the arbitrary engine and load torque/speed characteristics shown in Fig. 2.

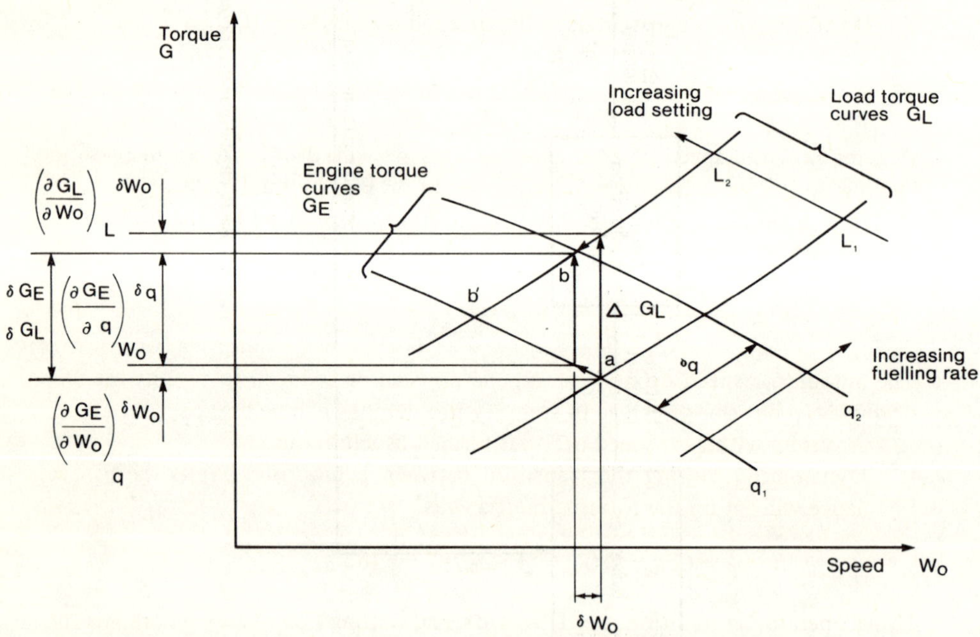

Fig. 2 – Torque/speed curves.

If when running at equilibrium point 'a' the load setting is changed by a small amount ΔG_L (e.g. by a small change in pitch of a marine propeller) and the engine is governed such that the fuelling rate is varied by an amount δ_q in response, then a new equilibrium point 'b' will be achieved associated with a small speed change $\delta\omega_0$.

The final changes in torque are given by:

$$\delta G_E = \left(\frac{\partial G_E}{\partial \omega_0}\right)_q \delta\omega_0 + \left(\frac{\partial G_E}{\partial q}\right)_{\omega_0} \delta q$$

and

$$\delta G_L = \Delta G_L + \left(\frac{\partial G_L}{\partial \omega_0}\right)_L \delta\omega_0$$

For equilibrium $\delta G_E = \delta G_L$ and writing

$$\left(\frac{\partial G_L}{\partial \omega_0}\right)_L = A, \left(\frac{\partial G_E}{\partial \omega_0}\right)_q = B, \left(\frac{\partial G_E}{\partial q}\right)_{\omega_0} = C$$

then

$$(A - B)\,\delta\omega_0 = C\delta q - \Delta G_L \tag{1}$$

This equation shows the steady state speed change $\delta\omega_0$ resulting from changes in fuelling rate δq and loading ΔG_L.

If fuelling rate is constant as in ungoverned engine, $\delta q = 0$, then

$$\delta\omega_0 = \frac{-\Delta G_L}{(A - B)} \tag{2}$$

showing how the speed of an ungoverned engine will 'droop' for an increase in loading. In this case the new equilibrium point on Fig. 2 would be 'b'.

In a governed engine this steady state droop is reduced by the increase in fuelling rate by δq then

$$\delta\omega_0 = \frac{C\delta q - \Delta G_L}{(A - B)} \tag{3}$$

The amount of speed droop with a governor can be reduced to zero by making $C\delta q = \Delta G_L$ but this gives rise to stability problems as will be discussed later.

A system with zero speed droop is termed 'isochronous'.

Dynamically, during the transition between equilibrium points the torque imbalance will act on the rotating inertia thus:

$$B\delta\omega_0 + C\delta q - \Delta G_L - A\delta\omega_0 = I\frac{\mathrm{d}}{\mathrm{d}t}(\delta\omega_0)$$

Using operator D notation, i.e. $D = \mathrm{d}/\mathrm{d}t$, and writing $C\delta q = K\delta y$ where y is the governor output position then:

$$\delta\omega_0 = \frac{K\delta y - \Delta G_L}{[I\mathrm{D} + (A - B)]} \tag{4}$$

This is the simple basic transfer function for a governed engine on load.

Note that if $A = B = 0$ the engine transfer function is effectively:

$$\frac{\delta\omega_0}{\delta y} = \frac{K}{ID} \qquad (5)$$

showing the engine to be nearly a pure integrator. In practical terms an integrator will continue to increase its output (i.e. speed in the case of an engine) as long as there is input (excess fuel). This explains why a diesel engine, in which normally $B \triangleq 0$, has a tendency to run away if not governed.

2.3 Firing Delay

A very significant feature of a diesel engine, from a control point of view, is the discontinuous manner in which power is produced owing to the sequential firing of a relatively small number of cylinders. This feature is important for two reasons.

First, it means that there is a time delay between the action of the governor in demanding a change in fuelling rate and the response of the engine to that change. Such a delay is classically destabilizing in a control system. Secondly, the engine crankshaft does not rotate at a uniform speed but rather experiences a cyclic variation in torque which gives rise to a cyclic variation in speed. A sensitive governor will respond to these cyclic speed variations and exhibit a small rythmic movement (jiggle) at its output lever. Such movements are very undesirable because of the wear induced in the governor and fuel pump mechanism, and normally have to be filtered out by a mechanism in the governor drive.

The *effective* firing delay has been found empirically to be the actual time between consecutive pistons arriving at the injection point plus approximately a quarter of a revolution of the crankshaft.

The effective firing delay is therefore approximately

$$T_F = \frac{60s}{2Nn} + \frac{60}{4N} \text{ seconds} \qquad (6)$$

where

$s = 2$ or 4 for two-stroke or four-stroke engine
$N =$ speed in rev/min.
$n =$ number of cylinders

This delay is included in the transfer function for the engine combustion system such that

$$\frac{\delta G_E}{\delta y} = K e^{-T_F D} \qquad (7)$$

See standard texts for transfer function of a delay.

The transfer function for a governed engine is therefore

$$\delta\omega_0 = \frac{Ke^{-T_F D}\,\delta y - \Delta G_L}{[ID + (A - B)]} \qquad (8)$$

for which a block diagram can be drawn as shown in Fig. 3.

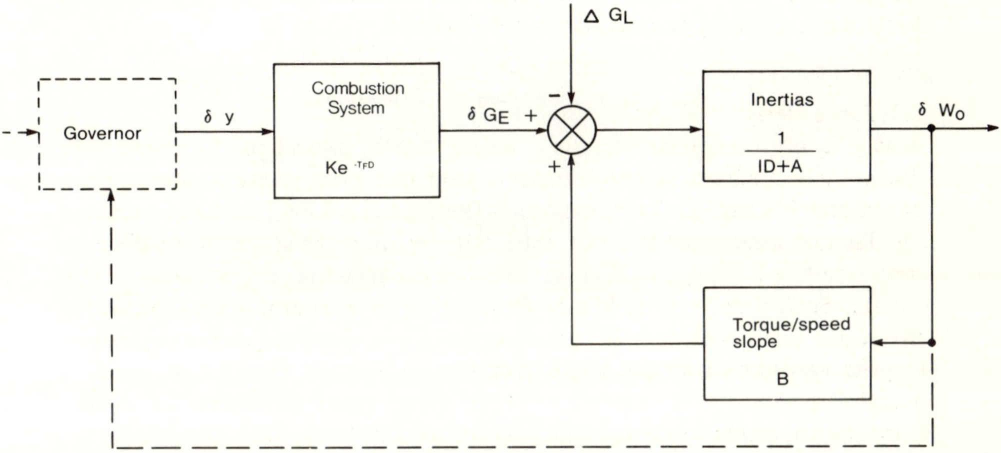

Fig. 3 – Engine block diagram.

2.4 Engine Acceleration Rate

The acceleration rate of the engine, taken as the percentage of full speed per second that the engine accelerates, at full speed and full load torque, is an important guide to the 'governability' of the engine. An engine capable of accelerating at above 50%/s. can be difficult to govern stably, although satisfactory governing above 100%/s. is commonly achieved. High acceleration rates are experienced in declutched marine propulsion engines which have small flywheels for weight and cost saving.

A method of calculating the acceleration rate is shown in Section 4.6 under Transient Response.

2.5 Typical Example of an Engine Model

Consider the following 16-cylinder diesel engine coupled to a typical generator.

Total inertia $I = 270$ kg m^2
Full load power $P = 1.5$ MW
Full (synchronous) speed $N = 1200$ rpm, $\omega_0 = 126$ rad/s.
16-cylinder four-stroke

No load–Full load governor movement – 15 mm

Firing delay equation (6) $T_F = \dfrac{60 \times 4}{2 \times 1200 \times 16} + \dfrac{60}{4 \times 1200}$

$= 0.00625 + 0.0125 = 18.75$ ms

Full load torque $G_E = \dfrac{P}{\omega_0} = \dfrac{1.5 \times 10^6}{126} = 11900$ NM

Acceleration rate $= \dfrac{G_E}{I\omega_0} = \dfrac{11900 \times 100}{270 \times 126} = 35\%/\text{s}.$

Torque/Rack gain K $= \dfrac{G_E}{y} = \dfrac{11900}{15} = 793$ Nm/mm

Torque/Speed slope B $= \left(\dfrac{\partial G_E}{\partial \omega_0}\right)_q$ approx. at full load –27 Nm/rad/s. from engine curves.

Generator load with automatic voltage regulator gives

$$A = \left(\frac{\partial G_L}{\partial \omega_0}\right)_L \simeq 0$$

Hence the engine transfer function at constant load ($\Delta G_L = 0$) is, from equation (8),

$$\frac{\delta \omega_0}{\delta y} = \frac{793 e^{-0.01875D}}{(270D + 27)} \text{ rad/s/mm}$$

In dimensionless form the gain

$$= \frac{793}{27} \times \frac{y}{\omega_0} = 3.5$$

$$\frac{\delta \omega_0}{\delta y} = \frac{3.5 e^{-0.01875D}}{(10D + 1)} \qquad (9)$$

The frequency response represented by this equation is shown in Fig. 4.

It can be seen that the engine is effectively an integrator except at very low frequencies. This means that the engine speed would be limited on run-away but in practice at a speed well above that at which the mechanism would fail.

The effect of the firing delay is also clearly to be seen. This is manifested as a rapidly increasing phase lag with frequency, above about 1.5 Hz.

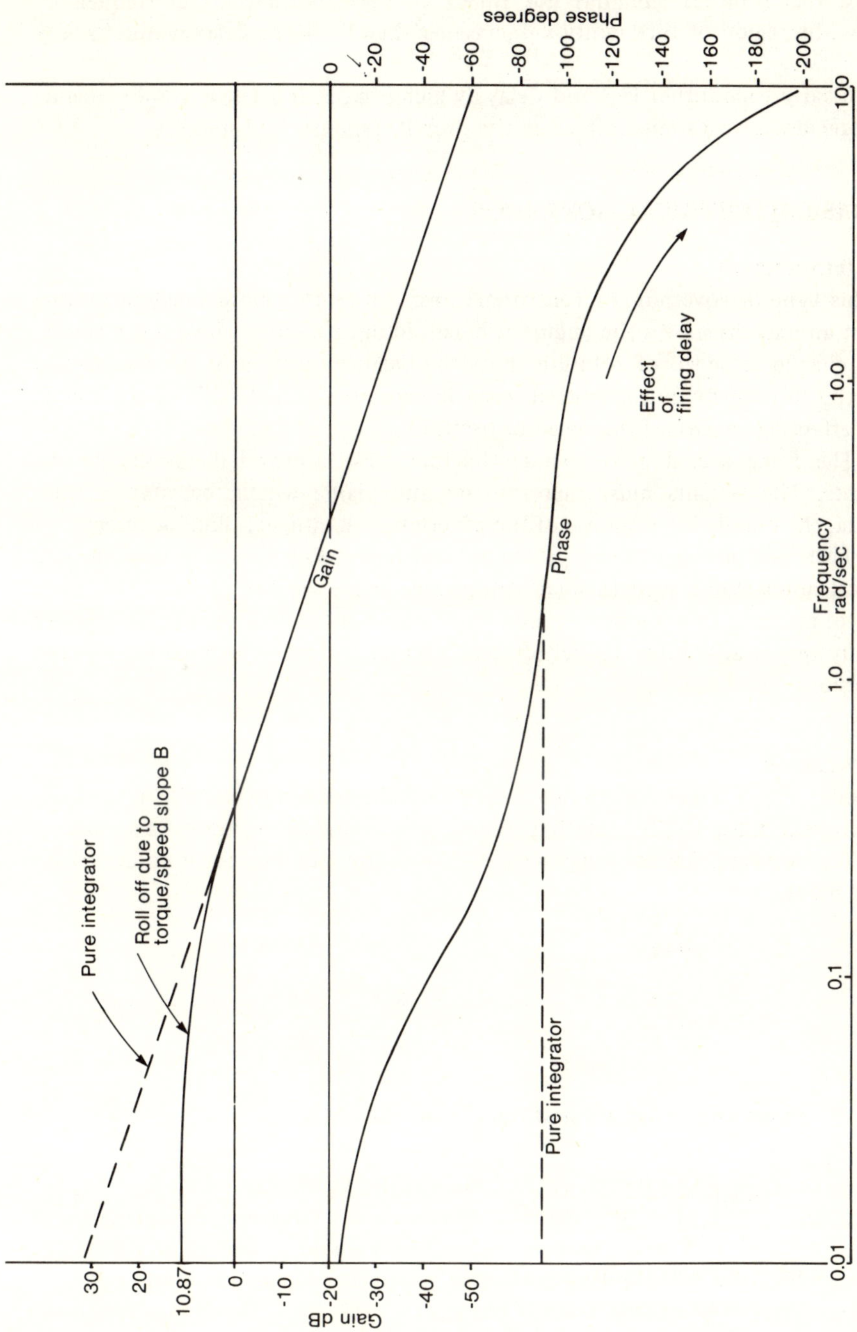

Fig. 4 – Engine frequency response.

If the governor gain has not rolled off very substantially at frequencies below the region of this additional phase lag then the firing delay would be very destablizing.

Clearly then, other lags and delays which can occur in the governor–engine linkage, etc. have a similar effect on the overall responses and stability.

3. BASIC MECHANICAL GOVERNOR

3.1 Introduction

In this type of governor the centrifugal force on two or more weights rotating about an axis driven by the engine is balanced against either gravity or a spring. Any change in speed of rotation shifts the balance position of the mechanism and the movement is transmitted via a linkage to the fuel valve of the engine thus effecting control of the speed of rotation.

The force available to operate the fuel valve is provided directly by the weights. The weights must therefore be quite large which, because of their inertia, then limits the response of the governor. Friction, etc. limit accuracy.

These simple mechanical principles are, however, still widely used in non-critical applications such as small automotive engines where precision of speed control and rate of response are not very important. Governors of this type are usually very neatly integrated with the fuel pump, and are termed pump-mounted governors.

3.2 Analysis

Consider Fig. 5 which shows two weights, usually termed flyweights in the U.K. and flyballs in the U.S.A., rotating about a spindle and acting against a spring.

The resolving forces along the axis of rotation (see Fig. 5 for notation) are given below.

Upwards Downwards

$$F = m\omega_0^2 \frac{q}{p} r \qquad = me \frac{d^2x}{dt^2} - f\frac{dx}{dt} + \beta(X - x) + F_L$$

where:

m_e = total effective mass referred to axis

$$= m\left(\frac{q}{p}\right)^2 + \frac{1}{3}m_s + m_v \tag{10}$$

m_s = mass of spring

m_v = mass of output rod, etc.

(and control valve in hydraulic governor, shown later)

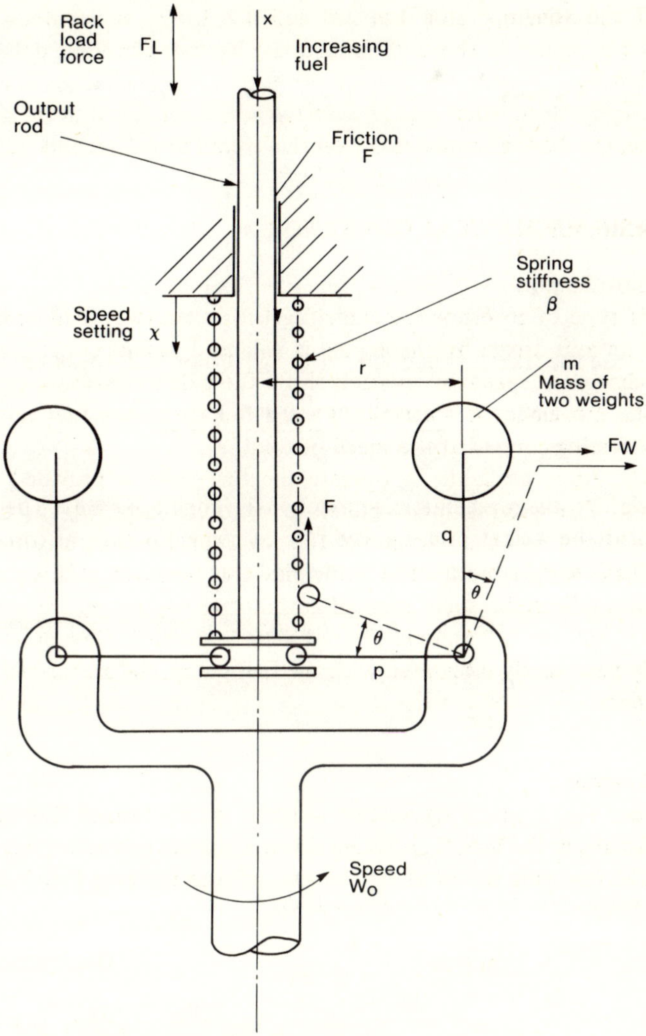

Fig. 5 – Simple mechanical governor.

See other standard texts for referred mass and effective spring mass.

and

f = viscous friction coefficient of moving parts
β = spring stiffness
F_L = load force due to fuel rack and linkage friction, etc.

Consider small changes around reference conditions * with X, the speed setting, constant and ignoring load forces.

Then, using operator D notation

$$2m\omega_0^* \frac{q}{p} r^* \,\delta\omega_0 + m\omega_0^{*2} \frac{q}{p}\, \delta r = -(m_e D^2 + fD + \beta)\delta x$$

$$\delta r = -\frac{q}{p}\,\delta x$$

Therefore

$$-2m\frac{q}{p}\, r^* \delta\omega_0 = \left[m_e D^2 + fD + \beta - m\left(\omega_0^* \frac{q}{p}\right)^2 \right] \delta x$$

Therefore the transfer function for the governor with speed setting constant is

$$\frac{\delta x}{\delta\omega_0} \quad \frac{-(2\,m\omega_0^*\, q/p\, r^*)}{[m_e D^2 + fD + \beta - m\,(\omega_0^*\, q/p)^2]} \tag{11}$$

To relate to the mathematical model of the engine, $\delta x \equiv \delta y$. The term $m(\omega_0^* q/p)^2$ is called the weight rate, i.e. the rate of change of weight force with radius, at speed ω_0^*, and may otherwise be derived thus:

Consider radial force $F_w = m\omega_0^{*2} r$ at reference speed ω_0^*. Resolving axially

$$F = F_w \frac{q}{p}$$

and

$$r = x\frac{q}{p}$$

$$F = m\left(\omega_0^* \frac{q}{p}\right)^2 x$$

The weight rate resolved axially is then

$$\frac{\partial F}{\partial x} = m\left(\omega_0^* \frac{q}{p}\right)^2 = R, \text{ say} \tag{12}$$

In the steady state, the governor transfer function is dimensionless

$$\frac{\delta x}{\delta\omega_0} = \frac{-2m\omega_0^* \dfrac{q}{p} r^* \dfrac{\omega_0^*}{x^*}}{(\beta - R)} = \frac{-K_0}{(\beta - R)} \tag{13}$$

showing the governor, in essence, to be a simple proportional device.

This term is the proportional gain of the governor and it determines the steady state change in fuelling rate for a given speed change. (Negative because increasing speed reduces fuel.)

Note that the gain is dependent on operating speed and radius; the higher the gain, the less the speed will droop as load is applied.

For a given size and geomtry of weights the gain can be increased by reducing the spring stiffness β so that in practical terms a greater output rod movement, equivalent to fuelling rate, occurs for a given speed change, i.e. the governor sensitivity is increased.

In the limit, if $\beta = R$ then the governor sensitivity will be infinite. In this case the governor will in principle increase the fuelling rate so as to maintain the speed droop zero for any load increase. Such a mode of operation is called isochronous governing. In practice this type of governor in the simple form under consideration would give instability. Any small deviation from the operating point would cause 100% change in fuelling rate and the engine would 'hunt' violently up and down in speed.

The effect of the other terms in transfer function equation (11) is to give a normal damped harmonic characteristic to the operation of the governor due simply to its being a spring/mass system with some damping due to friction.

Clearly, if m is large the natural frequency of the governor will be low, so limiting the response.

4. HYDRAULIC SERVO-ASSISTED GOVERNOR

4.1 Introduction

On large engines, where good governing is most likely to be needed, quite large forces are required to operate the fuel racks, but it is essential for accuracy and speed of response that the flyweight spring/mass system is made small and sensitive. Hence a servomechanism, usually comprising a hydraulic, spool-type pilot valve and power piston, is used to amplify the flyweight force–see Fig. 6. This basic arrangement is employed on, probably, 90% or more of the large industrial, marine and rail traction engines in the world.

Governors are usually classified in terms of the work available at the output shaft. The work output is defined as the stalled force multiplied by the full stroke of the power piston. In operation, only about 60% loading can be applied to the governor without significantly affecting its response.

Work levels range from 1 J for small engines, up to 350 J for very large engines. The most widely used mid-range sizes lie in the 10–80 J bracket for, say, 750 to 6000 kW engines.

4.2 Basic Servo Governor

In this case the flyweights operate at constant radius and any deviation from that position moves the pilot valve vertically such that the control land partially uncovers the control port passing oil to or from the power piston.

The flyweight/speeder spring arrangement is essentially identical to that described for the mechanical governor except that it is much smaller and more sensitive.

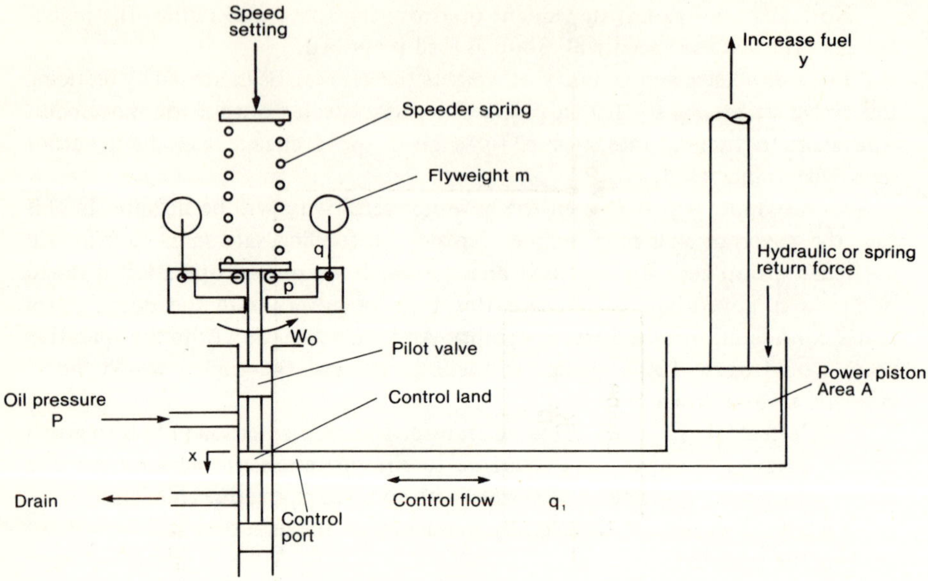

Fig. 6 – Basic hydraulic servomechanism.

Constant radius operation is advantageous because then the radius does not appear in the gain term although weight rate is still important as it detracts from the speeder spring stiffness.

For a given speeder spring force (speed setting) the weights will cause the valve to close only at one precise speed. Thus any diviation from this speed will open the valve and the power piston will continue to move, changing fuelling rate until such time as the speed is correct. This is isochronous operation and, as with a simple isochronous mechanical governor, the operation would be unstable. To see why it is unstable consider (in terms of small changes) the following.

Assume that the control flow q_1 is proportional to valve opening δx, i.e. $\delta q_1 = \alpha \delta x$, where α = valve flow coefficient $\partial q_1/\partial x$; assume also that the piston displacement rate is equivalent to the flow rate, i.e. $AD\delta y = \delta q_1$. Eliminating δq_1

$$\frac{\delta y}{\delta x} = \frac{\alpha}{AD} \tag{14}$$

Taking from the mechanical governor analysis that fundamentally $\delta x = K_g \delta \omega_0$, where K_g = governor gain = $K_0/(\beta - R)$, the simple servo governor is then

$$\frac{\delta y}{\delta \omega_0} = -\frac{K_g \alpha}{AD} = -\frac{1}{\tau D}, \text{ say} \tag{15}$$

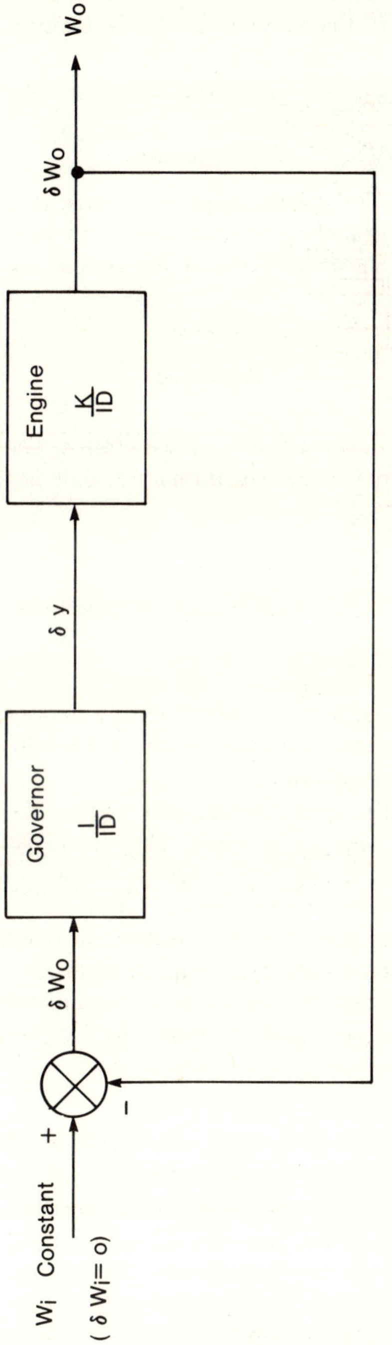

Fig. 7 – Engine/governor loop with integrating governor.

i.e. an *integrator* with integrating time τ. Note: for the isochronous case mechanical governor, when $\beta = R$ the alternative way of looking at it is to write from equation (11) thus:

$$\frac{\delta x}{\delta \omega_0} = \frac{-2m\omega_0^* \frac{q}{p} r^*}{D(m_e D + f)} \qquad (16)$$

i.e. an integrator also. The special property of integrators is one of infinite gain at zero frequency, as already argued from a different viewpoint.

It has also been shown earlier that the engine, too, is approximately an integrator, equation (5),

$$\frac{\delta \omega_0}{\delta y} = \frac{K}{ID}$$

If two integrators are connected in a closed loop, as shown in Fig. 7, then the closed loop transfer function, i.e. equation of motion, is, from standard texts,

$$\frac{\text{Output}}{\text{Input}} = \frac{G}{1 + G}$$

where

$$G = \frac{K}{ID} \cdot \frac{1}{\tau D} = \frac{K}{\tau I D^2}$$

thus

$$\left[\frac{\tau I}{K} D^2 + 1\right] \delta \omega_0 = 0 \qquad (17)$$

This is a second-order differential equation with constant coefficients and no term in D. The solution of this equation is an undamped harmonic oscillation of frequency:

$$\frac{1}{2\pi} \sqrt{\frac{K}{\tau I}} \text{ Hz} \qquad (18)$$

i.e. the arrangement would oscillate continuously up and down in speed.

4.3 Practical Servo Governor

A governor with this elementary form of hydraulic servomechanism is therefore of no use. In order to stabilize the system, some form of compensation is necessary. Two methods are normally employed involving closed loops within the governor itself.

First, a feedback of output position to speed setting, usually in the form of a mechanical linkage between power piston and the top of the speeder spring is used to impart proportional action to the servomechanism. The effect of this is slightly to decrease set speed as the output moves in the 'increase fuel' direction so deliberately causing speed droop with load to occur. This happens naturally in the simple mechanical governor owing to the change in weight radius. The feedback linkage ratio determines the amount of droop, i.e. the proportional band of the governor. For example, for good stable generating set operation a speed droop of 4% between no load and full load is most common.

Secondly, a force related to output (power piston) velocity is fed back to the pilot valve to provide a stabilizing and damping term. This function is quite complex in practice and has adjustable parameters by which governing performance can be optimized during engine tests. It is always used together with the position (droop) feedback but very importantly it does allow the position feedback ratio to be turned down to zero while retaining stability, i.e. stable isochronous operation is possible and in fact is frequently used. Two methods of 'rate' feedback are commonly employed:

(a) a pressure feedback directly onto the pilot valve;
(b) a mechanically applied force on the pilot valve via a dashpot.

Line diagrams of these two types of system are shown in Figs. 8(a) and 8(b) where the position feedback (droop) linkages are also shown.

4.4 Operation of Practical Servo Governor

The engine speed is controlled at a level determined by the balance between speeder spring force and the centrifugal force of the flyweights. Speed setting is changed by changing the compression of the spring. In practice this is achieved by various methods described later.

Referring to Fig. 8(a), consider the engine running at nominal speed such that the forces are in balance. The flyweights are vertical, the pilot valve is closed and the power piston is hydraulically locked in a position giving exactly the right fuelling rate to maintain the engine speed against the prevailing load.

Consider now a small increase in load. This will cause a small dip in speed, and owing to a consequent reduction in centrifugal force, the flyweights will move inwards slightly under the action of the speeder spring.

The pilot valve is thus moved downwards, opening the control port and allowing high pressure oil from the integral pump/accumulator system (or in simple examples, from an outside source) to pass to the power piston. The power piston in this case moves downwards against a constant returning force provided by oil pressure acting on a partial area of the underside of the power piston. A spring is sometimes used for the return force but this gives a force dependent on stroke.

Such a valve/piston arrangement is called a three-way valve system. A four-way system could be used whereby control oil is ported to either side of an

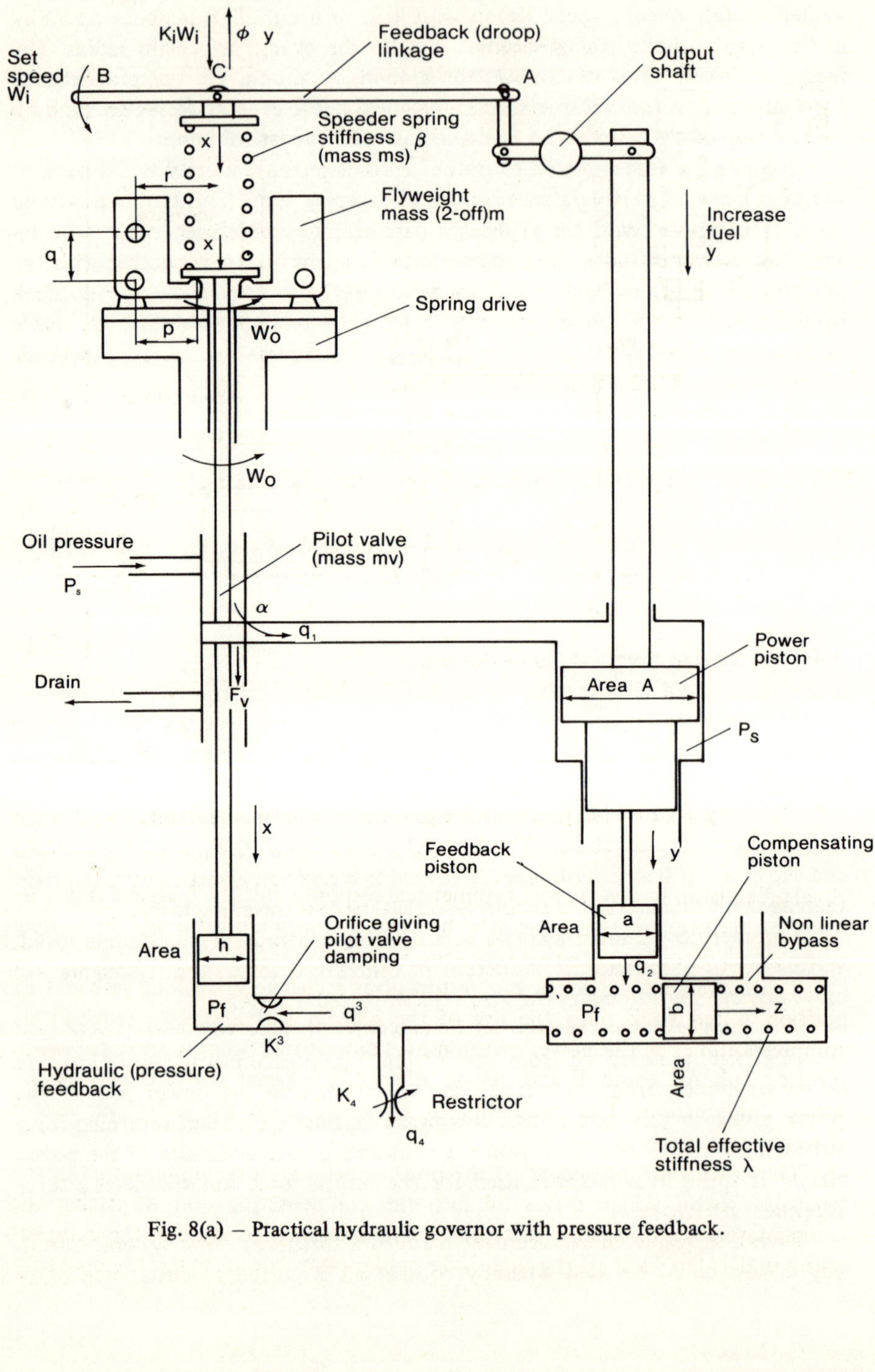

Fig. 8(a) – Practical hydraulic governor with pressure feedback.

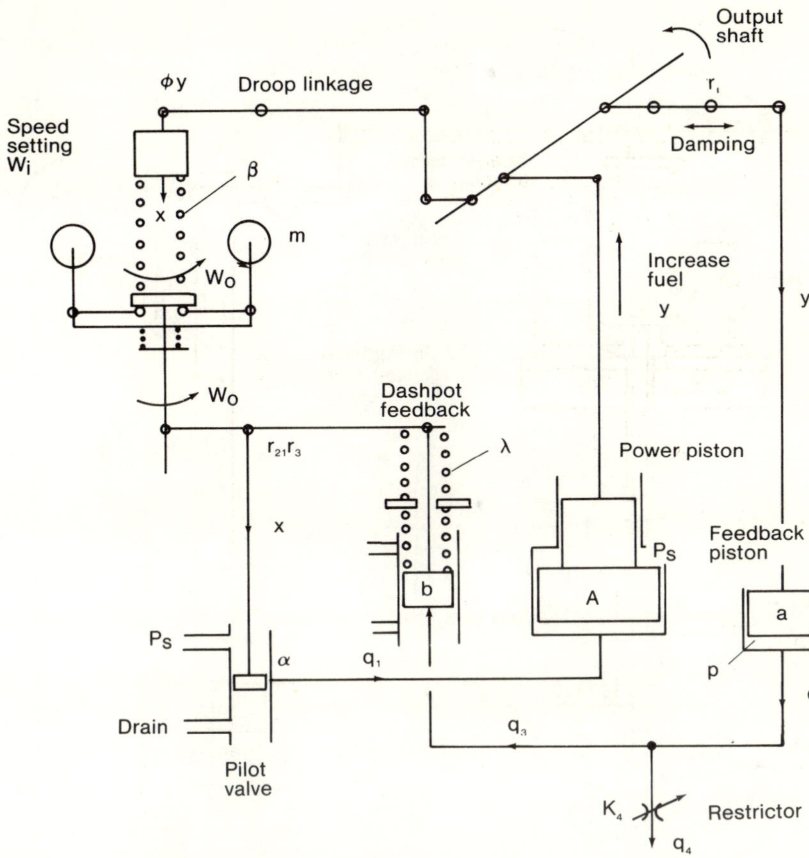

Fig. 8(b) – Practical hydraulic governor with dashpot feedback.

equal area piston to give a more symmetrical response, but the marginal improvement in performance is not worth the cost of the pilot valve and its mating sleeve which would require two very accurately positioned control lands.

The movement of the power piston gives a change in fuelling rate and its position is fed back onto the top of the speeder spring slightly reducing its compression, i.e. as the power piston moves downwards point A moves upwards pivoting link AB about B and raising intermediate point C which reduces the spring compression. This reduction in set speed gives the stabilizing droop characteristic.

The downward movement of the power piston is also communicated to the feedback piston which drives oil into the compensating unit displacing the compensating piston against the springs. The increased pressure in the compensating unit is fed back to the underside of the pilot valve forcing it temporarily

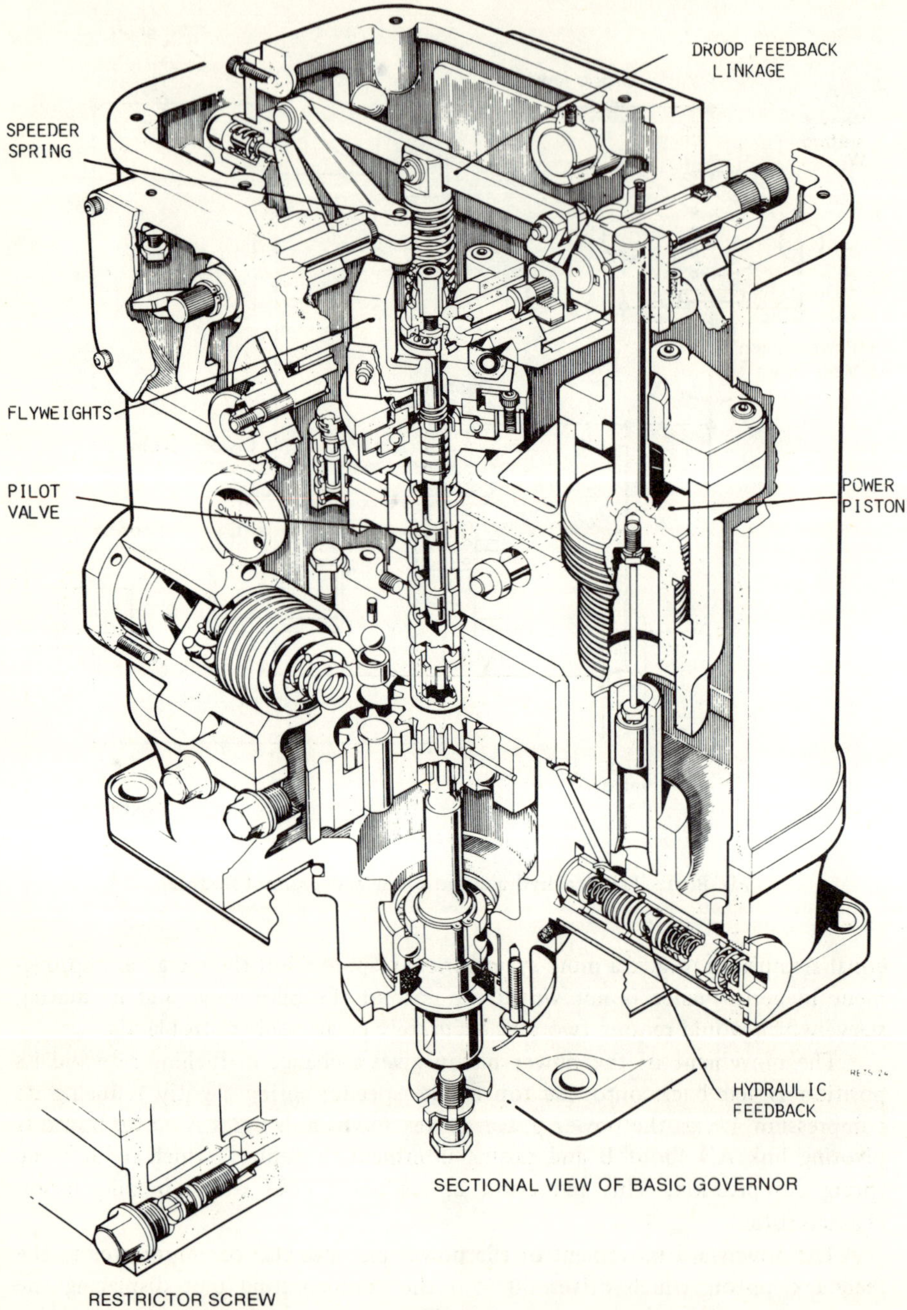

Fig. 9(a) – Sectional view of basic governor.

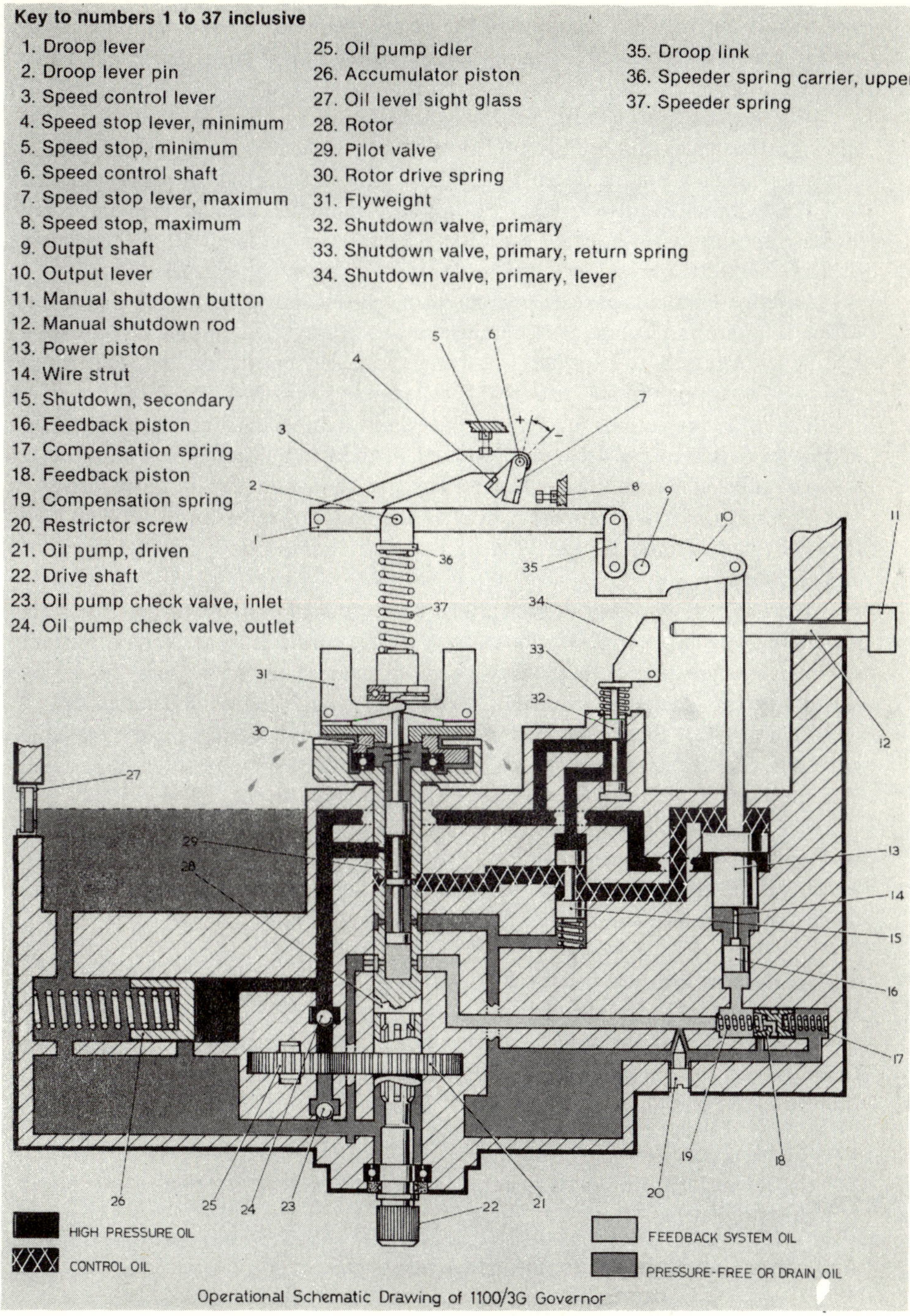

Fig. 9(b) – Operational schematic drawing of 1100/3G governor.

upwards to cut off the oil supply to the power piston. This has a damping effect on the transient action occurring in the governor and so enhances the stability of operation.

The feedback oil pressure gradually leaks away via the adjustable restrictor allowing the new equilibrium condition to be reached in a controlled fashion. The engine is now running against a slightly higher load at a slightly reduced speed.

The hydraulic feedback therefore only acts transiently and causes a temporary increase in droop to enhance stability. The rate of decay of its effectiveness through the restrictor, shapes the speed transient curve. See Fig. 13.

On large load changes the compensating piston moves further to a position where it uncovers a bypass port, momentarily negating the damping effect of the hydraulic feedback and initially allowing the speed correction to occur more rapidly so reducing the magnitude of the transient swing in engine speed.

The other very important feature necessary in diesel governors and indicated in Fig. 8(a) is a spring drive incorporated in the rotor head. This is in effect a low pass filter which attenuates the cyclic speed variations present in the drive

The spring drive comprises a light spring coupling between drive shaft and flyweight system and a vane type of hydraulic damper fed with oil by pilot valve leakage.

The natural frequency of the spring drive is selectable in the range 1.5 to 6 Hz depending on the application. Parameters of the spring drive are very important on engines operating with flexibly coupled loads, as in such cases the spring drive can either damp out or magnify instability around the flexible coupling.

Figures 8(a), 9(a) and 9(b) show a widely used commercially available governor, the Regulateurs Europa Limited type 1100 which operates as described. This unit will be used as the example for further analysis.

The operating principles of the dashpot governor shown in Fig. 8(b) are very similar, except that the pilot valve is acted upon by a mechanical linkage which senses the flyweight movement and the feedback pressure, the latter being converted into a movement by the dashpot.

4.5 Analysis of Pressure Feedback Governor

Refer to Fig. 8(a) for nomenclature.

This is a small signal analysis arranged to show the fundamentals of operation. In order not to lose sight of this objective, the analysis will be considered in three separate parts, i.e.

* the basic governor,
* the flyweight spring/mass system,
* the spring drive.

Also certain simplifying assumptions are made, i.e.

(a) effects of friction are ignored,

(b) oil compressibility is ignored,
(c) orifice characteristics are linearized (for small changes),
(d) pilot valve flow forces are ignored.

EQUATIONS OF MOTION

The basic governor

The force balance on the pilot valve is the same as that for the simple mechanical governor but with the addition of the feedback pressure P_f on the underside area h of the pilot valve.

The force balance on the pilot valve is given by

$$\text{Flyweight force } F_w + \text{Feedback force } F_f = \text{Spring force } F_s \tag{19}$$

Putting aside for the moment the mass/acceleration effects and viscous friction in the flyweight spring/mass system

$$F_w = m\,\omega_0'^2\,\frac{q}{p}\,r$$

Where ω_0' is the rotational speed of the flyweight system when loosely coupled to engine speed ω_0 by the spring drive. The feedback force, F_f is goven by

$$F_f = hP_f$$

In a modern governor the speeder spring is wound so that it has a square law relationship between force and compression to match the square law relationship between centrifugal force and speed of the flyweights. This gives constant sensitivity over the speed range. Then

$$F_s = K_s\,(X - x)^2$$

over the operating range. Also the speed setting varies with output y owing to the proportional feedback, i.e.

$$X = K_i\,\omega_i - \phi y$$

where $K_i\,\omega_i$ = speed setting and ϕ = feedback fraction. Hence

$$F_s = K_s\,(K_i\,\omega_i - \phi y - x)^2$$

Consider now small changes about reference conditions * and using operator D notation:

The flyweight force is

$$\delta F_w = K_0 \delta\omega_0' - R\delta x \tag{20}$$

where

$$K_0 = 2m\,\omega_0'^{*}\,\frac{q}{p}\,r^{*}$$

$$R = m\left(\omega_0'^{*}\,\frac{q}{p}\right)^2$$

as shown for the mechanical governor.

The feedback force

$$\delta F_\mathrm{f} = h\delta P_\mathrm{f} \tag{21}$$

$$\delta F_\mathrm{s} = \beta(K_i\,\delta\omega_i - \phi\delta y - \delta x)$$

where $\beta = 2K_\mathrm{s}\,(K_i\omega_i^* - \phi y^* - x^*)$, the spring stiffness at the operating point.

At constant speed setting $\delta\omega_i = 0$

The spring force, δF_s, is given by

$$\delta F_\mathrm{s} = -\,\beta(\phi\delta y + \delta x) \tag{22}$$

Equations (20), (21) and (22) in the force balance equation (19) give:

$$K_0\,\delta\omega_0' + (\beta - R)\,\delta x + h\delta P_\mathrm{f} + \beta\phi\delta y = 0 \tag{23}$$

The power piston motion is given by the simple flow/displacement relationship equation (14):

$$\delta x = \frac{A}{\alpha}\,\mathrm{D}\,\delta y \tag{24}$$

The feedback flow balance is given by

$$q_3 = q_2 - q_4 - b\frac{\mathrm{d}z}{\mathrm{d}t}$$

i.e.

$$-h\frac{\mathrm{d}x}{\mathrm{d}t} = a\frac{\mathrm{d}y}{\mathrm{d}t} - K_4P_\mathrm{f} - b\frac{\mathrm{d}z}{\mathrm{d}t} \tag{25}$$

where K_4 = the linearized orifice coefficient.

The force balance on the compensating piston gives

$$P_\mathrm{f}b = \lambda z$$

so

$$\frac{\mathrm{d}z}{\mathrm{d}t} = \frac{b}{\lambda}\,\frac{\mathrm{d}P_\mathrm{f}}{\mathrm{d}t} \tag{26}$$

Then equation (25) becomes, for small changes,

$$aD\,\delta y = -hD\delta x + K_4\delta P_f + \frac{b^2}{\lambda}D\,\delta P_f$$

$$aD\,\delta y + hD\delta x = \left(K_4 + \frac{b^2}{\lambda}D\right)\delta P_f \qquad (27)$$

Eliminating δx, δP_f and rearranging gives:
for $\phi \neq 0$

$$\frac{\delta y}{\delta\omega_0'} = \frac{-\dfrac{K_0}{\beta}\left(1 + \dfrac{b^2}{\lambda K_4}D\right)}{\dfrac{1}{K_4}\left[\left(\dfrac{\beta - R}{\beta}\dfrac{Ab^2}{\alpha\lambda} + \dfrac{h^2A}{\alpha\beta}\right)D^2 + \left(\dfrac{\beta - R}{\beta}\cdot\dfrac{AK_4}{\alpha} + \dfrac{ha}{\beta} + \dfrac{\phi b^2}{\lambda}\right)D + K_4\phi\right]} \qquad \text{Droop (28)}$$

for $\phi = 0$

$$\frac{\delta y}{\delta\omega_0'} = \frac{-\dfrac{K_0}{\beta}\left(1 + \dfrac{b^2}{\lambda K_4}D\right)}{\dfrac{D}{K_4}\left[\left(\dfrac{\beta - R}{\beta}\dfrac{Ab^2}{\alpha\lambda} + \dfrac{h^2A}{\alpha\beta}\right)D + \left(\dfrac{\beta - R}{\beta}\dfrac{AK_4}{\alpha} + \dfrac{ha}{\beta}\right)\right]} \qquad \text{Isochronous (29)}$$

Note that the droop governor is basically proportional, with a damped second-order response and a stabilizing phase advance term in the numerator. The isochronous governor is basically an integrator owing to 1/D term.

The adjustable restrictor in the hydraulic feedback path in effect varies the value of K_4. Closing the restrictor reduces K_4 thus increasing the integrating time and increasing the phase advance time constant, so slowing and stabilizing the response.

The steady state gain in the droop case is

$$\left|\frac{\delta y}{\delta\omega_0'}\right| = \frac{K_0}{\beta\phi} \qquad (30)$$

If this is normalized to be dimensionless using reference values y^* and $\omega_0'^*$

$$\frac{\delta y}{y^*}\,\frac{\omega_0'^*}{\delta\omega_0'} = \frac{K_0\omega_0'^*}{\beta\phi y^*} \qquad (31)$$

This is the dimensionless gain and is equal to 1/droop p.u. Thus

$$\text{droop}\,\% = \frac{\beta\phi y^*}{K_0\,\omega_0'^*} \times 100 \tag{32}$$

so for example, gain $K_0\,\omega_0'^*/\beta\phi y^* = 25$ for 4% droop.

These equations, then, describe the action of the basic governor. There are, however, two other important features in the governor which influence the overall response: the flyweight spring/mass system and the spring drive.

If we cheat slightly and consider these separately for clarity, then for the flyweight system a damped second-order dynamic term can be applied to the simple gain term thus:

$$\frac{\delta F_{\mathrm{w}}}{\delta\omega_0'} = \frac{K_0}{\dfrac{\mathrm{D}^2}{P_0^2} + \dfrac{2\zeta}{P_0}\mathrm{D} + 1} \tag{33}$$

Where p_0 = the natural frequency of flyweights against speeder spring, typically 25 Hz, and ζ = viscous damping coefficient on motion of pilot valve, etc., including effect of orifice K_3 in the feedback to the pilot valve.

Also, the spring drive comprises a mass consisting of the flyweights and carrier, etc. coupled by a spring to the drive shaft, with viscous damping between them. It is readily shown that the transfer function of such a device is of the form

$$\frac{\delta\omega_0'}{\delta\omega_0} = \frac{1 + \dfrac{2\upsilon}{\omega_{\mathrm{n}}}\mathrm{D}}{\dfrac{\mathrm{D}^2}{\omega_{\mathrm{n}}^2} + \dfrac{2\upsilon}{\omega_{\mathrm{n}}}\mathrm{D} + 1} \tag{34}$$

where

ω_0 is the actual input speed to the governor
ω_{n} is the natural frequency of the rotor head mass against the drive spring
υ is the viscous damping in the spring drive

The overall transfer function of the governor is then, approximately, for droop operation:

$$\frac{\delta y}{\delta\omega_0} = \frac{-\dfrac{K_0}{\beta}\left(1 + \dfrac{b^2}{\lambda K_4}\mathrm{D}\right)\left(1 + \dfrac{2\upsilon}{\omega_{\mathrm{n}}}\mathrm{D}\right)\dfrac{\omega_0^*}{y^*}}{\dfrac{1}{K_4}\left[\left(\dfrac{\beta - R}{\beta}\cdot\dfrac{Ab^2}{\alpha\lambda} + \dfrac{h^2A}{\alpha\beta}\right)\mathrm{D}^2 + \left(\dfrac{\beta - R}{\beta}\dfrac{AK_4}{\alpha} + \dfrac{ha}{\beta} + \dfrac{\phi b^2}{\lambda}\right)\mathrm{D} + K_4\phi\right]\left(\dfrac{\mathrm{D}^2}{\omega_{\mathrm{n}}^2} + \dfrac{2\upsilon}{\omega_{\mathrm{n}}}\mathrm{D} + 1\right)\left(\dfrac{\mathrm{D}^2}{p_0^2} + \dfrac{2\zeta}{p_0}\mathrm{D} + 1\right)} \tag{35}$$

where ϕ varies the droop (adjustable), K_4 varies the stability and response (adjustable) and spring changes λ, varying ω_n, can be made to match the governor to the engine.

Dashpot Governor

A similar equation can be written for the dashpot governor, viz. see Fig. 8(b) for nomenclature

$$\frac{\delta y}{\delta \omega_0} = \frac{-\dfrac{K_0}{\beta}\left(1+\dfrac{b^2}{K_4\lambda}\mathrm{D}\right)\left(1+\dfrac{2v}{\omega_n}\right)\dfrac{\omega_0^*}{y^*}}{\left[\dfrac{Ab^2}{r_2K_4\lambda\alpha}\mathrm{D}^2+\left(\dfrac{A}{\alpha}+\dfrac{r_3}{r_2}\dfrac{r_1ab}{K_4\lambda}+\dfrac{\phi b^2}{K_4\lambda}\right)\mathrm{D}+\phi\right]\left(\dfrac{\mathrm{D}^2}{\omega_n^2}+\dfrac{2v}{\omega_n}\mathrm{D}+1\right)\left(\dfrac{\mathrm{D}^2}{p_0^2}+\dfrac{2\zeta}{p_0}\mathrm{D}+1\right)} \tag{36}$$

It can be seen that this is a slightly simpler expression, with adjustments K_4 and r_1 having similar effects on stability and response, and ϕ adjusting droop.

Both types of governor are in widespread use, with comparable performances.

4.6 Performance

1. *Typical Governor Data*

Consider the case of a 20 J work output governor running at 1200 rev/min. with 4% droop and the feedback restrictor one turn open.

No Load–Full load stroke of power piston 15mm

ω_0^*		125 rad/s
A		$1535 \times 10^{-6}\,\mathrm{m}^2$
b		$201 \times 10^{-6}\,\mathrm{m}^2$
a		$127 \times 10^{-6}\,\mathrm{m}^2$
h		$71 \times 10^{-6}\,\mathrm{m}^2$
P_s	set for 20 J work output	$1.0\ \mathrm{MN/m^2}$
β	rate at 1200 rpm setting	2600 N/m
λ	selectable – taken as	350 N/m
m	2×0.061	0.122 kg
p		20 mm
q		10 mm

$$m_e = m\left(\frac{q}{p}\right)^2 + \frac{m_s}{10} + m_v$$ $m_s/10$ since square law spring

$$K_0 = 2m\,\omega_0'^*\,\frac{q}{p}\,r^* \quad (r^* = p) \ldots\ldots\ldots\ 0.3\ \mathrm{kg}\ \frac{\mathrm{m}}{\mathrm{sec}}$$

$$R = m\left(\omega_0'^*\,\frac{q}{p}\right)^2 \ldots\ldots\ldots\ 484\ \mathrm{N/m}$$

K_4 . . . difficult to enumerate but about $\dfrac{0.05}{10^9}$ m^5/s

$\phi = \dfrac{K_0 \omega_0'^*}{25\beta y^*}$ 0.0388

α . . . . for small openings . . . . $0.1\ m^2/s$

$p_0 = \sqrt{\dfrac{\beta}{m_e}}$, i.e. at op. point . . . . 168 rad/s (26.75 Hz)

ζ flyweight system damping coefficient, about 1.0, determined mostly by the orifice under the pilot valve in the hydraulic feedback system.

ω_n . . . . selected to be 4 Hz . . . . 25 rad/s

υ spring drive damping coefficient is about 1.0

The overall transfer function of the governor is then typically:

$$\frac{\delta y}{\delta \omega_0} = \frac{25(1 + 2.31D)}{\dfrac{D^2}{1.15^2} + \dfrac{5.07}{1.15} D + 1} \cdot \frac{(1 + 0.08D)}{\dfrac{D^2}{25^2} + \dfrac{2}{25} D + 1} \cdot \frac{1}{\dfrac{D^2}{168^2} + \dfrac{2}{168} D + 1} \tag{37}$$

Note that this 'simplified' function has three second-order lags with break points at 1.15 rad/s, 25 rad/s and 168 rad/s, and phase advance terms in the hydraulic servo with break points at 1/2.31 rad/s and in the spring drive at 1/0.08 rad/s.

The phase advance term in the spring drive is not really wanted. It arises because the rotating top is damped relative to the rotating drive whereas ideally the rotating top would be damped relative to 'earth'. This ideal is unfortunately not readily achievable in a practical mechanism of acceptable size.

In view of the complexity of this 'simple' model it is worth noting that governor manufacturers are often asked for 'the time constant' of the governor. The model shows why there is no valid answer to that question. The only simple indicator of governor response is the traverse time resulting from a large step speed change. In this case the pilot valve opens fully, taking a small time to achieve full opening, and the power piston then moves at saturation velocity. The traverse time of a typical modern governor is 0.1 s.

FREQUENCY RESPONSE

The gain and phase plots for the typical case chosen are shown in Fig. 10, where some relevant test results are also plotted as individual points for comparison. Reasonable agreement is shown with the test results considering that in practice there are very many minor factors which influence the actual performance and there are very many detail differences in governor build.

The intention here is to show the principal features and characteristics

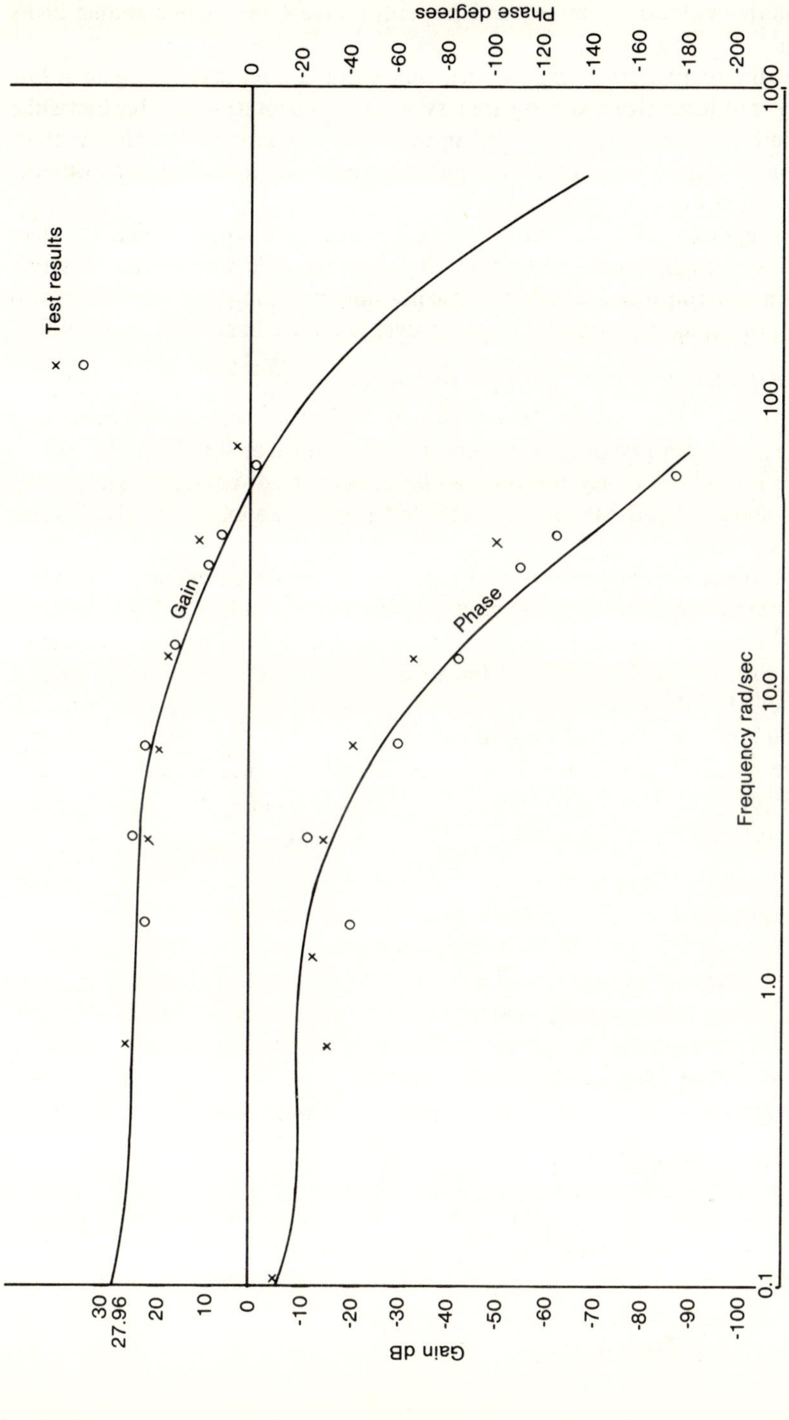

Fig. 10 – Hydraulic governor typical frequency response.

of a modern hydraulic governor and hopefully give some understanding of its operation.

Referring to Fig. 10, it can be seen that there is a reduction in gain at low frequency resulting from the lag/lead type of compensation in the hydraulic servo. There is a small degree of 'lift' in the phase characteristic to give stability in the middle (most important) frequency range and then the gain rolls off quite quickly at higher frequencies.

Note, lag/lead compensation is normally used in a regulator since it gives steady state accuracy but limits sensitivity at the middle frequencies. A servomechanism, on the other hand, will usually employ lead/lag compensation to give a flat, unity closed-loop gain response over as wide a bandwidth as possible.

ENGINE/GOVERNOR FREQUENCY RESPONSE

If Figs. 4 and 10 are added (hence use of dB's) then the overall open-loop frequency response of engine and governor can be drawn as shown in Fig. 11.

From this, the stability margins can be directly read, being, for the chosen example, about 16 dB gain margin and 48° phase margin, which should give adequate stability.

If the phase margin decreases owing, say to the development of backlash in the system or to cold oil in the governor (cold start stability being a common problem) or, alternatively, to an increase in gain due to incorrect lever ratios between governor and engine or due to air trapped in the governor, then it can be seen that the combination will tend to oscillate at around 1 Hz, which is the typical hunt frequency of most engines.

If we contrast this with a simple mechanical governor which might have flyweights of total mass 2 kg rotating at a radius of 50 mm, then typical figures might be, for 1200 rpm operation,

$$\omega_0^* = 126 \text{ rad/s}$$
$$\frac{q}{p} = 1$$
$$m = 2 \text{ kg total}$$
$$m_e = 5 \text{ kg, say}$$
$$r^* = 50 \text{ mm}$$
$$\left.\begin{array}{r} y^* = 15 \text{ mm stroke} \\ \text{Droop} = 4\% \end{array}\right\} \begin{array}{l} \text{for direct} \\ \text{comparison with hyd. unit} \end{array}$$

$$\text{Weight rate } R = m\omega_0^{*2} \cdot \left(\frac{q}{p}\right)^2 = 31752 \text{ N/m}$$

$$\text{Gain equation (13)} = \frac{2m\omega_0^* \left(\dfrac{q}{p}\right) r^* \left(\dfrac{\omega_0^*}{y^*}\right)}{(\beta - R)} = \frac{100}{\text{Droop } \%} = 25$$

hence

$$\beta = 40219 \text{ N/m}$$

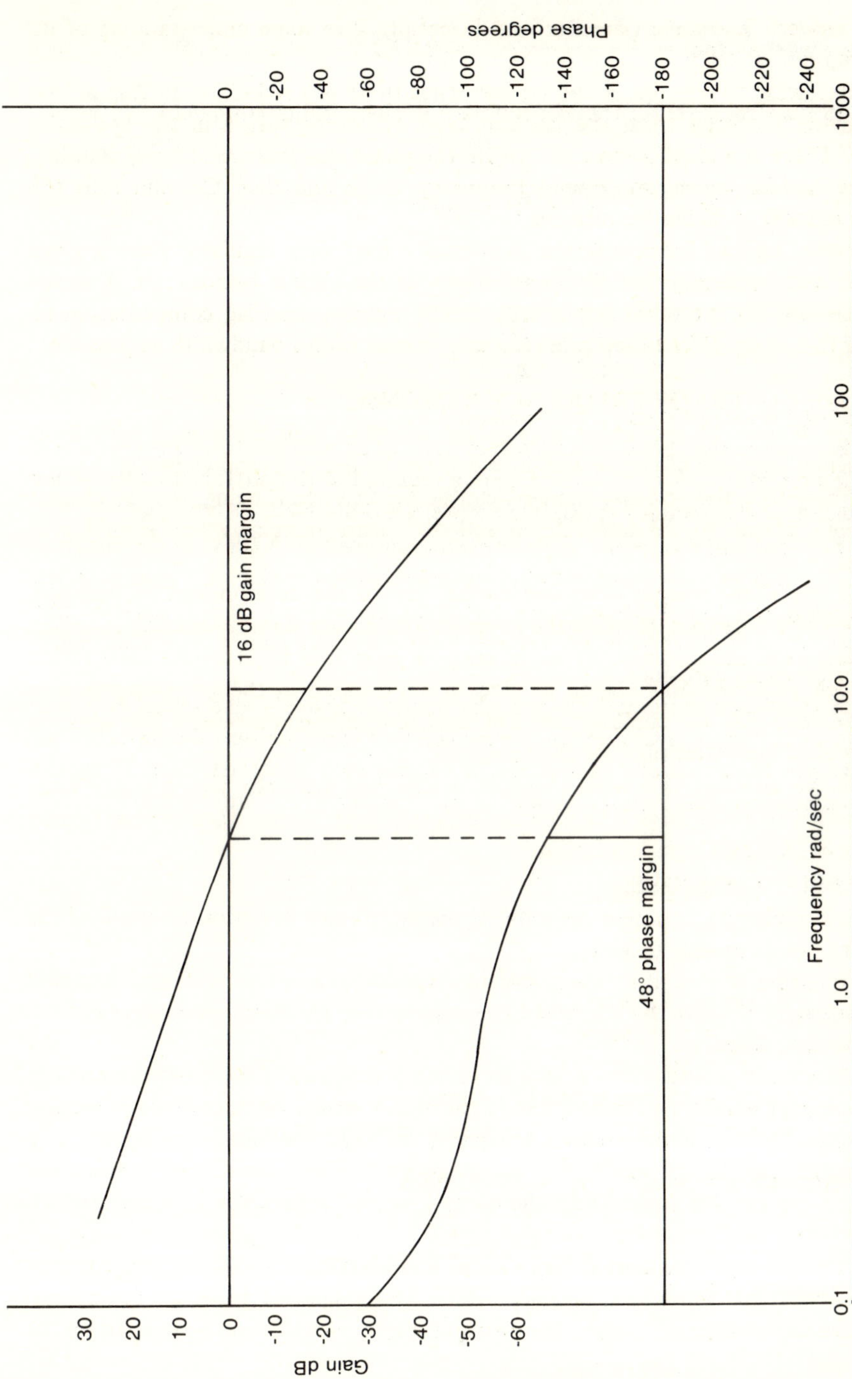

Fig. 11 – Overall engine/governor open loop frequency response.

$$\text{Stalled work output} = \frac{(\beta - R)y^*}{2} \cdot y^* = 0.95 \text{ J}$$

The mechanical governor transfer function is, then, in Equation (11)

$$\frac{\delta x}{\delta \omega_0} = \frac{-211680}{(5D^2 + fD + 8467)}$$

i.e.

$$\frac{\delta x}{\delta \omega_0} = \frac{-25}{\dfrac{D^2}{41^2} + \dfrac{2D}{41} + 1}$$

assuming f is such that damping overall is 1.0.

Because of the choice of a high gain case for comparison (droop = 4% instead of a more realistic 8%), the speeder spring required is very stiff, hence the natural frequency of the assumed spring/mass system is quite high.

The frequency response plots, shown combined with those of the engine in Fig. 12, indicate adequate stability margins, although the gain margin is less than with the hydraulic governor owing to the absence of the lag/head gain reduction at low frequencies. Overall loop 'hunt' frequency is a little higher in the arbitrary case chosen.

A significant difference between hydraulic and simple mechanical governors is the work output available to operate the fuel racks. Using a simple definition of average force × stroke as a guide, it can be seen to be less than 1J against an available capability on the market of up to about 350 J with hydraulics.

TRANSIENT RESPONSE

What is meant by transient response is the temporary deviation of speed of the engine due to change in load.

In order to calculate the transient response of a governed engine it would be necessary to solve the combined engine/governor closed-loop transfer function in the time domain.

Since this would mean solving at least a seventh-order differential equation, a computer would be needed and in the end it would be unrewarding because the calculated response would inevitably be very inaccurate owing to non-linearities and limitations in the real equipment. Extensive academic work has been done on this subject but the models derived tend to be too complex for day-to-day use.

To estimate transient performance, manufacturers resort to a mixture of experience and empirical methods which give sufficient accuracy for normal requirements since ultimately the performance is checked and optimized on the test bed.

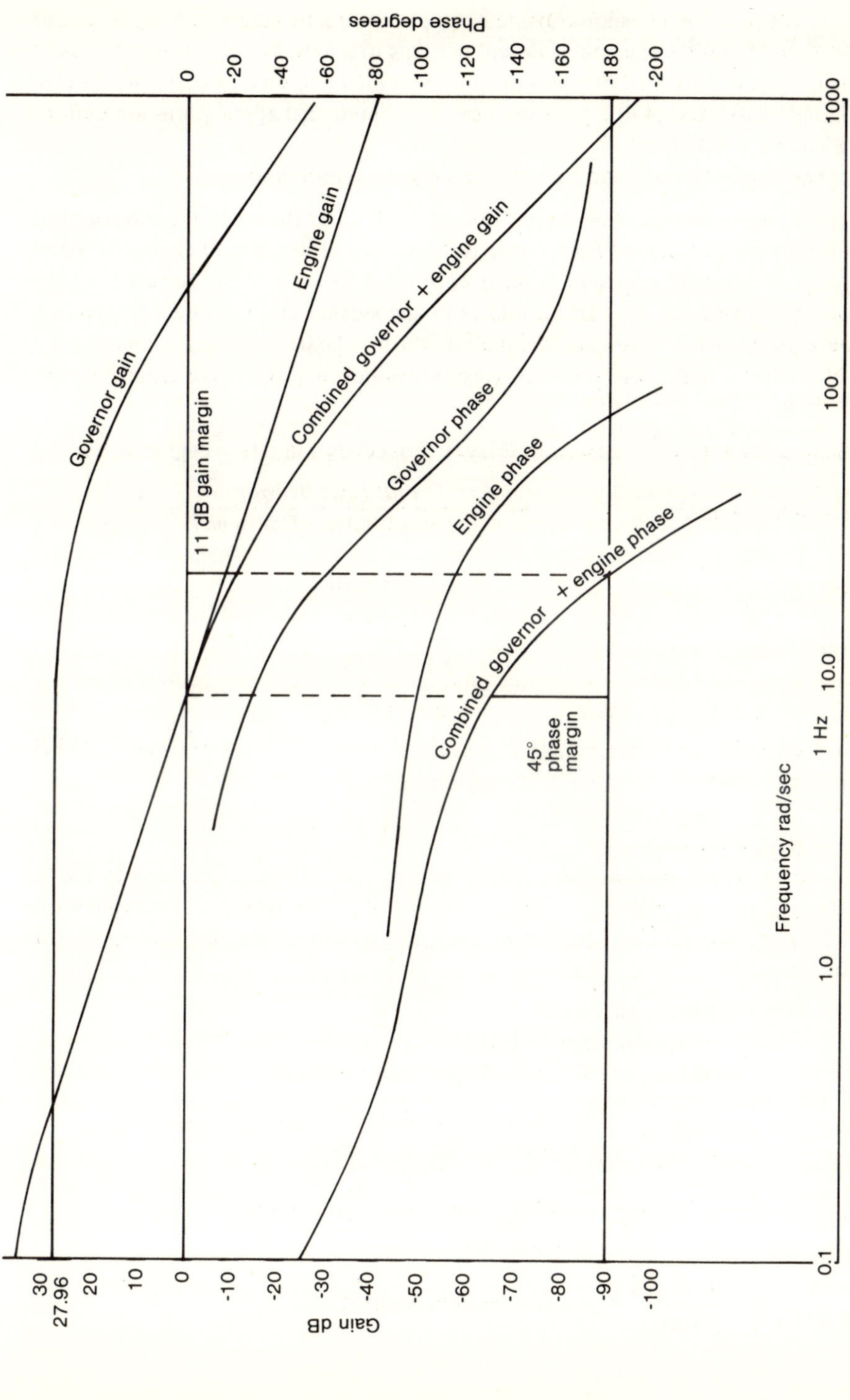

Fig. 12 – Mechanical governor frequency response, and combined with engine.

A 100% full load cannot normally be put onto a turbocharged engine in one step without stalling the engine down, perhaps to a halt. For this reason transient tests consist of part load onsets on top of zero load and on a range of pre-defined standing loads. Acceptable performances for commercial applications are defined in BS.5514. (ISO 3046).

The empirical methods are based mainly on commonsense.

1. If an engine has an acceleration rate of 50%/s and the governor traverse time from no load to full load fuel setting is 0.1 s, then clearly a load change of 100% is going to result in a speed transient of *at least* 50 x 0.1 = 5%. In fact it will be about 8% owing to the various delays in the system and because the governor is only progressively changing fuelling rate during the 0.1 s traverse time.

2. More thoroughly, the transient speed deviation is given quite closely by the following:

$$(\%\ \text{acceleration rate})^{\frac{1}{2}} \times \text{governor delay} + \%\ \text{acceleration rate} \times \text{engine delay}$$

% acceleration rate $\quad \gamma = \dfrac{\textit{Full Load Torque}}{\text{Total Inertia} \times \text{Full speed}} = \dfrac{G_E^*}{I\omega_0^*}$

In Imperial Units $\quad \gamma = \dfrac{5 \times 10^6 \times \text{BHP}}{IN^2}$, I in lb.ft.s^2

In S.I. Units $\quad \gamma = \dfrac{9.1 \times 10^6 \times \text{kW}}{IN^2}$, I in kg m^2

Note, sometimes in Europe, I is expressed in terms of diameter-squared; hence a factor of 4 must be applied to the equation for γ.

GOVERNOR DELAY

For the governor used as the example in this text, which is designed to run at 1500 rpm with a capability of 10, 20 or 34 J of work output and varying sensitivities with standard or small flyweights, the governor delay is made up of four factors.

$f = 1.0$ for 10 J build
or 1.1 for 20, 34 J build
(slower due to use of larger power piston)

$g = \sqrt{\dfrac{1500}{N_g}}$ where N_g rpm is governor nominal driven speed, e.g. if geared from the engine

$h = 0.75$ for standard flyweights (sensitive build)
or 1.0 for small flyweights

$k = \left(\dfrac{L}{100}\right)^{\frac{2}{3}}$ where L = percentage load change

ENGINE DELAY

This varies for different types of engine. If

N = engine speed – 1200 rpm
n = number of cylinders – 16
s = 4 for four-stroke engine
or 2 for two-stroke engine

then for a diesel engine from equation (6)

$$\text{Engine delay, } T_{\mathrm{f}} = \frac{60s}{2Nn} + \frac{60}{4N}$$

Thus the transient deviation is estimate to be

$$n_0\,\% = f.g.h.k.\sqrt{\gamma} + \gamma\, T_{\mathrm{f}}$$

For the example used, the transient response would approximately be, for say 50% load change and acceleration rate $\gamma = 50\%/\mathrm{s}$,

$$n_0\% = 1.1 \times \sqrt{\frac{1500}{1200}} \times 0.75 \times \left(\frac{50}{100}\right)^{\frac{2}{3}} \times \sqrt{50} + \left[\frac{60 \times 4}{2 \times 1200 \times 16} + \frac{60}{4 \times 1200}\right] \times 50$$

$$= 4.102 + 0.9375$$

i.e. approximately *5% for 50% load change* and *7.5% for 100%* load change. Such a performance is well within normal requirements.

Figure 13 shows the typical transient shapes and the effect of restrictor screw adjustment.

4.7 Practical Design Features

Special features to note in the design include (see Fig. 9(a)):

1. The rotor rotates about the pilot valve which remains stationary. This effectively negates static friction so that the pilot valve moves under very small speed changes without any stick/slip effect.
2. Most governors of this type are self-contained with in-built oil reservoir, bi-directional pump and accumulator/relief valve. The input drive then provides both speed sensing and hydraulic power. Typical figures are 10–26 bar oil supply pressure and power input 180–1100 W.
3. Pistons are grooved to ensure a pressure balance around the circumference, thus avoiding hydraulic locking effects of piston to cylinder wall.
4. The hydraulic feedback piston is connected to the power piston by a piano wire strut to alleviate alignment problems.
5. A two-way oil seal is incorporated in the drive so that the governor is sealed against ingress of oil from the engine as well as leakage from the governor.
6. Flyweights are pivoted on needle roller bearings to enhance sensitivity.

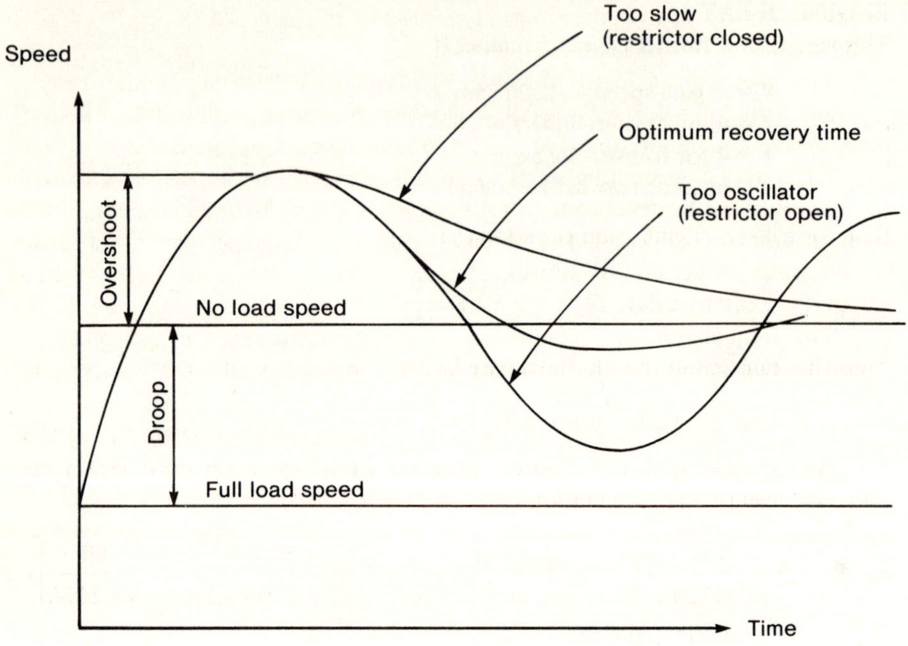

Fig. 13 – Sketch of transients.

7. Most moving parts are hardened for long life and the design in all respects has to take account of the very severe effects of engine vibration. Some larger governors are fitted with internal oil sprays to ensure lubrication to the upper linkage to ease problems of fretting corrosion.

5. ANCILLARY FUNCTIONS

The basic governor as so far described and analysed is but a minor part of many modern governors in that numerous ancillary functions may also be required to protect the engine and manipulate its characteristics to suit the application.

The main variables are:

Speed setting method
Fuel limitation
Load control
Shutdown
De-rating
Protection

5.1 Speed Setting Methods

Probably the most common form of speed setting is a small electric motor mounted on top of the governor and connected to the speed setting linkage via a nut and screw or similar simple mechanism.

This method of speed setting is used on generating sets which usually operate with some sort of overall control system often with automatic start/stop/protection circuits and load sharing/frequency control functions.

When an a.c. generating set is to be run in parallel with other sets, or with the grid, it must be synchronized to the existing supply before being connected. This is done by adjusting the speed (frequency) of the set to get the phase relationship correct and by adjusting the generated voltage to match the running supply.

The frequency/phase relationship is adjusted through the agency of the speed setting motor on the governor and the voltage is adjusted via the automatic voltage regulator (A.V.R.), not discussed here.

Once in parallel, the role of the governor changes somewhat in that it then acts as a proportional power controller such that raising the speed setting of one set in a paralleled group will cause that set to take on more load to the relief of the others. Thus the electric motor speed setting also provides a ready means for adjusting the load sharing between paralleled sets.

For marine propulsion use, a much more rapid response is required to speed setting changes than is provided by the normal electric motor arrangement, and so either a simple direct cable/lever system (on small boats) or a pneumatic pressure signal (on larger vessels, where air is usually available for engine starting, etc.) is used.

The ship's command lever incorporates an air pressure regulator giving a pressure, over typical range 0–6 bar, from neutral to full-ahead. This signal is fed to a small servomechanism within the governor which moves to adjust the speed setting. The required response times will be perhaps 10 s neutral to full-ahead and 3 s back to neutral for emergency stop.

Pneumatic systems are favoured on ships from reliability and maintainability viewpoints.

A third, fairly common speed setting method is the electric step controller and is mostly found on rail traction, although pneumatic systems find wide use on rail traction also.

In this case the locomotive control handle gives a coded output of five, eight or sixteen steps to a cluster of solenoid operated valves in the governor which, via small hydraulic valves and pistons and a summing mechanism, select preset speed settings.

5.2 Fuel Limitation

Most large diesel engines are turbocharged and it is very desirable to limit the fuelling rate when the charging pressure (boost pressure) is low, to maintain a reasonable air/fuel ratio and avoid the emission of black smoke.

To achieve this the air inlet manifold pressure is fed to a small servomechanism within the governor which acts directly on the pilot valve, limiting the amount of governor output movement in proportion to the charge pressure. A one-way mechanism is used in that low charge pressure limits fuelling rate but high pressure does not cause high fuelling rate.

Thus, rate of fuelling increase is limited during transients to that which can be properly burnt with the air available. This system is usually called boost fuel limitation or boost bias.

Simple manual fuel limitation in the form of a knob on the governor is sometimes employed to put a preset limit on governor output position (fuelling rate) during starting. This operates on the pilot valve in the same way as the boost method except that the setting is fixed until manually removed.

5.3 Load Control

In applications such as marine propulsion with controllable pitch propellers (CPP) or rail traction with field controlled alternators, the governor includes a load control valve which provides a control signal to the CPP or alternator.

In the marine propulsion case a dual output command lever, usually called a combinator, commands engine speed and propeller pitch according to some predetermined relationship.

If, owing to sea conditions, ship's loading, etc., the engine becomes overloaded the load control valve opens so as to reduce the propeller pitch, overriding the command lever. When the loading decreases the governor allows the command lever signal to again increase pitch – note the governor does not actively cause an increase in pitch.

On rail traction the driver's control lever commands engine speed, as already described, and the governor sets the alternator power via a field control system which may also be inside the governor. Thus the driver of the train is given a certain power level for each step of lever setting. The train then accelerates or retards depending on gradient, etc., and the driver maintains speed by operation of his control lever between steps.

Rail traction governors can be very complex incorporating:

stepped speed control
controlled rate of speed change between steps
load control valve
vane motor controlled field resistance
boost fuel limit
altitude de-rating
weak field start facility
wheel spin correction
engine low oil pressure protection

Marine governors for use with fixed pitch propellers (FPP) can likewise be complex for applications such as fast patrol boats where it is necessary to have in-built torque control to protect the engine but with the facility for sprint override for emergency use.

Each of these special governors could be a major study in itself.

The initial setting and adjustment of most governors outlined above can only be carried out with sufficient precision on purpose-designed test rigs (hydraulic engine simulators) since it is not possible to create the necessary conditions safely on a running engine.

The important matter to appreciate is that governors are no longer romantic devices in polished brass, mounted atop grand machines but are complex high technology devices performing many vital functions on highly rated modern engines.

6. GOVERNING OF SMALL DIESEL ENGINES

Large engines invariably have separate fuel pumping elements for each cylinder. Sometimes up to six to eight elements are housed in one block. For the larger and higher performance engines, totally separate pumps may be used distributed around the engine, each close to the cylinder. The governor normally employed with these arrangements of fuel pumps is a further separate unit as already described in detail, and manufactured usually by a completely different company.

On small engines, a monobloc pump with separate pumping elements may be used, but for reasons of simplicity, compactness, ease of installation and hence cost, a single element distributor type of pump is most common with integral governor, timing control and sometimes boost fuel limit for turbo-charged engines.

Small engine governors are mostly simple mechanical hydraulic devices and may be all-speed or max-min. devices.

An all-speed governor controls engine speed over the whole operating range as in the case of the large engine types already analysed.

A max–min. governor is inoperative over the normal speed range and only comes into operation when the speed demand is too high or too low. For example, the fueling rate on a truck engine with a max–min. governor would be under the direct control of the driver until such time as he either tries to exceed the permitted top speed or removes his foot from the pedal when stationary. Then the governor takes over and limits the top speed or holds a preset idling speed, as appropriate.

The mechanical type of governor is in principle exactly as described in section 3 except that the axis of rotation is usually horizontal for compact integration with the horizontally driven fuel pump.

The hydraulic type of governor as used in this context is worth a little extra consideration as it is an example of a different principle used for speed measurement and is neatly integrated into the fuelling rate control function.

Fuel is fed into the injection pump by a transfer pump. The delivery of the transfer pump is proportional to engine speed and this is converted into a pressure proportional to engine speed by a regulating valve. This pressure is then applied to a metering valve plunger where it is balanced against a spring. The spring is pre-loaded by the speed setting command provided, for instance, by a vehicle driver's foot pressure on the accelerator pedal.

Referring to Fig. 14 and ensuring linearity by considering small changes δ as before:

Transfer pressure is proportional to engine speed, i.e.

$$\delta P = K_a \, \delta\omega_0 \tag{38}$$

Force balance on the metering valve gives

$$\beta(\delta X - \delta x) = A\delta P \tag{39}$$

Displacement δX is proportional to set speed $\delta\omega_i$, i.e.

$$\delta X = K_i \delta\omega_i \tag{40}$$

Fuelling rate is proportional to metering valve displacement,

$$\delta q = K_b \delta x \tag{41}$$

Combining equations (38), (39), (40) and (41) gives

$$\delta q = K_b \left(K_i \delta\omega_i - \frac{K_a A}{\beta} \delta\omega_0\right)$$

For constant set speed

$$\frac{\delta q}{\delta\omega_0} = -\frac{K_a K_b A}{\beta}$$

This shows the governor to be basically a simple proportional device as is the mechanical governor.

A more thorough analysis would show the presence of lags in the dynamic performance due to the responses of the regulating and metering valves.

A dashpot type of damping device would be incorporated with the metering valve in practice to prevent rapid movement of the metering valve causing instability.

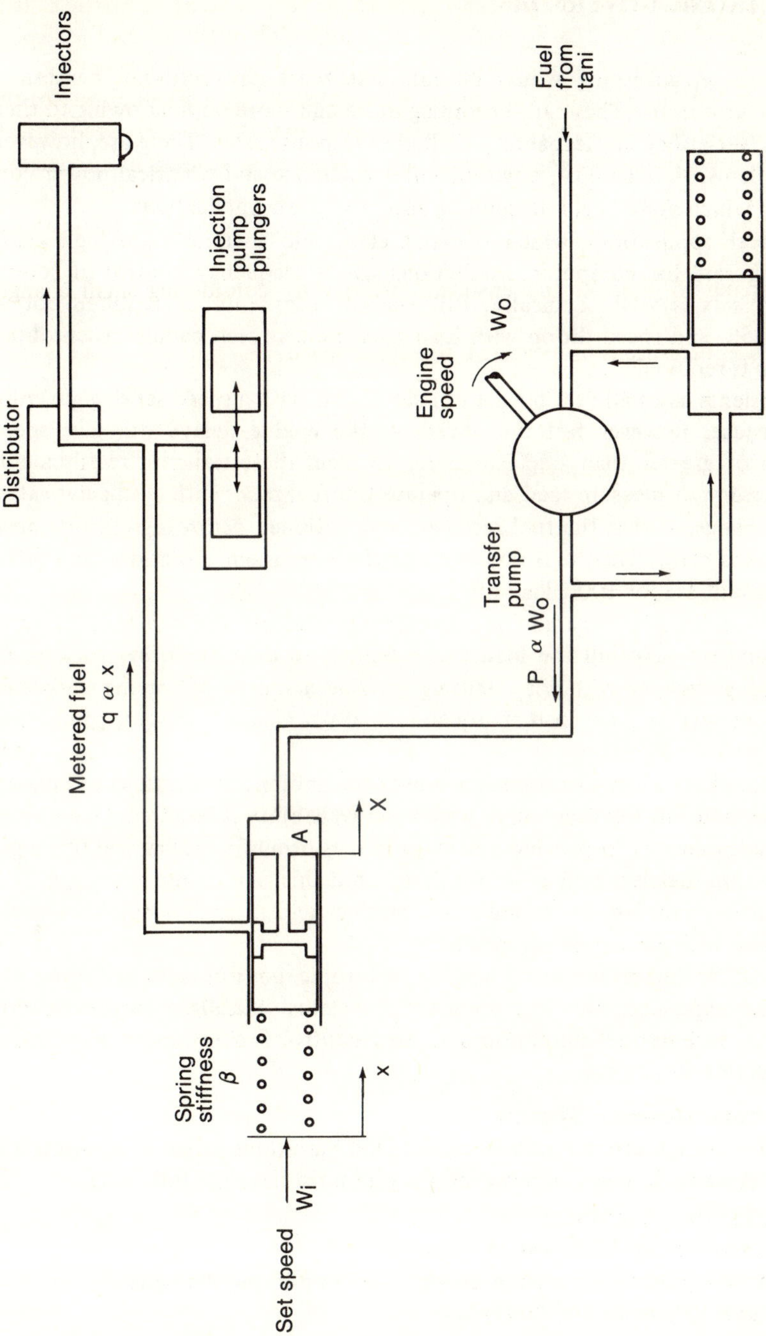

Fig. 14 — Simple pump-mounted hydraulic governor system.

7. ELECTRONIC GOVERNORS

7.1 General

No treatise on governing would be complete these days without mentioning electronic governors. These are becoming more and more popular owing to their accuracy, versatility and capability of higher response rates. There are, however, also problems of reliability, environmental tolerance and electrical power consumption which make them difficult to accept in some applications.

Typical applications which demand electronic governors are high grade generating sets for computer installations where frequency control of several paralleled sets is vital. Typically, full load-off transients are required not to exceed 1.5% and speed droop with load must be zero, i.e. parallel isochronous operation is required.

Transients as small as 1.5% are not achievable with a speed sensing governor only because, however fast the governor, the engine delays will give speed deviation of greater than 1.5%. In order to meet these transient requirements it is necessary to measure load and operate the fuel racks with a time derivative of load change so that the fuel racks are re-positioned *before* significant speed change takes place. Then it is possible to reduce transient deviations to a minimum of about 1% for 100% load off.

Isochronous sets in parallel are divergently unstable and will tend to move so that one set takes all the load and attempts to drive the other backwards. Electronic governors with load sensing provide a means of forcing paralleled isochronous sets to share load stably and equally, or share it in some proportion with the grid.

Applications where electronic governors are finding only cautious acceptance are marine and rail traction where very high availability is required. One way of getting acceptance is to provide a backup in the form of a mechanical flyweight system within the electronic governor design, but this is of course expensive.

A further problem in the merchant marine field is lack of suitably trained staff to deal with electronic equipment.

The other important attraction of electronic governors is that they can readily be expanded into comprehensive systems, including automatic synchronizing, various fuel limitation and load control functions and monitoring and protection functions.

7.2 Electronic Governor Systems

As in all things there are numerous detailed variations possible in electronic governor systems. In essence, however, they all must have the following:

1. Speed sensing transducer.
2. Load sensing capability (when required).
3. Electronic module for mixing, shaping and amplifying the signals.
4. Actuator to operate the fuel racks.
5. Electrical power source.

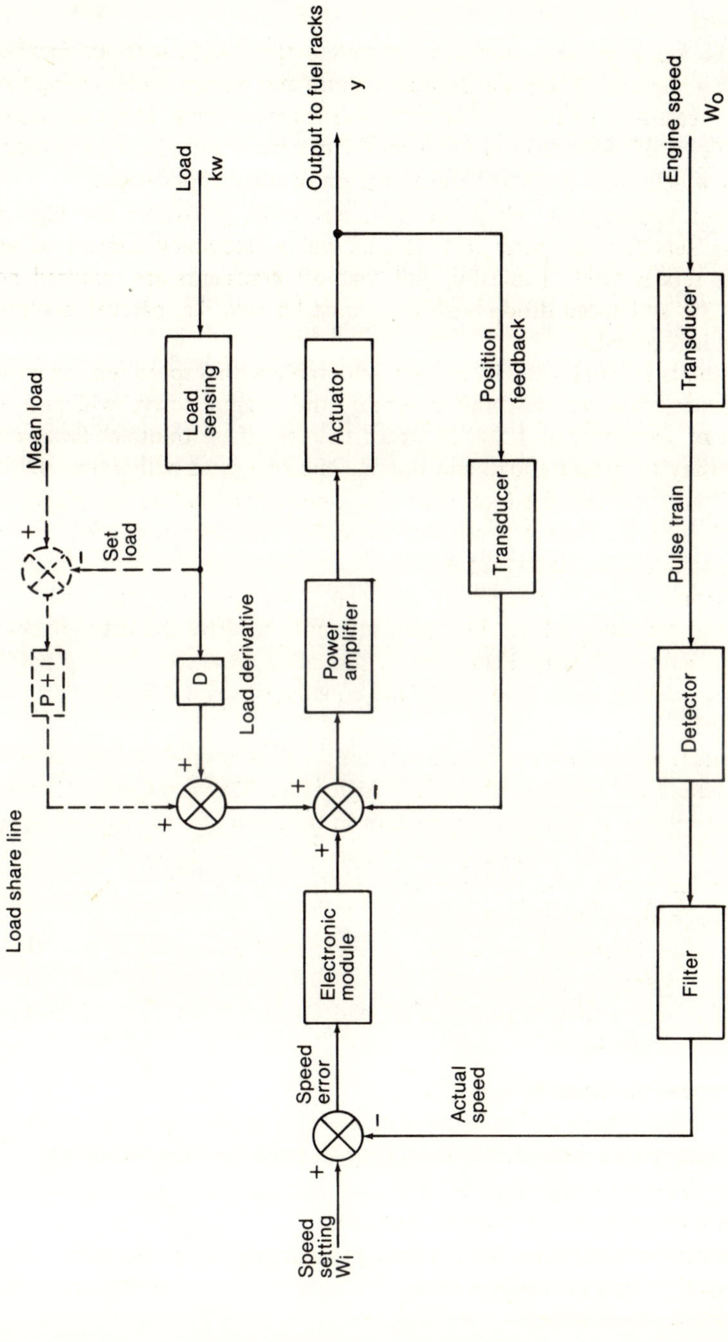

Fig. 15 – An electronic governor system.

In addition, they usually must also include filters to remove cyclic variation on the speed signal (e.g. a spring drive equivalent) and a position feedback transducer (droop linkage equivalent). Systems can be designed without this latter item but good accurate performance is difficult to achieve. Unfortunately the position feedback transducer can be a costly and troublesome device.

The electronic governor system is shown in Fig. 15.

The speed sensing transducer is normally an inductive perception head detecting teeth on, say, the starter ring of the engine and giving a pulse train signal in the range 1000–5000 Hz. This signal is then integrated in a detector circuit to produce a voltage level proportional to speed.

After filtering, the speed signal is compared with a set speed reference to produce a speed error signal which is then shaped in the electronic module. The shaping is typically proportional plus integral or lag/lead as in the mechanical governor to give a frequency response curve of the same general shape.

Note, no derivative in the forward path as this, although possibly beneficial in enhancing stability, would amplify the high frequency content of the signal and produce output 'jiggle'.

The shaped error signal is then amplified in power level so as to drive the actuator. This power amplification may be quite modest if a hydraulic actuator is used with a small electro/hydraulic force motor or it may be quite substantial if an all-electric actuator based on an electric motor or large solenoid is employed.

The actuator position feedback transducer can be a potentiometer but this normally has a very short life on a vibrating engine. More commonly, an L.V.D.T. (linear variable differential transformer) is used, this being a contactless device of potentially much longer life. The main drawback of the L.V.D.T. is that it needs a fair amount of associated electronics which can give rise to accuracy and temperature drift problems.

Load sensing can be taken from the fuel rack position, also by an L.V.D.T., for load sharing duty on coupled marine propulsion engines but the signal from this source cannot, of course, be used for transient improvement since it comes in effect from the actuator movement and so is 'after the event'. True load measure is only practical on generator sets whereby means of current and voltage transformers on all three phases of alternator output, an accurate measure of kW can be achieved. The load measure circuit produces a voltage

$$v \propto VI \cos \Phi$$

For transient improvement this voltage is a capacitor-coupled to give a load derivative signal. For load sharing it is summed with similar outputs from all sets to produce a mean load signal and then each individual set load is compared with the mean to produce a load error signal. This load error signal is shaped for stability, probably with $P + I$, and fed into the electronic module to effect a balance of load between sets.

8. GLOSSARY OF TERMS

Boost	– pressurized air supply to the engine combustion chambers produced by a turbocharger.
Damping	– rate dependent force which 'damps' out oscillations in a control system or element.
Delay	– a finite dead time between output of a system element and its input.
Droop	– drop in engine speed when load is applied; usually expressed as percentage of nominal full speed.
Firing Delay	– delay between consecutive cylinders of an engine firing.
Frequency Response	– solution of the transfer function in the frequency domain (harmonic response).
Hunting	– colloquial term applied to an unstable engine/governor system.
Integrator	– a device whose output continues to move so long as there is an input, e.g. a hydraulic piston, electric motor.
Isochronous	– operation with zero droop.
Lag	– a system element whose output follows input with an exponential response.
Parallel Operation	– two or more engines coupled together to jointly provide the required power output.
Regulator	– a device or system for maintaining some variable sensibly constant, e.g. a governor is a regulator for maintaining engine speed.
Servomechanism	– a system for enhancing power, response rate, accuracy, of some variable.
Stability Margins	– the amount by which the open loop gain and phase lag of a system could be increased before it becomes unstable (oscillates) in closed loop.
Synchronism	– electrical alternators running together in phase and voltage correspondence.
Transfer Function	– a differential equation (in operator D or Laplace form) relating the dynamic behaviour of output to input of a device or system.
Transients	– the temporary overshoot of engine speed due to sudden application or removal of load – solution of the transfer function in the time domain.
Work Output	– stall torque at governor output shaft multiplied by full stroke.

5

Supercharging and turbocharging

N. WATSON, Reader in Mechanical Engineering, Imperial College of Science and Technology London

---

1. INTRODUCTION TO SUPERCHARGING

The principal objective of supercharging is to increase the power output of a given size of engine. Subsidiary objectives may be to improve efficiency and/or reduce exhaust emissions and combustion induced noise.

Supercharging means increasing the density of the air entering the cylinders in order that more fuel may be burnt. Density is increased by a compressor. Many types of compressor are available but only positive displacement rotary, or rotodynamic compressors can deliver a sufficiently large quantity of air from a small compressor to be practical. Usually the positive displacement rotary compressor will be limited in rotational speed, relative to a rotodynamic compressor and therefore will be larger. Both machines can be driven mechanically by the engine but the rotodynamic compressor is suitable for a turbocharger, where the driving power comes from exhaust gas energy via a turbine.

The density of the air delivered to the cylinders depends on the ambient air density, the pressure ratio and efficiency of the compressor and any heating (for example in the inlet manifold) or cooling (for example from a charge air cooler) of the air.

As far as the engine is concerned, supercharging influences the engine through three parameters only:

1. The inlet manifold pressure } hence density.
2. The inlet manifold temperature }
3. The exhaust manifold pressure, if turbocharged; or the power requirement to drive the supercharger, if mechanically supercharged.

Usually an engine manufacturer is concerned with what a particular supercharger can do for his engine, but it is more instructive to reverse this process and ask the question 'what does the engine require from the supercharger?'.

In this way, targets can be set for an 'ideal' supercharger to meet the requirements of a particular engine and its application. Here we will use a much simplified analysis and try to answer this question. The starting point will be how the three parameters listed above influence the airflow through an engine and therefore how much fuel can be burnt and its power output.

2. AIRFLOW THROUGH AN ENGINE

2.1 Four-stroke engine

Consider first the simple case of a four-stroke engine with no valve overlap, or at least a sufficiently small overlap period for scavenge flow to be negligible. The airflow rate can be expressed as

$$\dot{m} = \frac{N}{2} \rho_m V_{sw} \eta_{vol} \tag{1}$$

where

N = engine speed
ρ_m = inlet manifold density
V_{sw} = swept volume of engine
η_{vol} = volumetric efficiency

The inlet manifold density will depend on ambient air conditions and compressor pressure ratio and efficiency. Using the notation of Fig. 1:

$$T_2 = \frac{T_1}{\eta_c}\left[\left(\frac{P_2}{P_1}\right)^{\frac{\gamma-1}{\gamma}} - 1\right] + T_1 \tag{2}$$

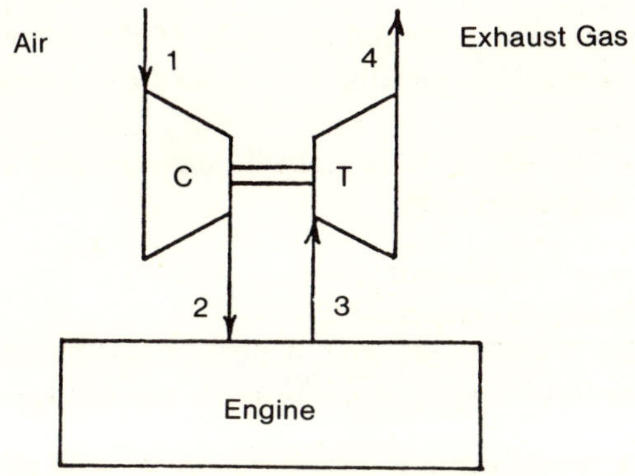

Fig. 1 – Schematic of engine and turbocharger.

where η_c = isentropic efficiency of the compressor. Treating air as a perfect gas, and assuming no charge air cooling, the inlet manifold density is given by

$$\rho_m = \rho_2 = \frac{P_2}{RT_2} \tag{3}$$

$$\rho_2 = P_2/RT_1 \left[\frac{1}{\eta_c}\left\{\left(\frac{P_2}{P_1}\right)^{\frac{\gamma-1}{\gamma}} - 1\right\} + 1\right] \tag{4}$$

Equations (1) and (4) may be combined to calculate the airflow through the engine. Note that the exhaust manifold pressure of the turbocharged engine does not influence the airflow, as zero scavenge flow has been assumed for simplicity.

Fig. 2 shows a typical volumetric efficiency curve of a truck engine. For simplicity, a constant volumetric efficiency of 0.9 will be assumed and a constant compressor isentropic efficiency of 0.7. Fig. 3 shows the effect of compressor pressure ratio (P_2/P_1) and engine speed (N) on the airflow characteristics of a four-stroke engine with ambient conditions of 1 bar, 23°C.

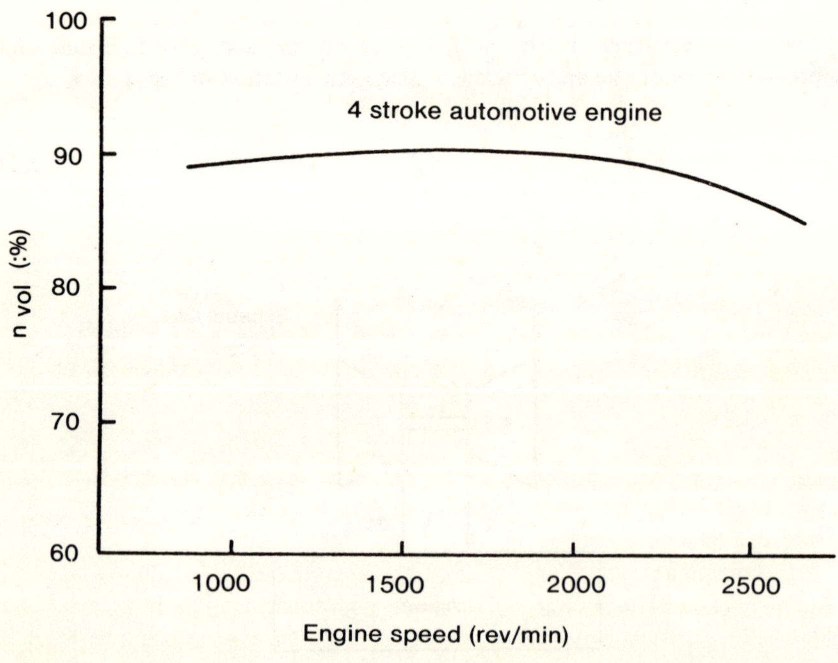

Fig. 2 – Volumetric efficiency curve.

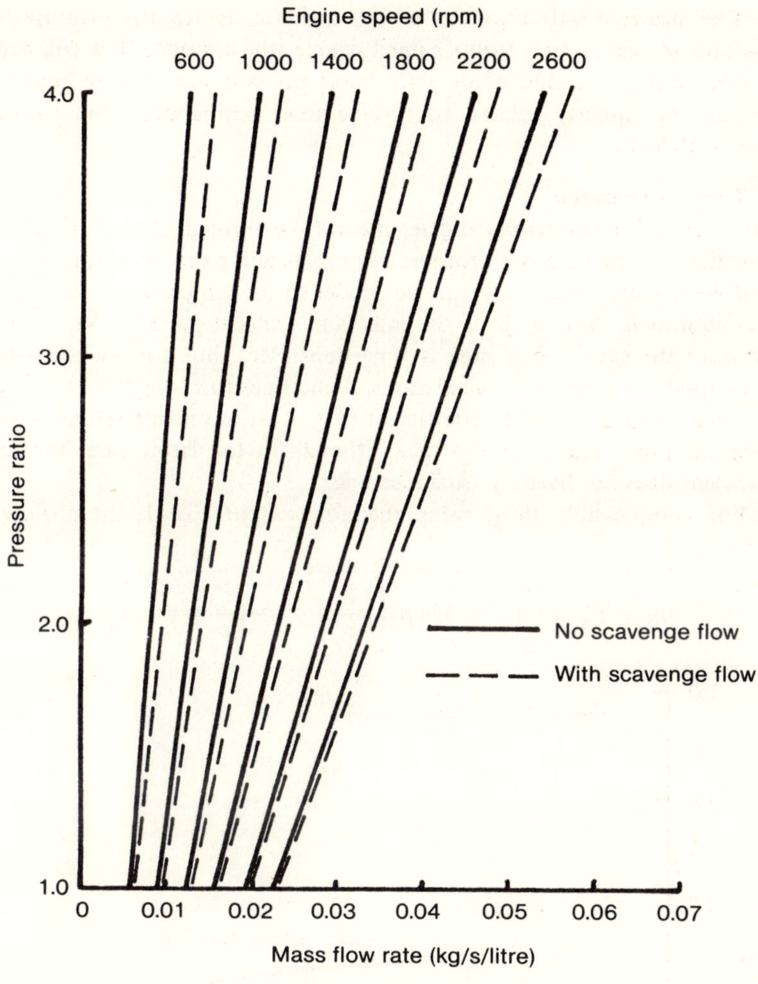

Fig. 3 – Engine airflow rate as a function of speed and compressor pressure ratio (four-stroke, no charge cooling).

Obviously airflow increases with engine speed, since the engine behaves as a positive displacement pump. In addition, airflow increases with pressure ratio—the objective of supercharging.

If the engine has significant valve overlap, then the airflow will be influenced by the difference between inlet and exhaust manifold pressures. In the case of a mechanically supercharged engine there will be a positive pressure difference between inlet and exhaust manifolds, creating a positive scavenge airflow (Fig. 3). The scavenge airflow will increase as the boost pressure (P_2) rises since exhaust manifold pressure will be close to atmospheric, hence the pressure between manifolds increases. In a turbocharged engine, the scavenge flow will depend on the relationship between compressor boost pressure and turbine inlet pressure. This can be directly related to turbine inlet temperature and turbocharger overall efficiency.

2.2 Two-stroke engine

In the case of a two-stroke engine, the airflow through the cylinders is largely a function of the pressure drop between inlet and exhaust manifolds and inlet manifold density. The flow can be modelled as compressible flow through a mean equivalent flow area of the inlet and exhaust ports (or valves) in series. Obviously the actual flow area is time dependent, but a mean equivalent area can be used for simplified calculations. Sometimes this is related to a 'reference' flow area using a discharge coefficient (Fig. 4). Convenient reference areas are maximum port area or piston area although in the latter case, the discharge coefficient does not have its usual meaning.

For compressible flow, using the notation of Fig. 1, the airflow rate is given by:

$$\dot{m} = C_\mathrm{d}\, A \sqrt{\left\{\left(\frac{2\gamma}{\gamma-1}\right) P_2 \rho_2 \left[\left(\frac{P_3}{P_2}\right)^{2/\gamma} - \left(\frac{P_3}{P_2}\right)^{\frac{\gamma+1}{\gamma}}\right]\right\}} \tag{5}$$

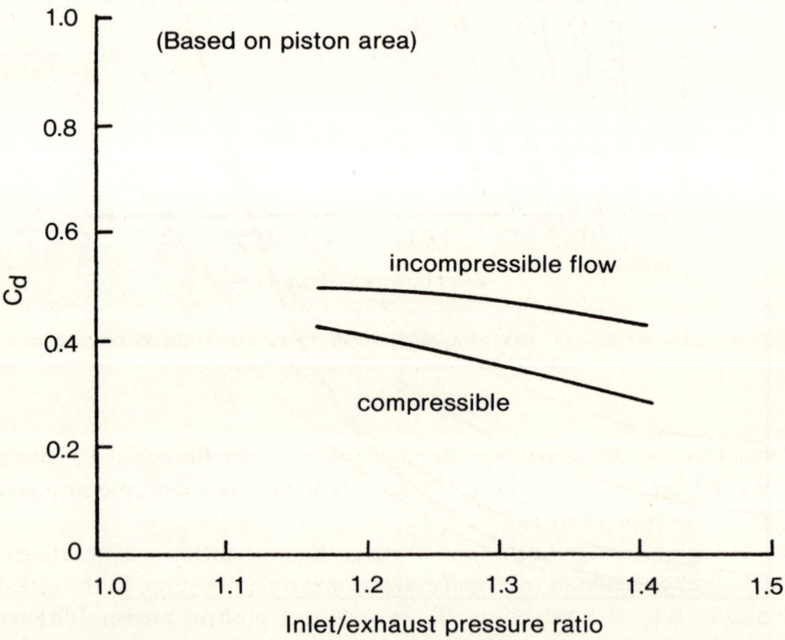

Fig. 4 – Overall valve/port discharge coefficient for an opposed piston 2-stroke engine.

If the pressure drop between inlet and exhaust manifolds is small, as is likely to be the case, then the simplification of incompressible flow can be made. Equation (5) then becomes

$$\dot{m} = C_d A \sqrt{[2 \rho_2 (P_2 - P_3)]} \tag{6}$$

In the case of a mechanically driven supercharger, the exhaust manifold pressure (P_3) is close to ambient, hence the mass flow rate is a function of boost pressure and temperature only. The airflow characteristic is a single line on Fig. 5. For a turbocharged engine, the expansion ratio of the turbine (P_3/P_4) is mass flow dependent, thus P_3 varies with engine speed and load in a manner determined by the flow characteristic of the turbine. Fig. 5 shows how the airflow characteristics of the engine vary with the exhaust manifold pressure P_3. In practice, P_3 may not be constant at steady engine conditions (depending on the turbocharging system selected) but this complication will be ignored here.

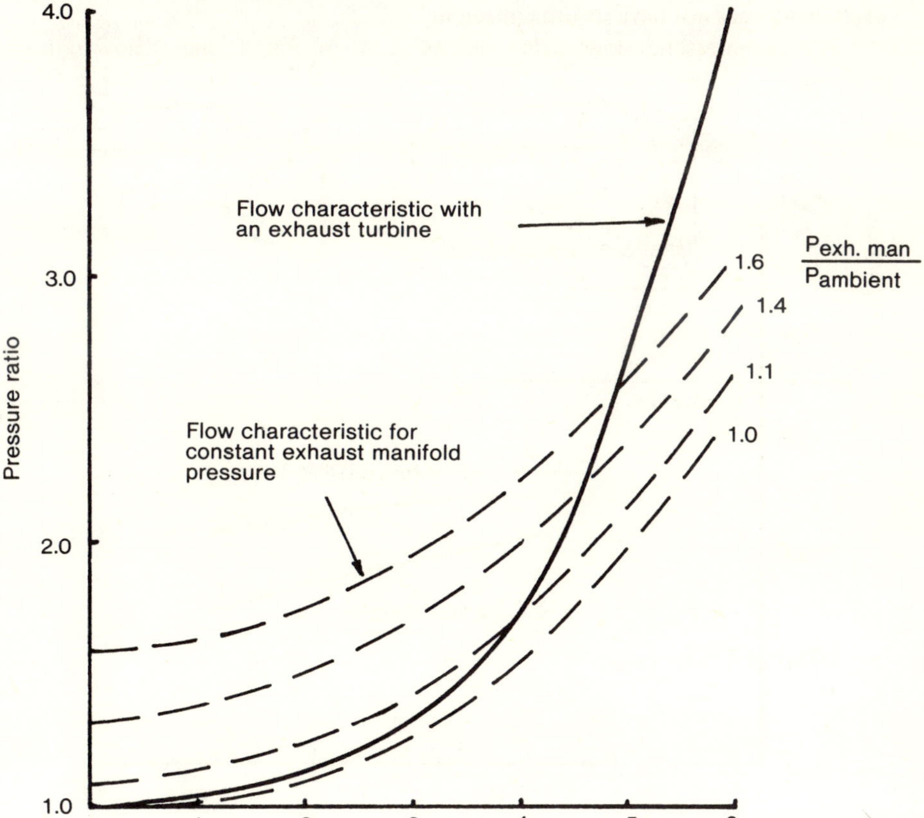

Fig. 5 – Two-stroke engine flow characteristics as a function of exhaust and inlet manifold pressures.

2.3 Charge cooling

Charge air cooling is used on many supercharged engines to increase the effectiveness of the supercharging system. Reducing air temperature by charge cooling increases air density, for a given compressor pressure ratio, and hence increases mass flow rate.

The effectiveness of the charge cooler is given by the actual heat transfer divided by the maximum possible heat transfer, which, for a perfect gas, reduces to

$$\epsilon = \frac{T_{2c} - T_2}{T_{2c} - T_{cool}} \tag{7}$$

where subscript 2c denotes compressor exit; '2' denotes charge cooler exit (inlet manifold conditions); and 'cool' denotes the coolant (Fig. 6).

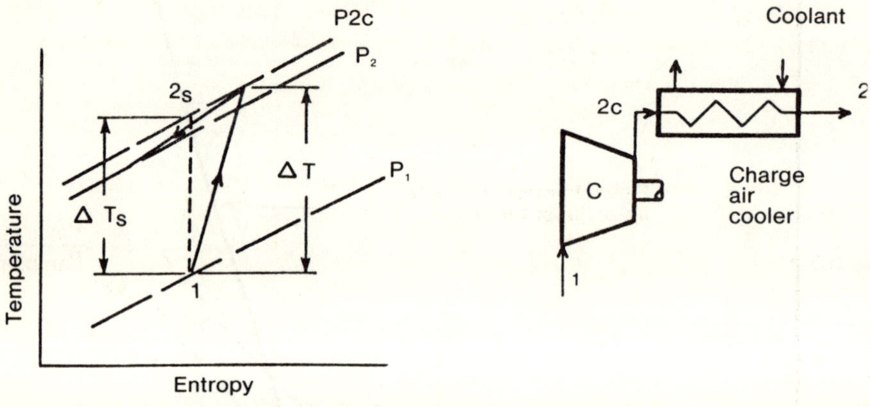

Fig. 6 – Charge air cooling, with a pressure loss.

Thus

$$T_2 = T_{2c}\,(1 - \epsilon) + \epsilon\, T_{cool} \tag{8}$$

Equation (2) becomes

$$T_3 = T_1 \left[\frac{1}{\eta_c}\left\{\left(\frac{P_2}{P_1}\right)^{\frac{\gamma-1}{\gamma}} - 1\right\} + 1\right]\left[1 - \epsilon\right] + \epsilon\, T_{cool} \tag{9}$$

Equation (9) is plotted for a compressor efficiency of 0.7 and T_1 and $T_{cool} = 20°$ in Fig. 7. As important is the density ratio ($\rho_2/\rho_{ambient}$), plotted in Fig. 8 showing the benefit of increased effectiveness of charge air cooling.

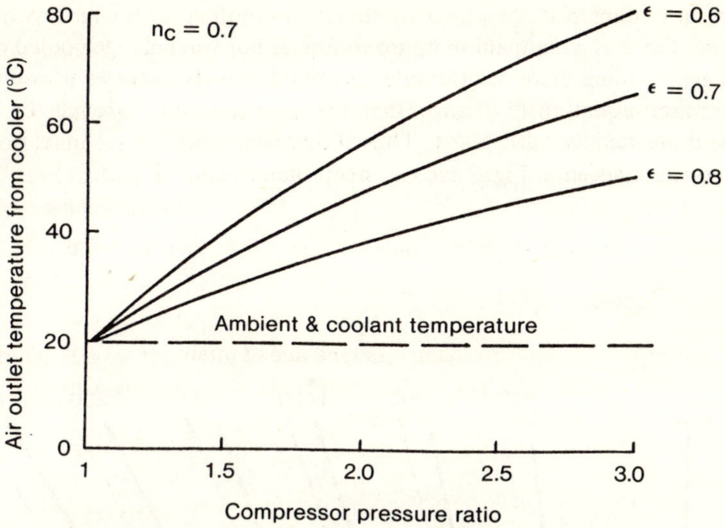

Fig. 7 – Intercooler effectiveness and inlet manifold temperature.

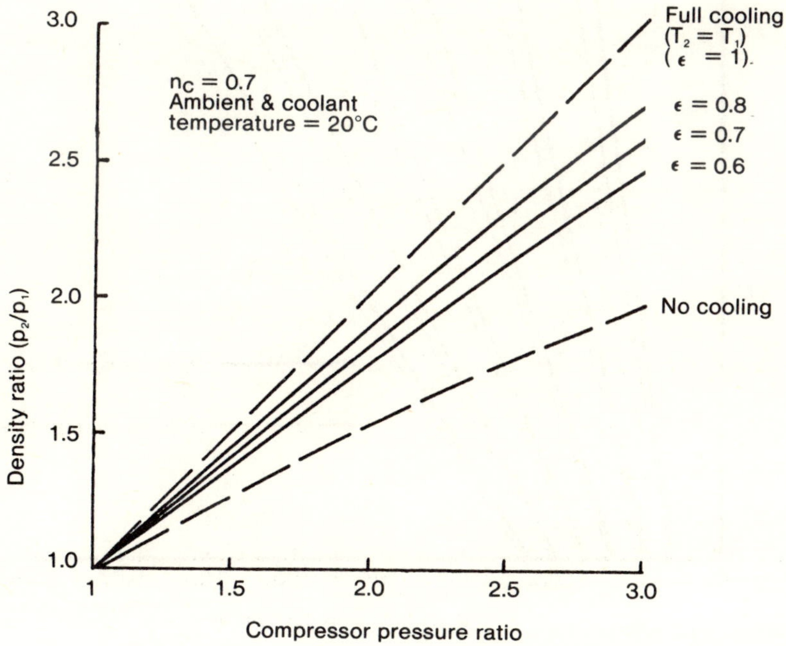

Fig. 8 – Effect of charge cooling on inlet air density.

Here our concern is the effect of charge air cooling on the airflow through the engine. Fig. 9 is a companion figure to Fig. 3, but for a charge cooled engine. Since charge cooling increases the inlet manifold density more as pressure ratio increases, then equation (1) shows that the mass flow rate through the engine increases more rapidly with boost. Thus the pressure ratio versus mass flow rate lines at constant speed in Fig. 9 are less steep than those of Fig. 3.

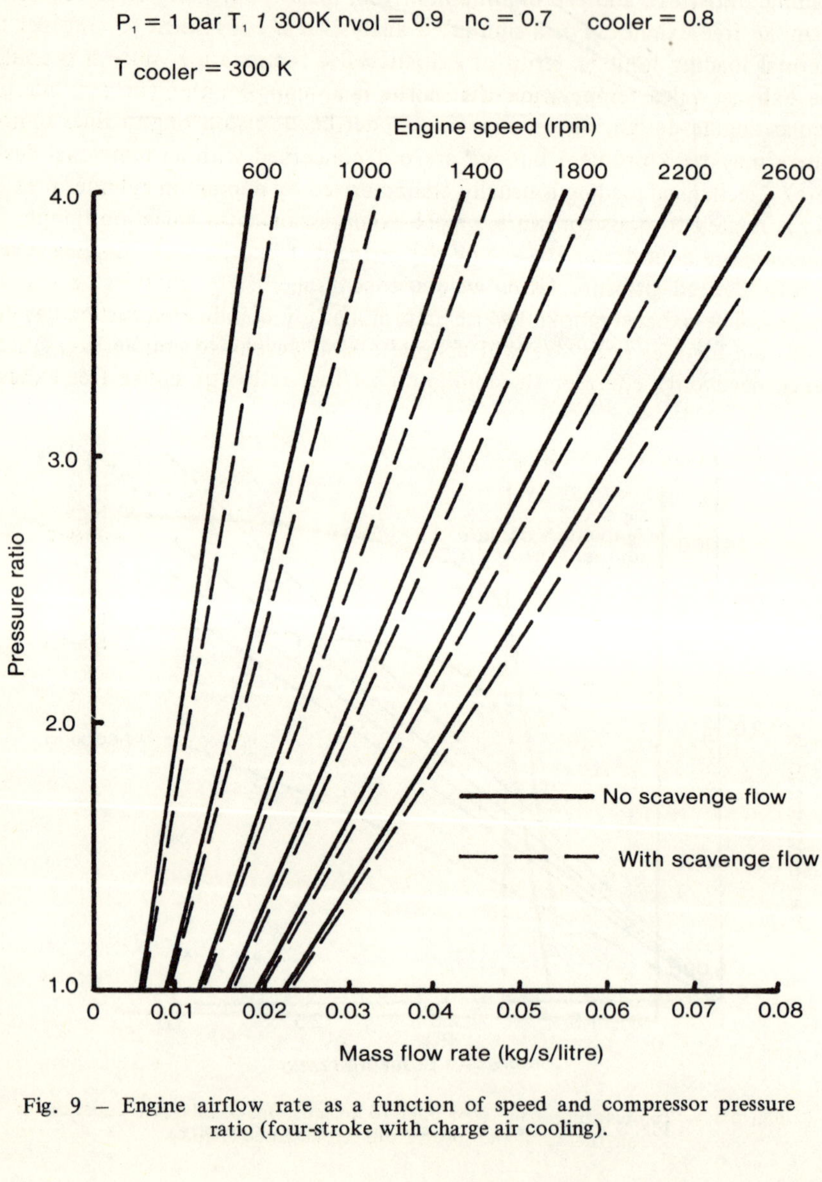

Fig. 9 – Engine airflow rate as a function of speed and compressor pressure ratio (four-stroke with charge air cooling).

3. ENGINE PERFORMANCE LIMITATIONS

Now that the basic airflow characteristics of four and two-stroke engines have been established, we need to link the airflow to fuel delivery and power output. In doing this account must be taken of various engine performance limitations imposed by the engine rather than the supercharging system.

The key limitations are mechanical loading and resultant stress, thermal loading and stress, and the minimum air/fuel ratio for efficient combustion and a smoke free exhaust. For a simplified analysis it is convenient to consider the thermal loading limit in terms of exhaust valve temperature, since it is usually the exhaust valve temperature that is the main consideration (on any one particular engine design, piston crown, cylinder head, piston ring or liner temperatures may be a problem but we are not concerned with an individual design here). Mechanical loading is usually characterized by maximum cylinder pressure and is related to boost pressure, engine compression ratio, injection timing, etc. Since we are considering the engine requirements from the supercharging system at any desired pressure ratio, we need not specifically consider mechanical loading. The exhaust smoke level is determined by combustion system development and air/fuel ratio. We will assume a well developed combustion system, hence need only consider the minimum air/fuel ratio for smoke free exhaust.

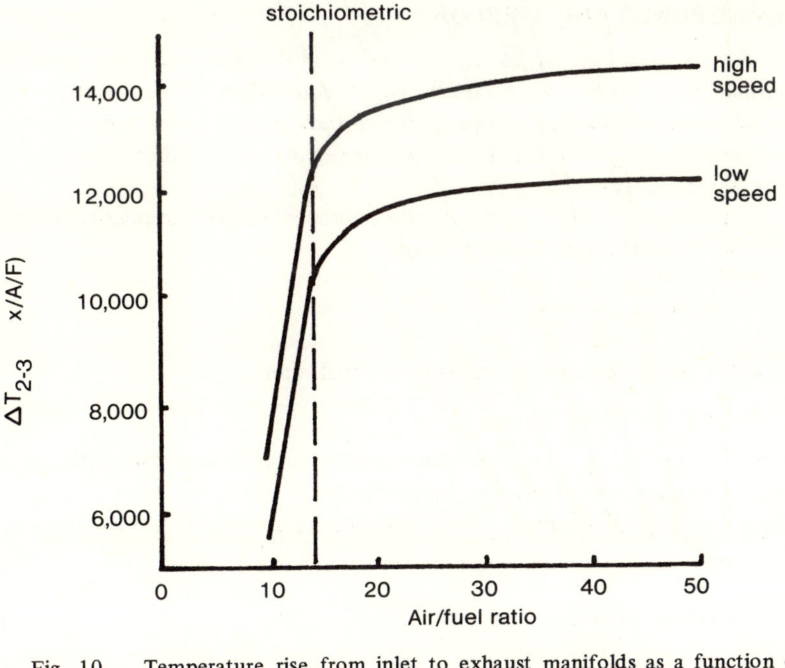

Fig. 10 – Temperature rise from inlet to exhaust manifolds as a function of air/fuel ratio and speed (four-stroke truck engine).

However, this may not be the air/fuel ratio for most efficient engine operation at any particular speed and load, so the 'optimum' air/fuel ratio must also be considered.

Compressor efficiency and charge cooler effectiveness influence the inlet manifold temperature and therefore engine thermal loading directly as well as through airflow rate. However, this factor will be ignored in the current analysis. In addition the exhaust manifold pressure is strongly influenced by overall turbocharger efficiency and in turn influences engine efficiency through piston pumping work. This will not be considered since its influence is relatively small, except on very highly rated engines.

In conclusion, for a simplified analysis, the most important limitations are exhaust (or turbine inlet) temperature and air/fuel ratio. In practice these are strongly linked, particularly on a four-stroke engine with little scavenge airflow. Fig. 10 shows the temperature rise ($T_3 - T_2$, Fig. 1) across a turbocharged four-stroke truck engine as a function of air/fuel ratio, but plotted as temperature rise × air/fuel ratio versus air/fuel ratio. It can be seen that the result is almost constant except near stoichiometric air/fuel ratios when dissociation becomes significant. For a large, slower speed engine, with a quiescent combustion chamber, the curve would become non-linear below a greater air/fuel ratio.

4. ENGINE POWER AND AIRFLOW

The power requirements of an engine need to be linked to airflow, within the limitations given above, to establish the requirements of the supercharging system. Obviously this link is unique to a particular engine and depends primarily on the combustion system but is also influenced by many other factors such as mechanical efficiency etc.

It is convenient to start with the indicated specific fuel consumption (ISFC) or indicated thermal efficiency (η_i), where:

$$\text{ISFC} = \frac{1}{\eta_i \times \text{CV}} \tag{10}$$

(where CV is the lower calorific value of the fuel) and

$$\text{BSFC} = \text{ISFC}/\eta_{\text{mech}} \tag{11}$$

where BSFC and η_{mech} are the brake engine specific fuel consumption and mechanical efficiency of the engine, respectively.

Also the fuel consumption ($\dot{m}_f$) relates brake specific fuel consumption to power output

$$\text{BSFC} = \left(\frac{\dot{m}_f}{\dot{W}}\right) \tag{12}$$

where $\dot{W}$ = power output.

The air/fuel ratio is conveniently considered in terms of the stoichiometric air/fuel ratio divided by the equivalence ratio (F)

$$\frac{\dot{m}_a}{\dot{m}_f} = \left(\frac{\dot{m}_a}{\dot{m}_f}\right)_{st} / F \tag{13}$$

Figure 11 shows typical variations of indicated specific fuel consumption with equivalence ratio.

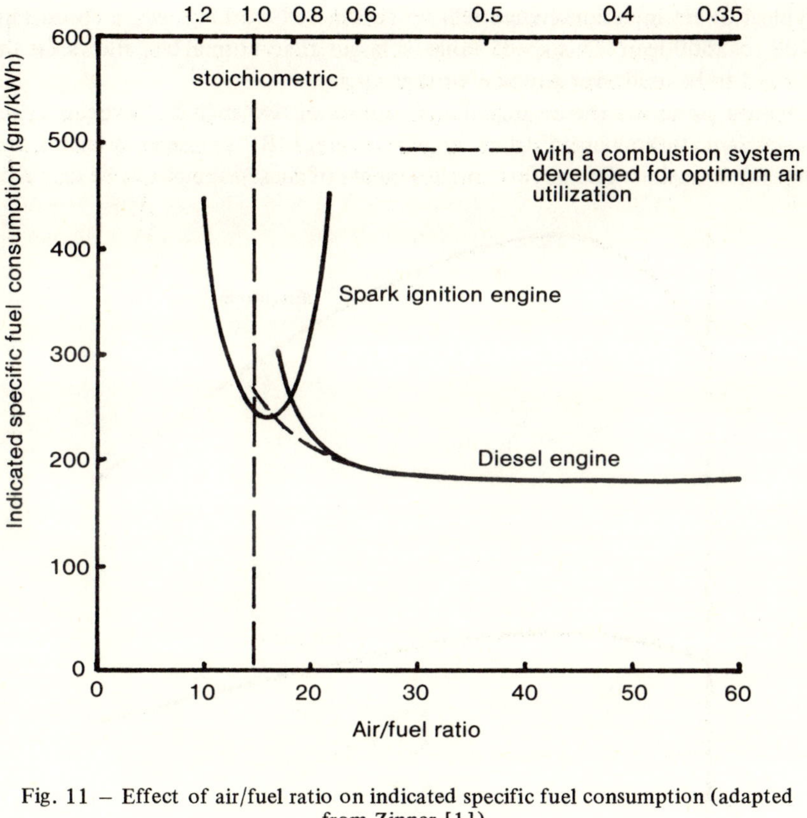

Fig. 11 – Effect of air/fuel ratio on indicated specific fuel consumption (adapted from Zinner [1]).

From equations (10) to (13), power can be linked to airflow as follows:

$$\dot{m}_a = \dot{W} \times \text{IFSC} \times (\dot{m}_a/\dot{m}_f)_{st}/(F \times \eta_m) \tag{14}$$

Equation (14) enables the air requirement for any particular power requirement to be established. Equations (1) and (4) (or (2) and (9) for a charge cooled engine) link the airflow requirement to the supercharger.

5. ENGINE POWER AND SUPERCHARGER REQUIREMENTS

The requirements placed on a supercharger vary substantially with engine type and its application. Three types of engine will be considered, as representative of widely different types, namely a four-stroke truck diesel engine, a four-stroke medium speed marine diesel engine and a four-stroke passenger car diesel engine. In addition the effects of different engine ratings and charge air cooling are considered.

5.1 Four-stroke truck diesel engine

A typical direct injection engine will be considered working over a speed range of 600 to 2600 rpm. This speed range is larger than normal but allows the data presented to be used over any part of that range.

Figure 12 shows the desired torque curves of two engines, 'A' being typical of a current turbocharged truck engine whereas 'B' represents a much more highly rated engine. The airflow requirements of these engines can be calculated

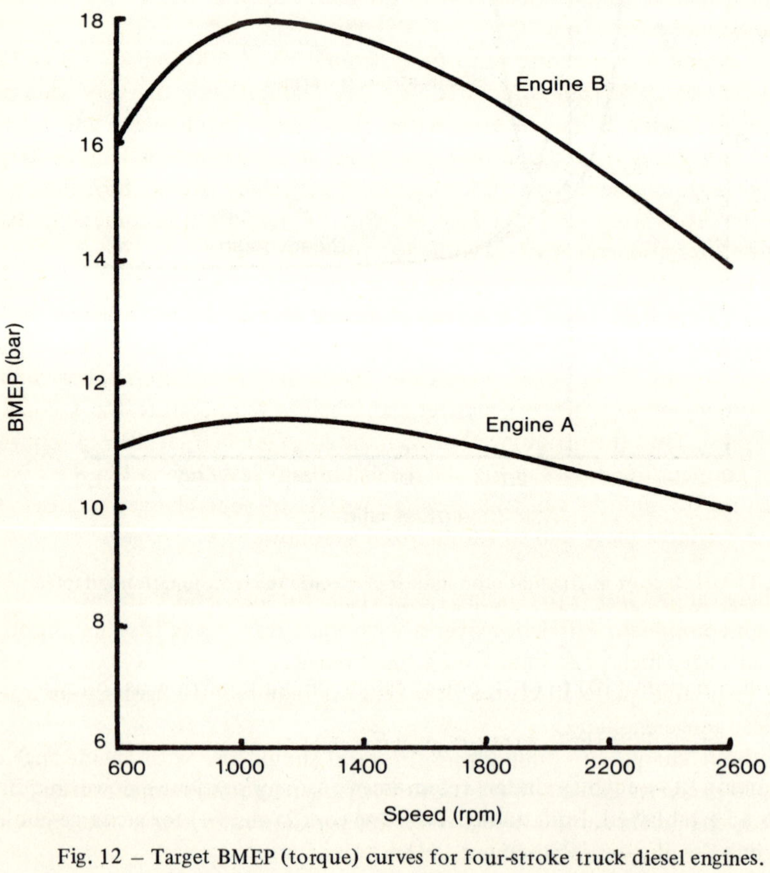

Fig. 12 – Target BMEP (torque) curves for four-stroke truck diesel engines.

from equation (14), once the air/fuel ratio has been established. Experience shows that a normal swirl-type direct injection diesel engine requires an air/fuel ratio of around 24:1 for optimum fuel consumption *for a specified BMEP.* Figure 11 shows that the benefit of higher air/fuel ratios is marginal. The fact that many engines run with much lower full-load air/fuel ratios is largely due to the limitations of existing superchargers and maximum cylinder pressure constraints that must be ignored herein.

The airflow required from equation (14) must match that calculated from equation (1). This defines the supercharger pressure ratio required as shown on Fig. 13, which is the pressure ratio versus mass flow rate characteristic for a four-stroke engine per litre engine capacity with a volumetric efficiency of 0.9 and a compressor isentropic efficiency of 0.7. Superimposed are the airflow requirements of engines A and B (assuming no charge air cooling), with an air/fuel ratio of 24:1. In practice, the exhaust temperature rises with engine speed at constant air/fuel ratio and this may impose a requirement for additional air at high speeds. This is illustrated by the lines shown in Fig. 13 for a constant exhaust temperature of 650°C.

For engine A, a pressure ratio rising from 1.85 at 600 rpm to 2.0 at 1000 rpm to 2.33 at 2600 rpm is required, the key feature being the very wide mass flow range. Engine B has an even wider flow range requirement and needs a pressure ratio rising from 3.5 at 600 rpm to 4.2 at 1000 rpm to 4.75 at 2600 rpm. Although a requirement for a 70% compressor efficiency over an 80% flow range at pressure ratios from 1.85 to 2.33 is difficult, but feasible, it becomes impossible if the pressure ratio exceeds 4:1, with current technology.

Intercooling reduces the pressure ratio required to achieve the desired mass flow rate. Thus Fig. 14 is a companion figure to Fig. 13 but assumes air–air charge cooling with an effectiveness of 0.8 (equation (7)). The target performance of engine B now looks more reasonable. Note that since intercooling reduces inlet air temperature, it also reduces exhaust gas temperature for a constant air/fuel ratio. Thus the difference between the constant air/fuel ratio and constant exhaust temperature curves in Fig. 14 is substantially less than in Fig. 13.

Figures 13 and 14 illustrate several aspects of supercharger requirements for truck engines. These can be summarized as follows:

(a) Charge air cooling (intercooling) is essential for high BMEP engines.
(b) High compressor efficiency over a wide mass flow range is more important than a very high peak value over a small range.
(c) A very large surge to choke flow range is required, particularly on a wide speed range engine.
(d) Without charge air cooling, pressure ratio should rise with engine speed to avoid excessive exhaust valve (or turbocharger turbine) temperatures.
(e) With charge air cooling a relatively flat pressure ratio versus mass flow rate and speed characteristic is desirable.

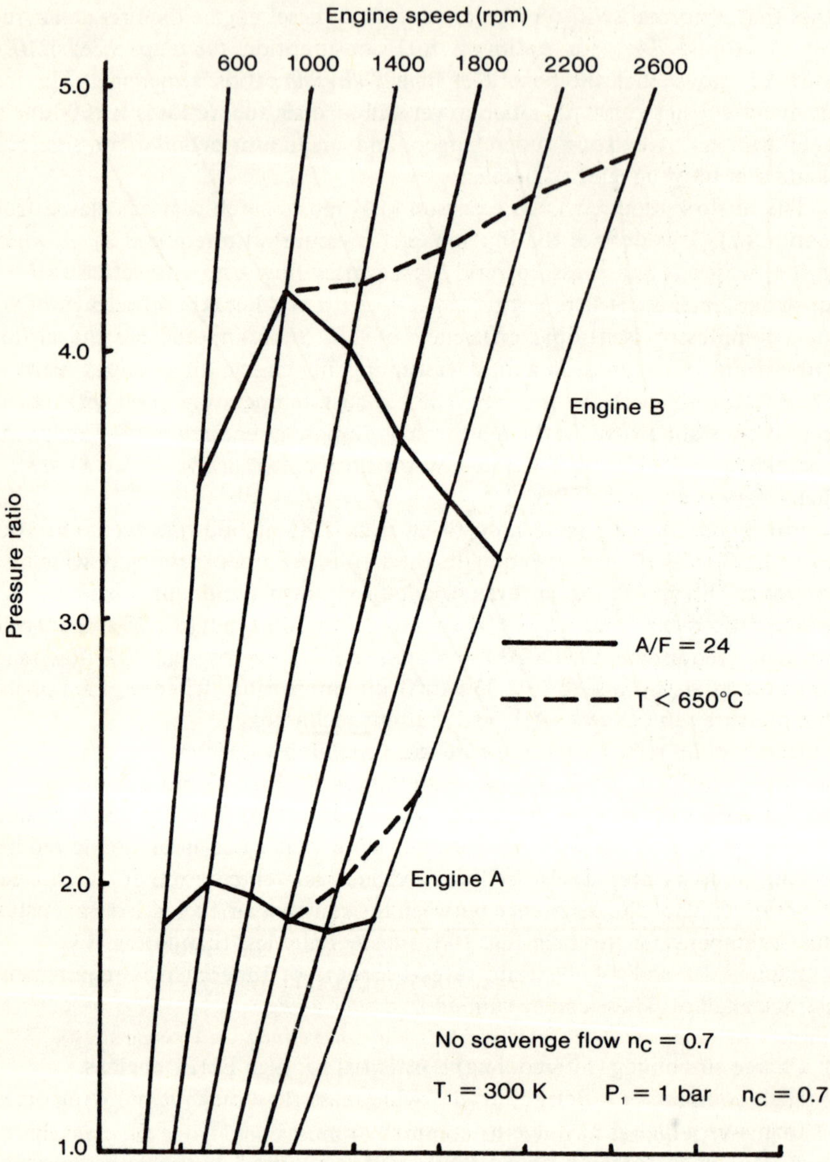

Fig. 13 – Compressor requirements for a four-stroke truck diesel engine without charge air cooling.

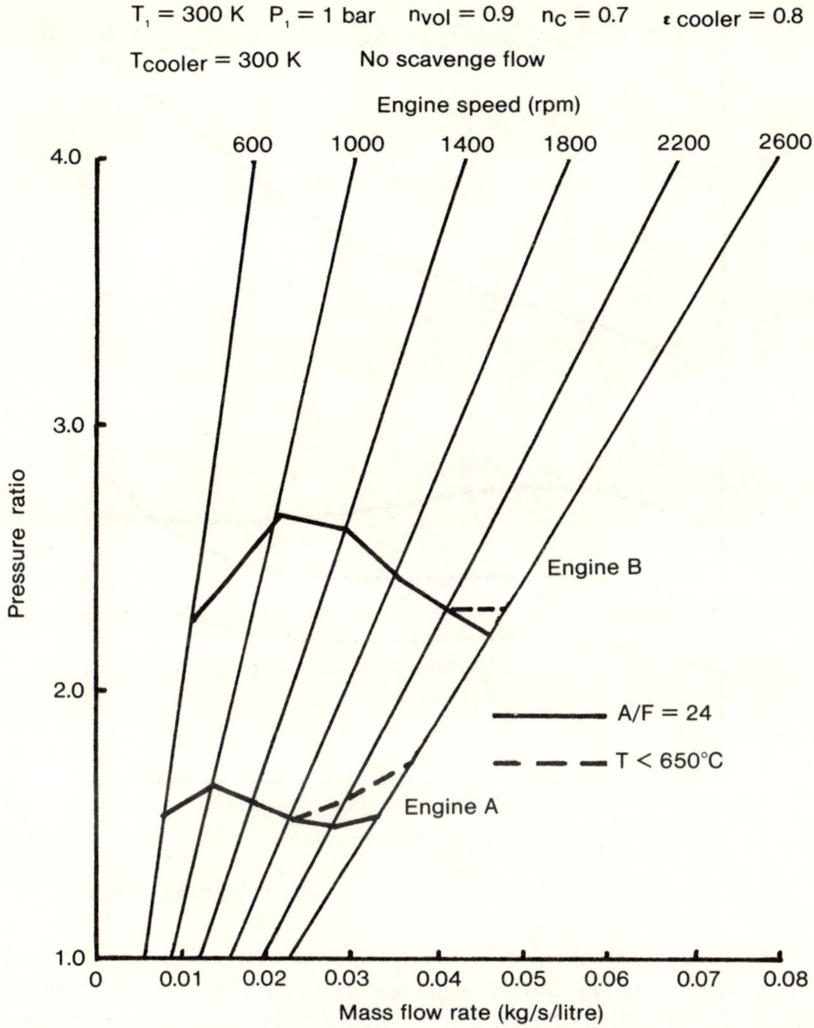

Fig. 14 – Compressor requirements for four-stroke truck diesel engines with charge air cooling.

The power required to drive the compressor must come from the crankshaft of the engine or an exhaust driven turbine. Obviously the latter system is potentially the more efficient but a mechanically driven supercharger is attractive if its power requirement is small. Figure 15 shows the power required to provide the airflows of Figs. 13 and 14, expressed as a fraction of engine power (with no allowance for mechanically driving the supercharger). With the exception of the non-intercooled engine B (which we have said was impractical), the superchargers

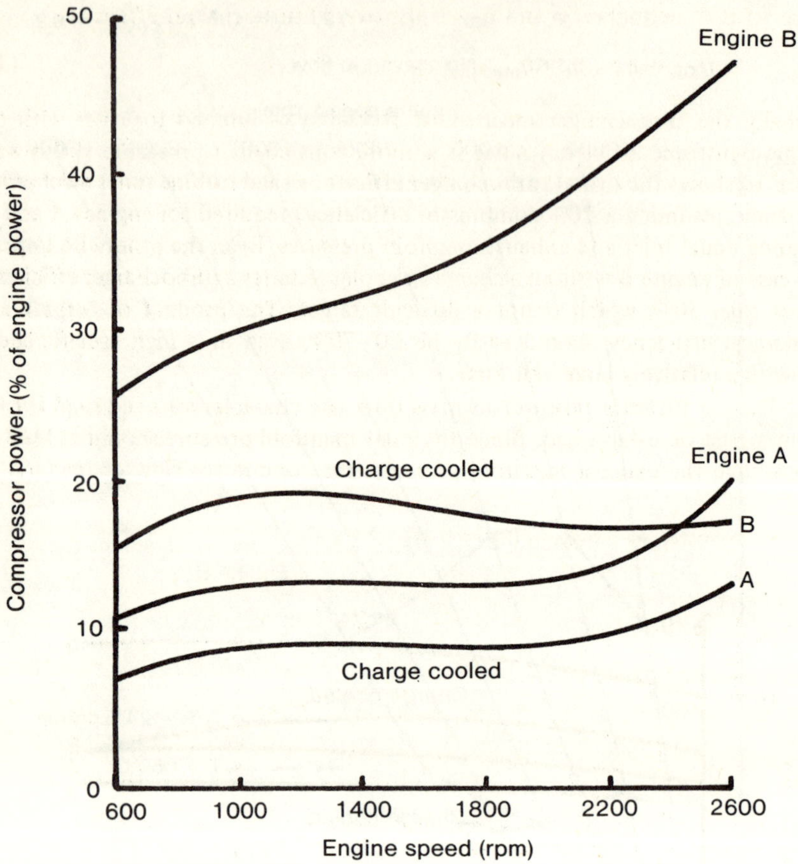

Fig. 15 – Compressor power requirements for four-stroke truck diesel engines.

require between 9 and 20% of the engine power output (less supercharger). In view of the fact that most truck engines spend much of their operational time at high load where the supercharger is needed, this rules out a mechanical supercharger on fuel economy grounds. Thus both engines A and B would be turbocharged.

In the case of turbocharged engines, the overall turbocharger efficiency required to achieve the pressure ratios of Figs. 13 and 14 is of interest. On the basis of a constant pressure turbine it is easy to calculate an energy balance between compressor and turbine. For a diesel engine, in which fuel is added in the cylinders, this becomes:

$$\left[\left(\frac{P_2}{P_1}\right)^{\frac{\gamma-1}{\gamma}} - 1\right]_{\mathrm{c}} T_1 C_{P1} \dot{m}_1 = \left[1 - \left(\frac{P_4}{P_3}\right)^{\frac{\gamma-1}{\gamma}}\right]_{\mathrm{t}} T_3 C_{P3}\left(1 + \frac{1}{\mathrm{AFR}}\right)\eta_{\mathrm{TC}} \tag{15}$$

where AFR = air/fuel ratio and η_{TC} is the overall turbocharger efficiency

$$\eta_{TC} = \eta_c \times \eta_t \times \eta_{mech(tc)} \tag{16}$$

Normally the turbocharger mechanical efficiency is lumped together with the turbine isentropic efficiency since it is more convenient to measure it this way. Figure 16 shows the overal turbocharger efficiency (and turbine times mechanical efficiency, assuming a 70% compressor efficiency) required for engines A and B, assuming equal inlet and exhaust manifold pressures. Even the otherwise impractical case of engine B without a charge air cooler requires a turbocharger efficiency of less than 50%, which is not a difficult target. The product of turbine and mechanical efficiency should easily be 60–70%, even in a high specific speed unit with a relatively large exit loss.

Next the pressure ratio versus mass flow rate characteristics required for the turbine must be established. Since the inlet manifold pressure should at least be no less than the exhaust manifold pressure, the compressor flow characteristics

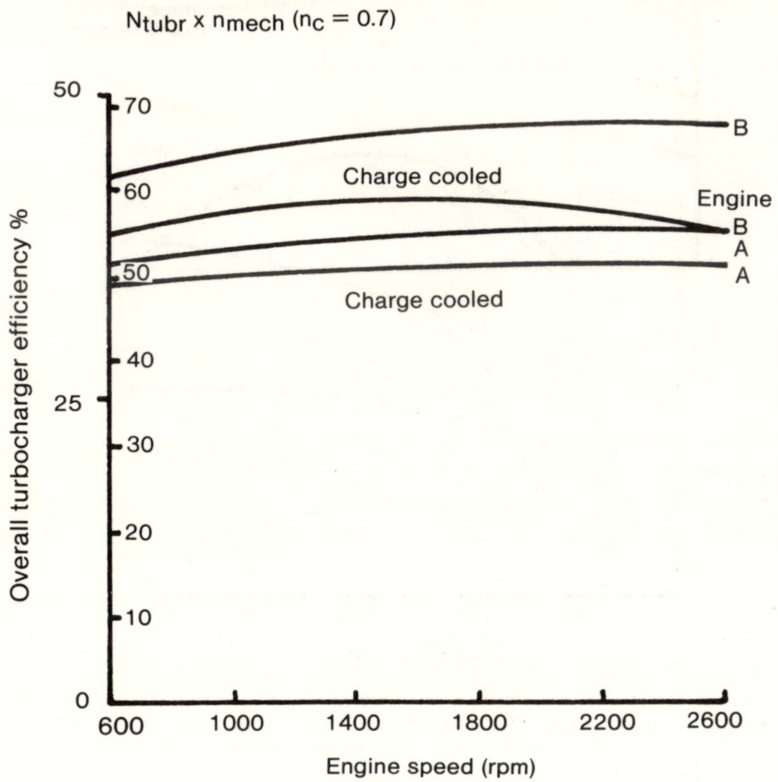

Fig. 16 – Turbocharger and turbine efficiency required for equal inlet and exhaust manifold pressures ($T < 650°C$).

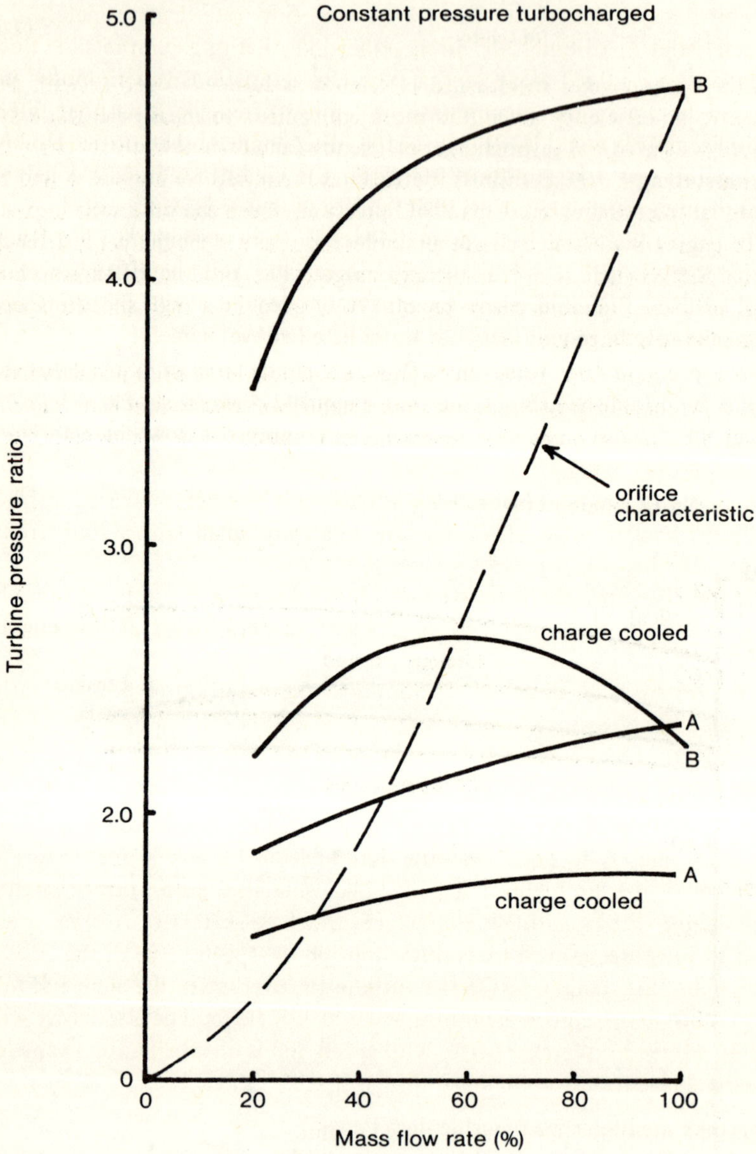

Fig. 17 – Turbine flow characteristics required for four-stroke truck engines, with equal inlet and exhaust manifold pressures ($T < 650°C$).

must virtually be duplicated by the turbine (Fig. 17). The requirement is a pressure ratio that is relatively insensitive to mass flow rate. Unfortunately the flow characteristics of turbines are often rather like that of a compressible flow through an orifice of fixed area. Such a flow characteristic is superimposed on Fig. 17, sized to match the maximum power point of engine B without inter-cooling (similar curves could be drawn for the other engines). The orifice chokes at a pressure ratio of 1.8 (approx.); thereafter the volumetric flow is constant and mass flow increases linearly with inlet density at higher pressure ratios.

Clearly the orifice type flow characteristic is quite unsuitable for a truck engine turbocharger turbine. Turbine expansion ratio and therefore power is insufficient to drive the compressor to achieve the desired engine performance! We can now add several requirements to those listed above, namely:

(f) The power required to drive the compressor is too large for a mechanically driven supercharger to be used, without a significant loss in fuel economy.
(g) The overall efficiency required from the turbocharger is not high–less than 50% (but see (k) below).
(h) The benefits of higher turbocharger efficiencies through increased airflow and better engine cycle efficiency are relatively small (for a four-stroke engine) – see Fig. 11.
(i) The orifice like flow characteristics of a normal turbine are quite unsuitable for this application, resulting in insufficient turbine power at low engine speeds.
(j) The turbine must be designed to minimize the variation of pressure ratio with mass flow rate.
(k) Very high turbine efficiency at low flow rates will partly offset the potential turbine power deficit.

It can be concluded that the turbine power deficit implied by Fig. 17 at low mass flow rates is so large that the combination of several measures are required to solve the problem. These include using the 'pulse' turbocharging system, designing for a turbine characteristic whose effective 'orifice area' increases with pressure ratio or mass flow, limiting the speed range of the engine (and hence mass flow range), sizing the turbine to 'overboost' the engine at full speed and finally designing for high efficiency at low flows. The alternative will be an engine whose torque curve rises with speed and is unsuitable for a vehicle, and/or greatly excessive boost pressure at high speeds.

5.2 Four-stroke medium-speed marine diesel engine

The requirements for this type of engine are totally different from the truck engine of section 5.1, and are dominated by the characteristics of the ship's propeller. Propeller power increases approximately as the cube of speed (the well known 'propeller law'). Thus the torque or BMEP required increases with speed squared. Fig. 18 shows the power curve required for a medium-speed engine having a rating of 25 bar BMEP at full speed (1000 rpm) and load.

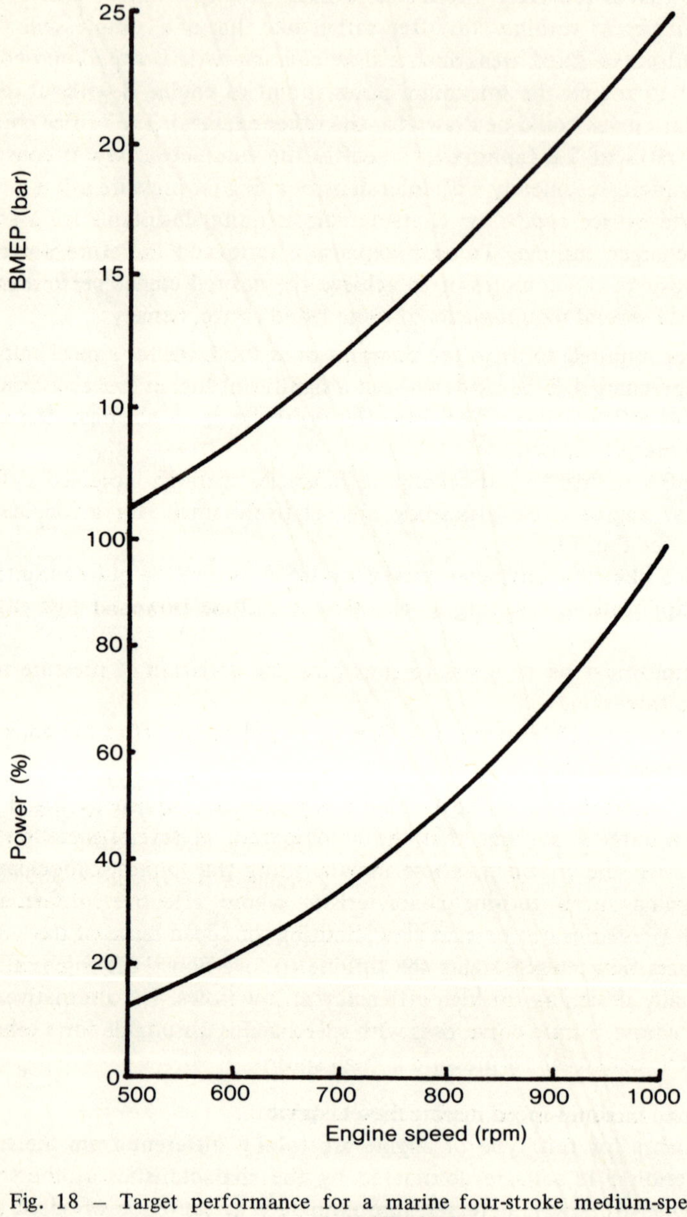

Fig. 18 – Target performance for a marine four-stroke medium-speed engine (propeller law).

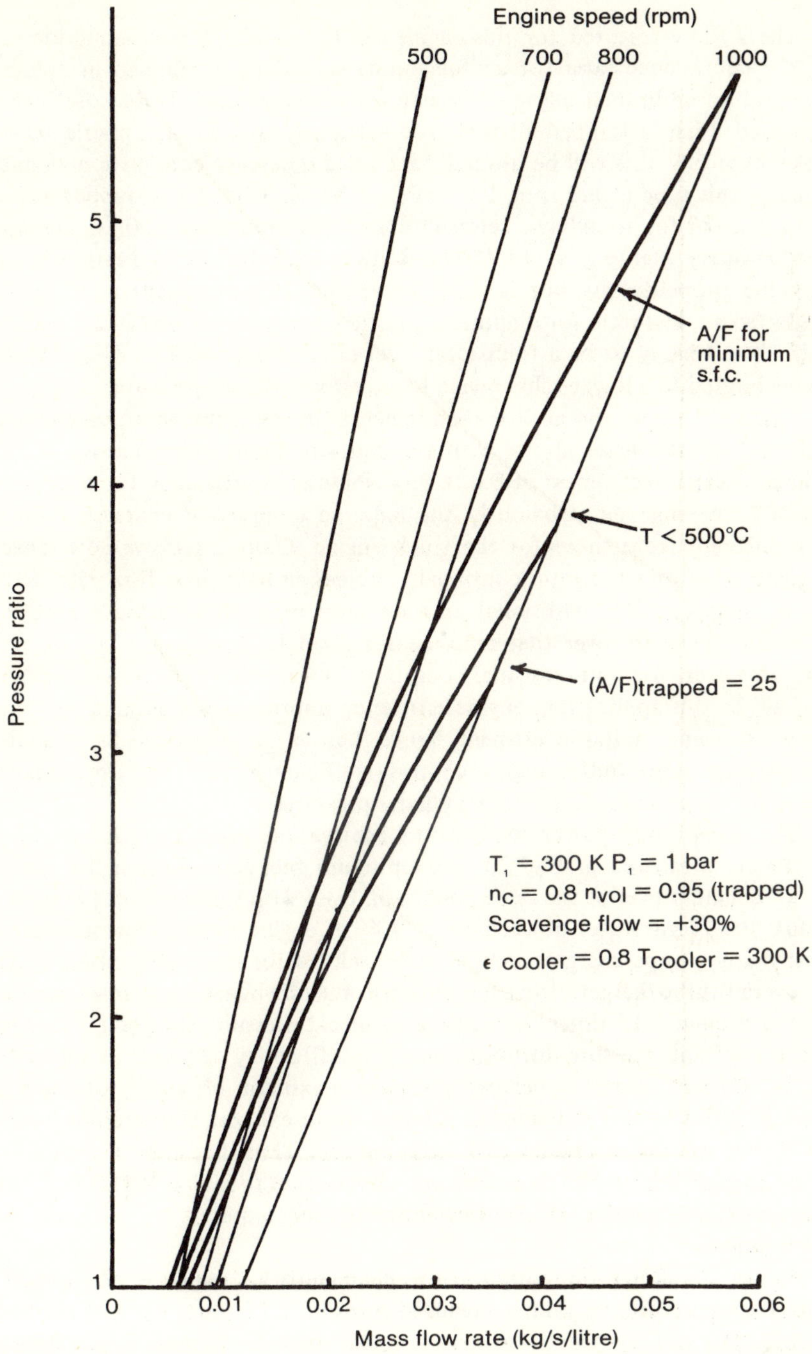

Fig. 19 – Compressor requirements for marine medium-speed, four-stroke engine (propeller law).

The airflow required for this engine can be calculated from equation (14) but the equivalence ratio (or air/fuel ratio) should be a 'trapped in cylinder' value and an additional allowance made for scavenge flow. Three cases can be considered. First, a trapped air/fuel ratio sufficient only for acceptable exhaust smoke; typically this will be around 25:1 for a quiescent combustion chamber engine of this type (1000 rpm, 0.2 to 0.3 m bore). Secondly, a trapped air/fuel ratio sufficient for an exhaust temperature of not more than 500 °C, for operation on heavy fuel (e.g. up to 3500 s Redwood I.). The air/fuel ratio will vary along the propeller law but is likely to exceed 25:1 (trapped) at all speeds. Thirdly, an air/fuel ratio for minimum specific fuel consumption (at the specified load). This is likely to be a trapped/air fuel ratio of around 40:1 (Fig. 11), but may be limited to a lower value owing to excessive cylinder pressure.

Figure 19 shows the airflow requirements for the compressor, based on the three air/fuel ratio cases above, with a compressor isentropic efficiency of 80%, an intercooler effectiveness of 0.8, a 95% volumetric efficiency for trapped air plus 30% scavenge air. Obviously the required compressor characteristics are quite different from those for the truck engine. Quite a narrow flow range is adequate, with pressure ratio continually increasing with mass flow rate. It can also be seen that the additional pressure ratio required to achieve a 500 °C exhaust temperature over that required for a smoke limit is not unreasonably large. However, for optimum fuel consumption a much higher pressure ratio is required. In this application, engine efficiency improvements as small as 1% are well worth some capital investment, hence there is a strong incentive to achieve high pressure ratio and/or high compressor efficiency, but the former may be limited by the allowable maximum cylinder pressure.

The compressor power required to achieve the pressure ratios shown in Fig. 19 are presented in Fig. 20. For optimum fuel consumption this rises to 67% of engine power at the rated condition. Even with sufficient air for burning residual fuel, compressor power rises to 40% of engine power. Obviously mechanically driven superchargers cannot be considered for this highly rated engine. The overall turbocharger efficiency required must be based on an inlet manifold pressure around 1.15 times the exhaust manifold pressure, for good scavenging with a constant pressure turbocharger. The efficiency required is plotted in Fig. 21. Current large turbochargers achieve maximum efficiency of 63–65%, which is sufficient for this engine running to an exhaust temperature limit of 500°C, but not for optimum fuel consumption. Thus if ratings of four-stroke marine engines are to rise above current values of up to 20 bar BMEP to 25 bar, compressor and turbine efficiencies over 85% are required for optimum fuel consumption.

Figure 22 shows the turbine flow characteristics required, for a 500°C exhaust temperature, with inlet manifold pressure 1.15 times exhaust manifold pressure. The requirement is an almost linear increase in mass flow rate with pressure, which suits the characteristics of a choked orifice quite well. It can be

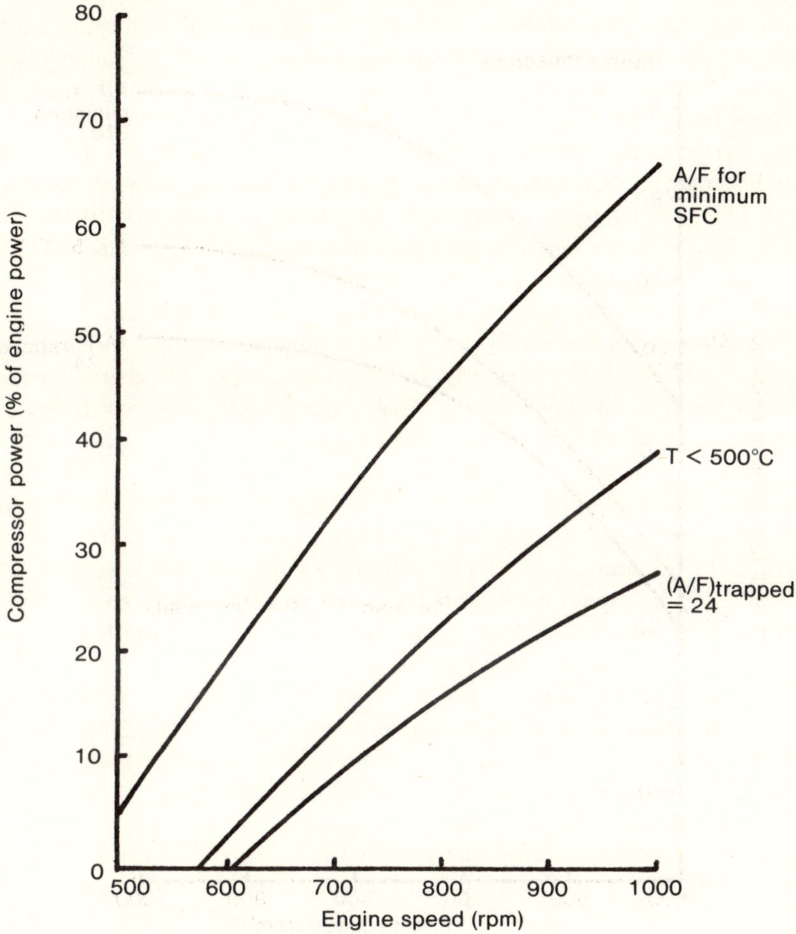

Fig. 20 – Compressor power requirement for marine four-stroke medium-speed engine (propeller law).

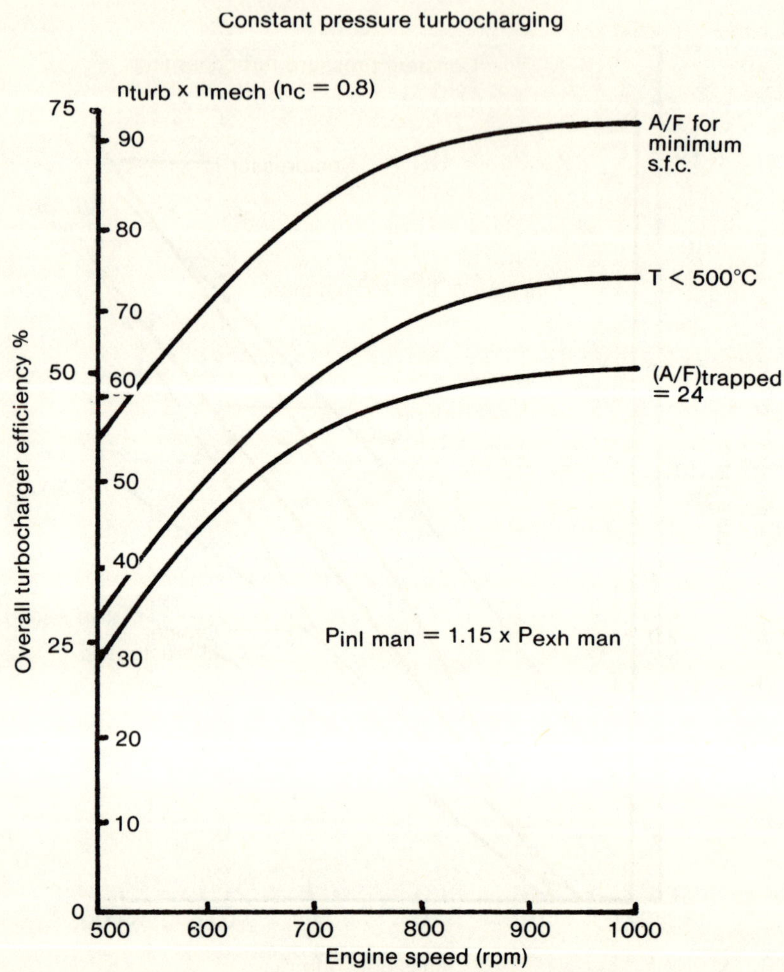

Fig. 21 – Turbocharger and turbine efficiency required for marine four-stroke diesel engine (propeller law).

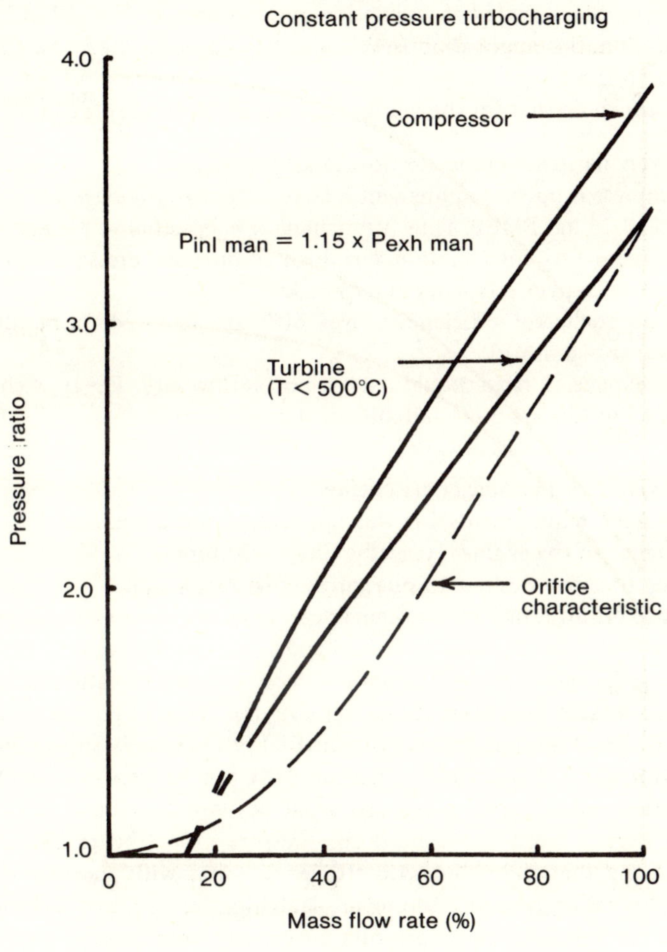

Fig. 22 – Turbine flow characteristics for marine four-stroke medium-speed engine (propeller law).

seen that the orifice flow characteristic is much closer to the required curve than in the case of the truck engine. The efficiency required to achieve the necessary boost with this turbine expansion ratio curve (Fig. 21) falls with reducing engine speed, hence higher efficiencies are likely to be available to compensate for the shortfall in expansion ratio shown in Fig. 22 at less than rated speed. Thus this application is well suited to an orifice like flow characteristic from the turbine.

We can conclude that for the marine four-stroke medium-speed diesel engine:

(a) A relatively narrow compressor flow range is adequate.
(b) The compressor power requirement is large relative to engine power, typically 40–60% at 25 bar BMEP. Thus turbocharging is essential.
(c) For optimum fuel consumption very high compressor pressure ratio and/or overall turbocharger efficiency is required.
(d) Overall turbocharger efficiencies over 60% are required for residual fuel operation at high BMEP.
(e) Turbine expansion ratio should rise with mass flow rate. The flow characteristics of an orifice are quite suitable for this application.

5.3 Four-stroke diesel passenger car engine

The supercharger requirements for this application are dominated by the very wide speed range of the engine (typically 1000–5000 rpm).

Consider first the desired torque curve. Figure 23 shows the power curve for an advanced European 2 litre passenger car diesel engine in naturally aspirated form and its BMEP (or torque) curve. 1.5 and 1 litre mechanically supercharged or turbocharged engines might be considered to replace the naturally aspirated 2 litre. Their BMEP requirements to achieve the same power curve are also shown in Fig. 23. Assuming a minimum air/fuel ratio (usually smoke limited to 17–18:1 with indirect injection), equation (14) can be used to calculate the required airflow rate. Equation (4) can again be used to plot mass flow as a function of engine speed and compressor pressure ratio. Figure 24 shows the results, assuming that the volumetric efficiency varies with speed in the same manner as the naturally aspirated engine. Matching the required airflows at all speeds gives the required boost pressure curve. Not surprisingly, since we are trying to match the torque curve of the naturally aspirated engine, the required compressor pressure ratio is almost constant, at about 1.75 for the 1.5 litre engine and 3.1 for the 1 litre engine. In both cases an extremely wide compressor flow range is required – virtually constant pressure ratio from 15 to 100% mass flow.

If a charge air cooler (air–air) is used, then the required pressure ratio reduces. Figure 25 is a companion to figure 24, an intercooled engine with a cooler effectiveness of 0.7, but negligible pressure loss (in practice a small charge cooler will have a significant pressure loss). The charge cooler reduces

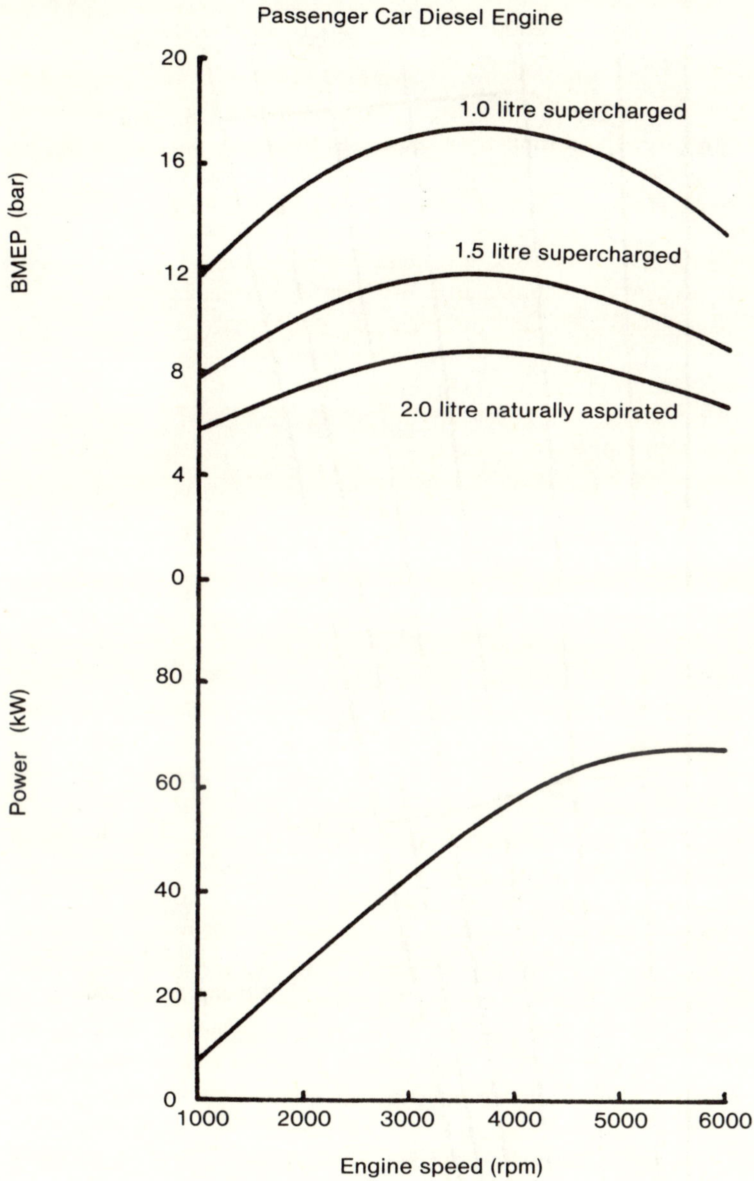

Fig. 23 – Performance target of 2 litre (naturally aspirated) and equivalent 1.5 and 1.0 litre supercharged diesel engines.

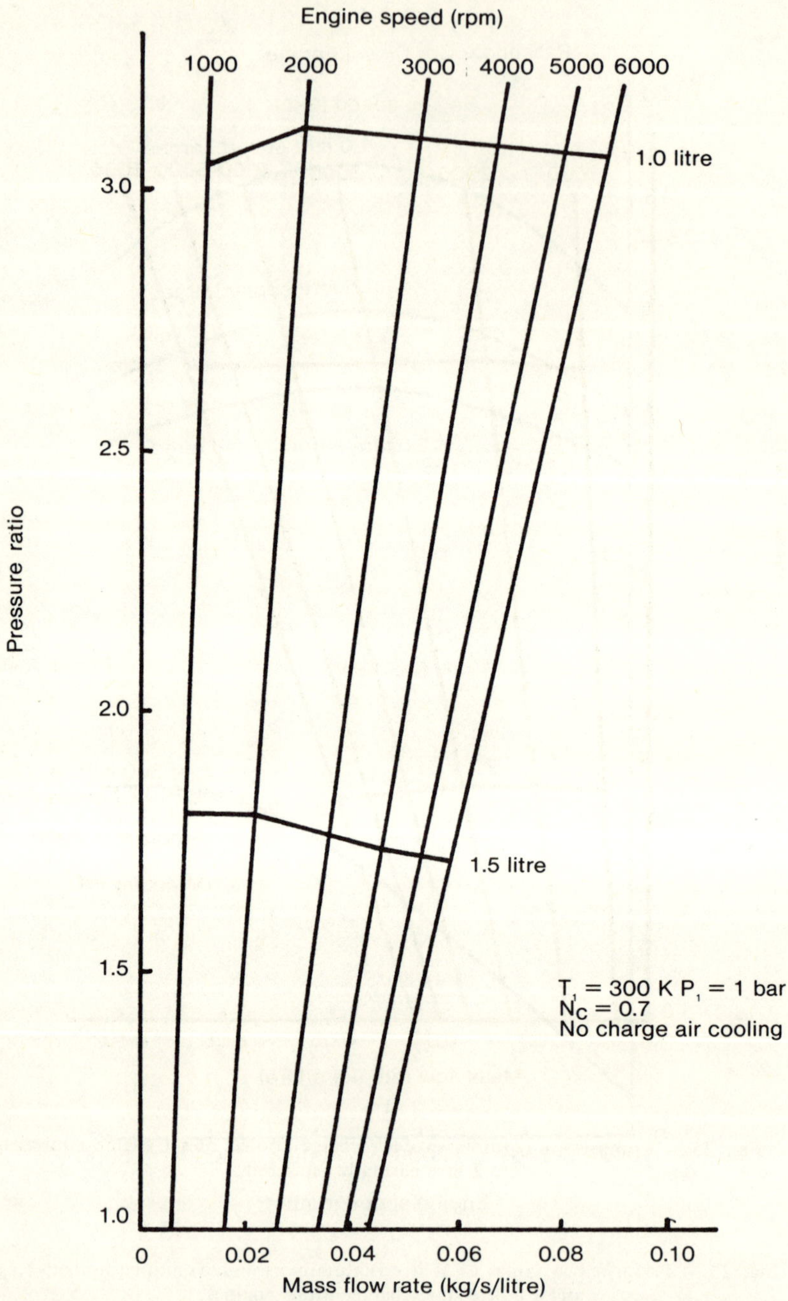

Fig. 24 – Compressor requirements for supercharged diesel engine equivalent to 2 litre naturally aspirated.

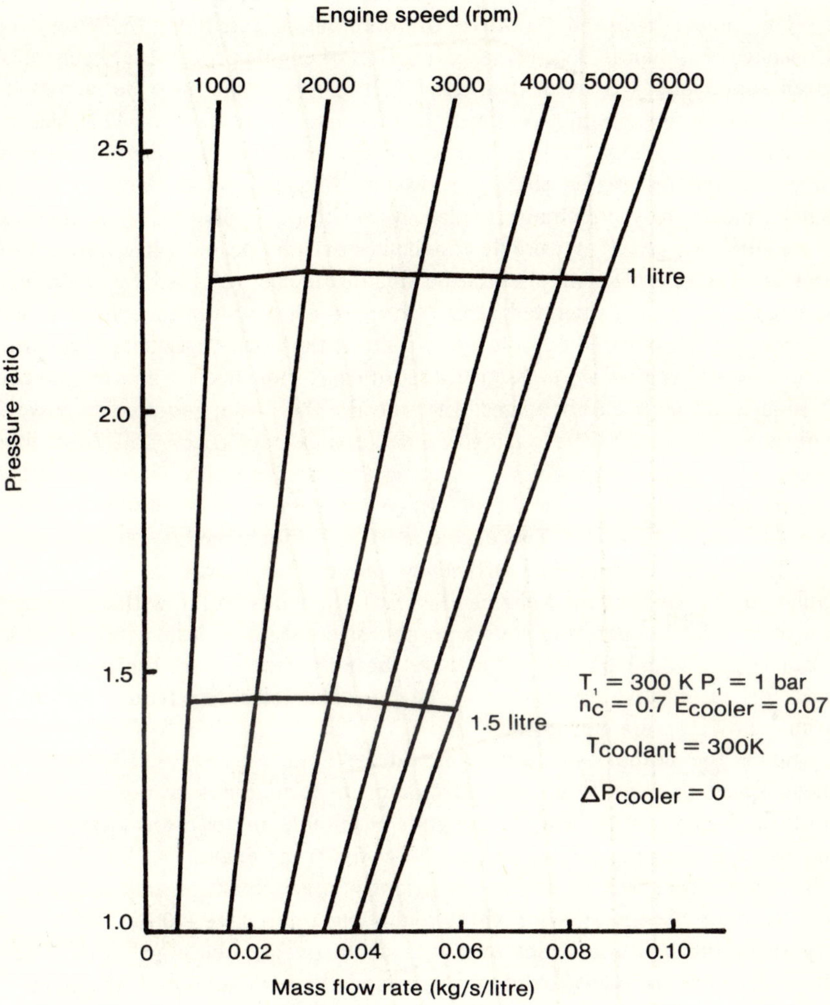

Fig. 25 – Compressor requirements for charge cooled diesel engine equivalent to 2 litre naturally aspirated.

the required pressure ratio from 1.75 to 1.45 (1.5 litre engine) and 3.1 to 2.25 (1 litre engine). The 1.5 litre engine looks a practical proposition, especially when charge cooled but the 1 litre still needs a high pressure compressor.

The power required to dirve the compressor (assuming 70% isentropic efficiency) is shown in Fig. 26, as a function of engine power. If a mechanically driven supercharger is used, then the IMEP of the engine must be increased to provide the power, hence fuel and airflow must increase also. This has been ignored for simplicity in Fig. 26, hence the power requirement for a supercharger is underestimated slightly. However, Fig. 26 reveals that the 1.5 litre engine, particularly in charge cooled form, could easily use a mechanically driven supercharger. If it were de-clutched or even by passed when not required, the loss of engine efficiency would be small except at full load. Since passenger car engines are rarely operated at full 'throttle' for any duration, this loss might be offset by the reduced friction and weight of the smaller engine, provided that the engine and vehicle were designed accordingly. Thus the 1.5 litre engine could be supercharged or turbocharged. In contrast, the power required to drive the compressor of the 1 litre engine is 14% (charge cooled) or 20% (not charge cooled) of engine power, which is excessive for a mechanically driven compressor. This is particularly true since the smaller engine will require boost for significantly more of its operating time. This engine must be turbocharged.

The overall turbocharger efficiency required is shown in Fig. 27 and is similar to that of the truck engine (Fig. 16) but tends to fall with engine speed as turbine inlet temperature rises with almost constant air/fuel ratio. A modern turbocharger should be able to achieve these figures, even in the case of small units for automotive applications. The flow characteristics required from the turbine, however, are a problem.

Since the compressor boost should be almost constant, irrespective of engine speed or mass flow rate, so should the turbine expansion ratio. Orifice like flow characteristics (Fig. 28) are quite unsuitable. In this respect the passenger car engine is even more of a problem than the truck engine. As Fig. 28 shows, it becomes impossible to achieve sufficiently large expansion ratio at low flow rates without becoming too high at high flow rates. One solution is to use a very small turbine, equivalent to a very small orifice, combined with a wastegate that opens to bypass exhaust gas around the turbine at high speeds (recognized by high boost pressure). Even so, it is difficult to achieve the required boost pressure at very low engine speeds without a significant loss of engine efficiency at high speed owing to high exhaust manifold pressure.

In summary:

(a) a 1.5 litre supercharged engine could replace a 2 litre naturally aspirated engine, particularly if charge cooled;
(b) the engine could be supercharged or turbocharged, but the supercharger should be bypassed or de-clutched when not required;

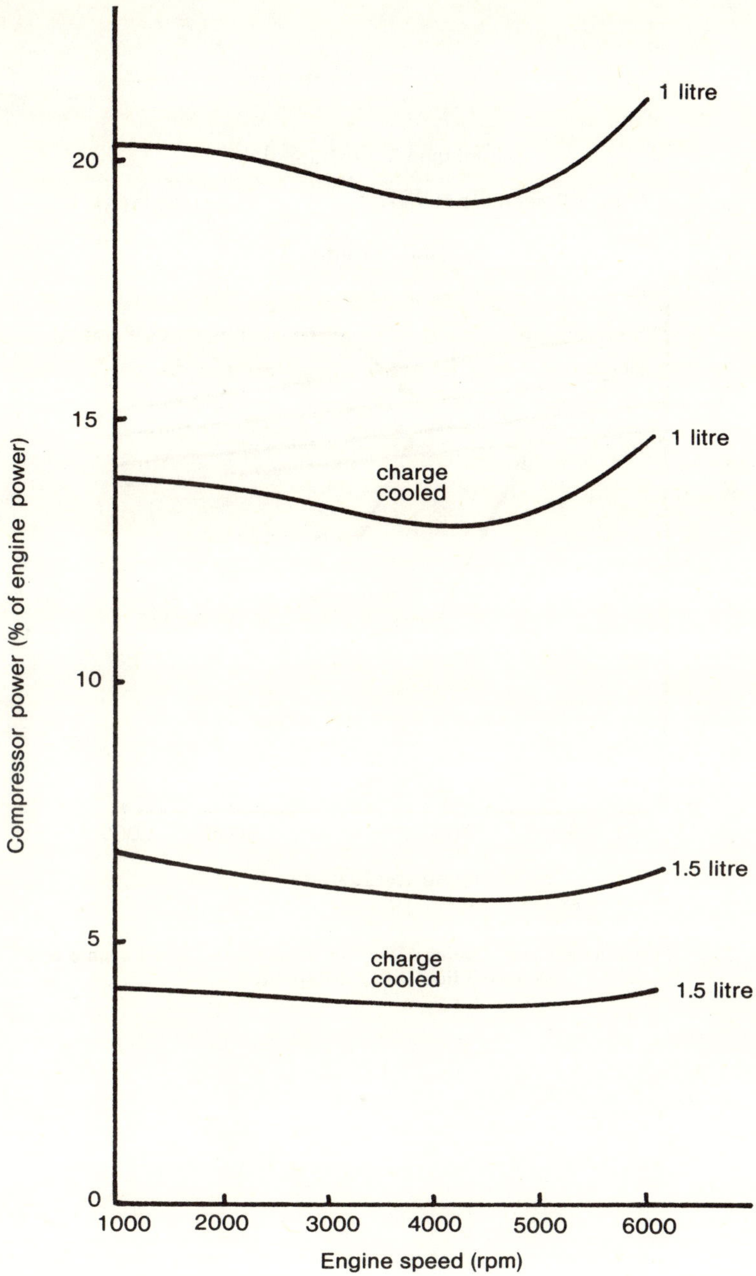

Fig. 26 – Compressor power requirement for diesel engine equivalent to 2 litre naturally aspirated.

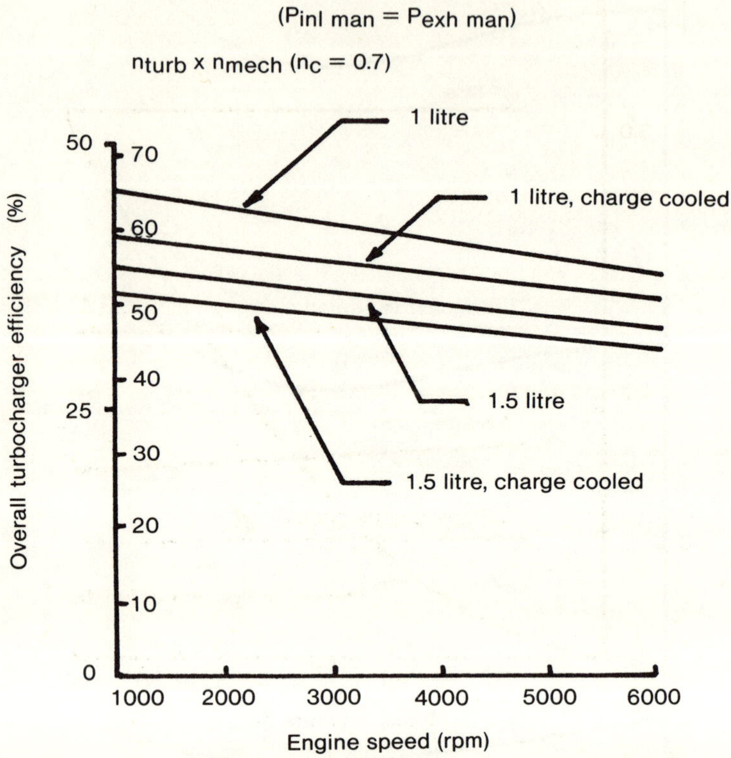

Fig. 27 – Turbocharger and turbine efficiency required for diesel engine equivalent to 2 litre naturally aspirated.

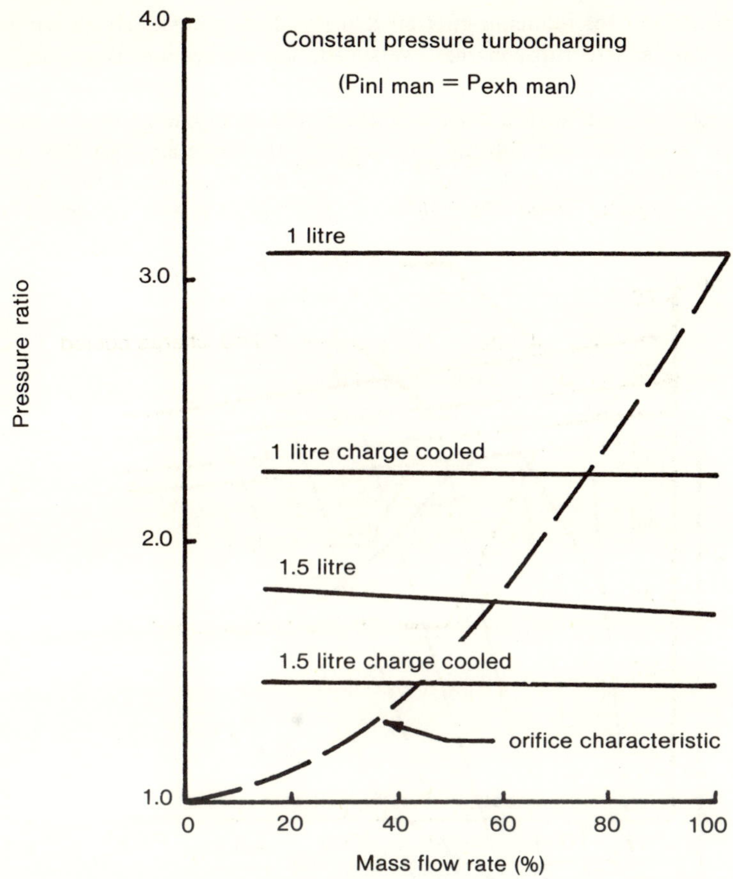

Fig. 28 – Turbine flow characteristic required for diesel engines equivalent to 2 litre naturally aspirated.

(c) the compressor power required for a 1 litre supercharged engine to replace a 2 litre naturally aspirated engine can only sensibly be provided by a turbocharger with charge cooling;

(d) both 1.5 and 1 litre engines require very wide flow range compressors, but overall turbocharger efficiency is not excessive;

(e) turbine flow characteristics are the most difficult problem; either a waste-gate or variable geometry is essential.

6. SUPERCHARGER REQUIREMENTS AT ALTITUDE

All the data so far presented have shown performance at typical sea level ambient conditions of 1.0 bar and 300 K (23°C). At altitude, the turbocharger offers partial compensation for reducing inlet air density. This is obviously a desirable attribute but can lead to turbocharger overspeed, and always means an increase in pressure ratio.

The air requirements of the truck engines A and B, in charge cooled form, will be taken as an example. Ideally they require the same air mass flow rate regardless of altitude. Figure 29 shows this same air requirement, but superimposed on compressor/engine characteristics calculated for inlet conditions of

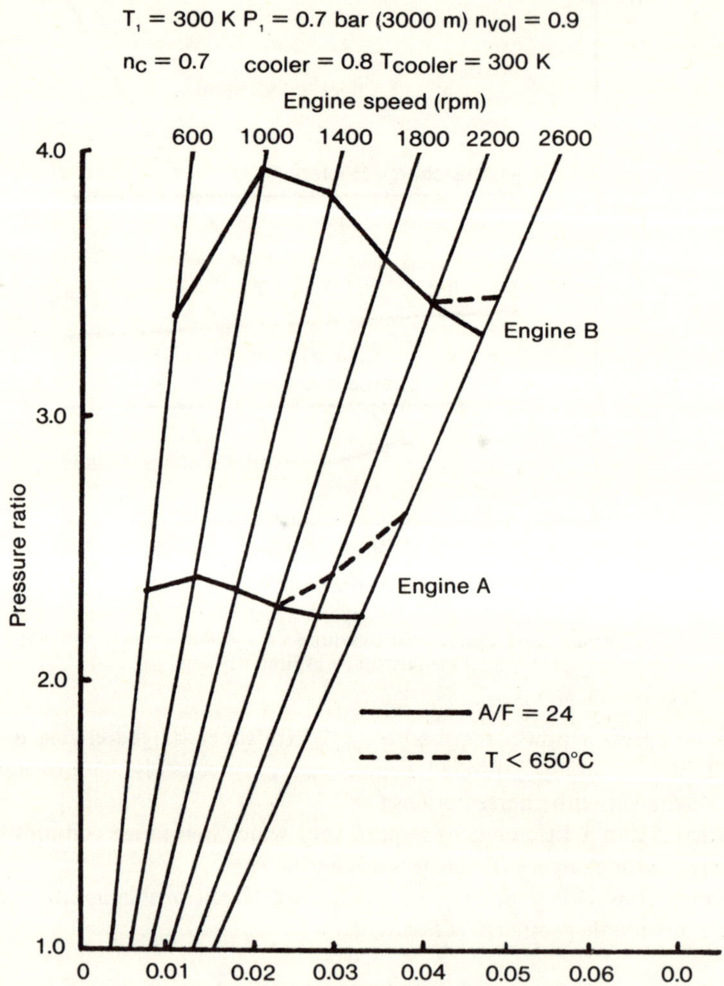

Fig. 29 – Compressor requirement at 3000 m altitude, four-stroke truck diesel engine.

0.7 bar, 300 K (i.e. approx. 3000 m altitude). Comparing Figs. 14 and 29 shows the need for a substantial increase in pressure ratio, for example, from 2.65 at 1000 rpm (engine B) to 3.93. Re-casting Fig. 29 in the more normal form of a compressor map, with volumetric flow or $\dot{m}\sqrt{(T/P)}$ as the horizontal axis, produces Fig. 30. This shows that operation at altitude amplifies the complete pressure ratio and volumetric flow rate requirement. In particular, the need for a very wide flow range is emphasized.

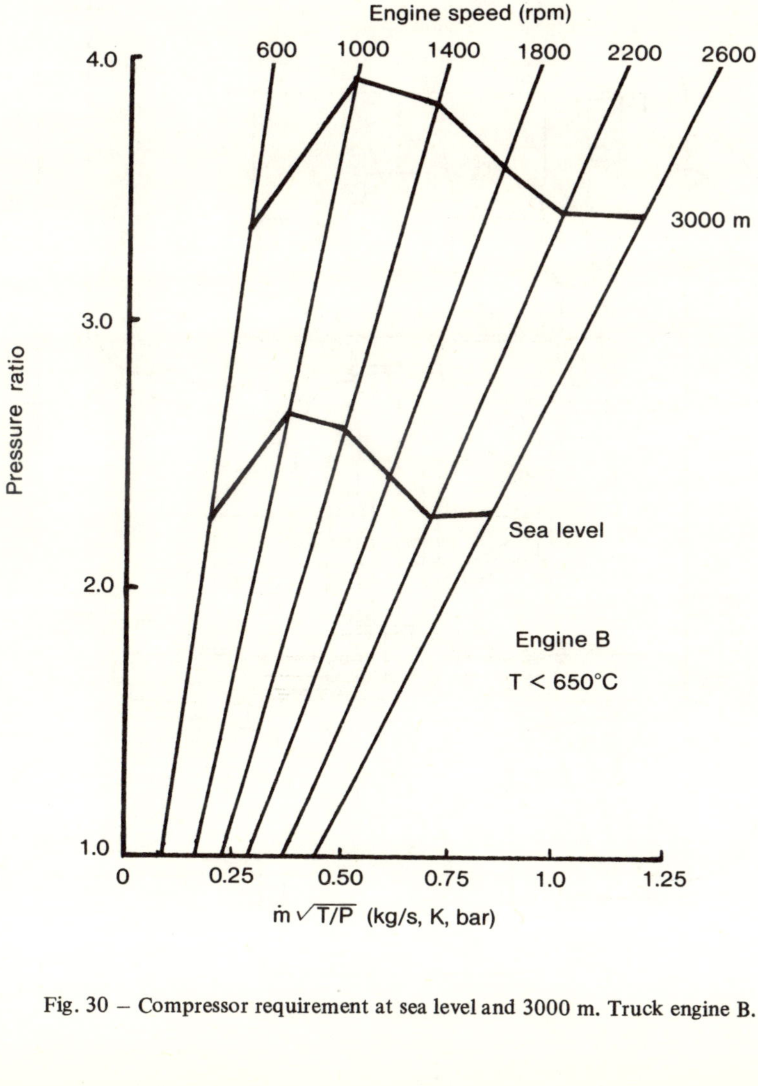

Fig. 30 – Compressor requirement at sea level and 3000 m. Truck engine B.

7. TURBOCHARGERS FOR VEHICLE DIESEL ENGINES

Since turbocharging is more efficient than supercharging, the remainder of this chapter will concentrate on the former system.

Turbochargers are used for passenger car diesel engines rated at 45 kW upwards to larger special heavy truck and construction vehicles rated at up to 600 kW. The most important design factors are cost, reliability and performance. To keep cost low, the design must be simple, hence a single stage radial flow compressor, and a radial flow turbine are mounted on a common shaft with an inboard bearing system (Fig. 31). This arrangement simplifies the design of inlet and exhaust casing and reduces the total weight of the turbocharger.

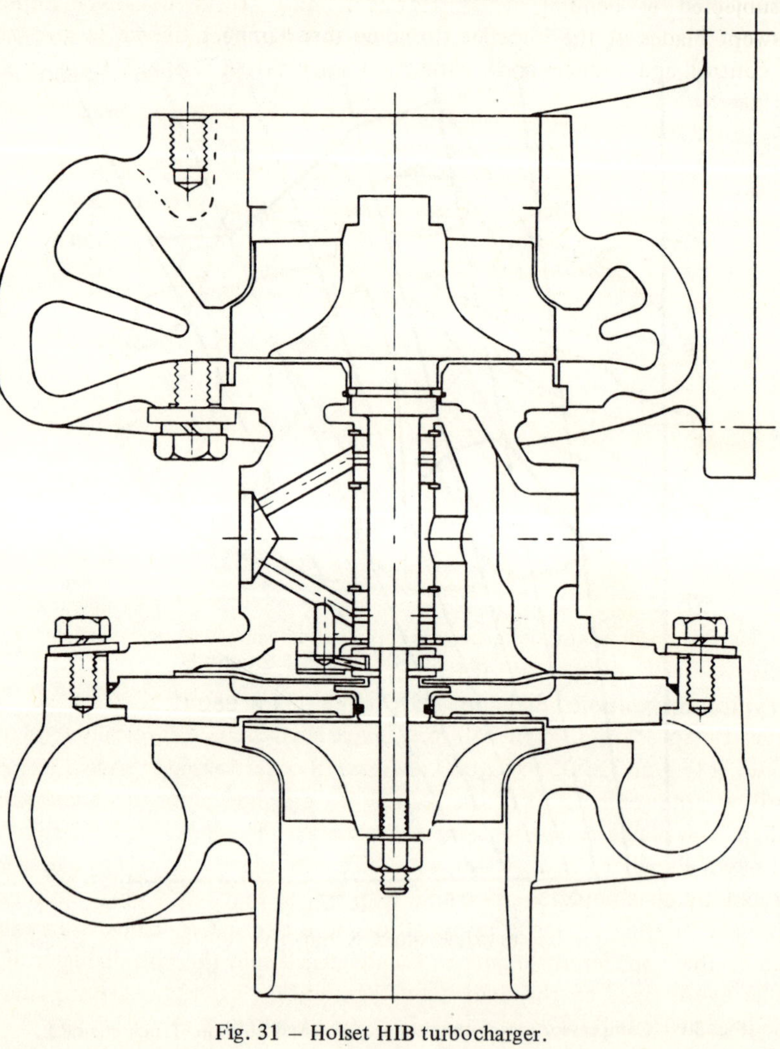

Fig. 31 – Holset HIB turbocharger.

7.1 Compressor

The compressor impeller is an aluminium alloy (LM-16-WP or C-356T61) investment casting, with a gravity die-cast aluminium housing (LM-27-M). The design of the impeller is a compromise between aerodynamic requirements, mechanical strength and foundry capabilities. To achieve high efficiency, and minimum flow blockage, very thin and sharp impeller vanes are required, thickening at the root (impeller hub) for stress reasons. It is common practice to use splitter blades (Fig. 32) that start part way through the inducer, in order to maintain good flow guidance near the impeller tip without excessive flow blockage at the eye. Until recentely the impeller vanes have been purely radial so that blades were not subjected to bending stress. However most recent designs incorporate backswept blades at the impeller tip since this has been shown to give better flow control and reduces flow distortion transmitted through from impeller to diffuser.

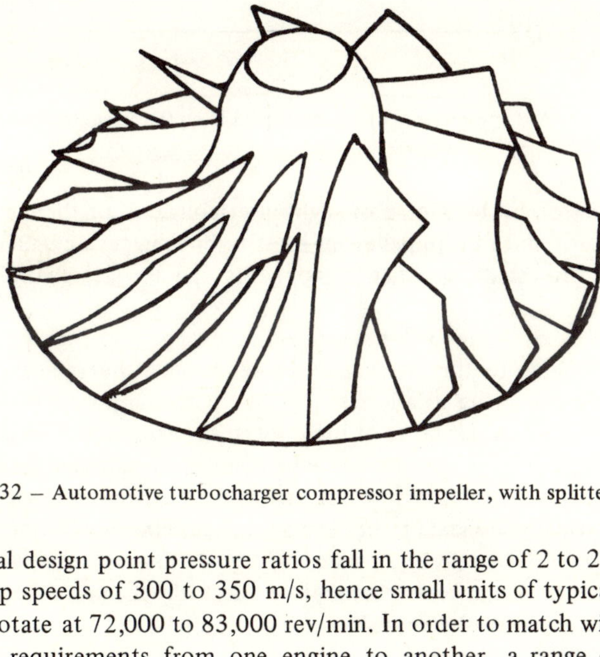

Fig. 32 – Automotive turbocharger compressor impeller, with splitter blades.

Typical design point pressure ratios fall in the range of 2 to 2.5:1, requiring impeller tip speeds of 300 to 350 m/s, hence small units of typically 0.08 m tip diameter rotate at 72,000 to 83,000 rev/min. In order to match wide differences in airflow requirements from one engine to another, a range of compressor impellers is available to fit the same turbocharger. These will be produced from one or two impeller castings, but with different tip widths and eye diameters generated by machining as shown in Fig. 33, and matched with appropriate compressor housings. Usually up to 10 or more alternative 'trims' are available but since the impeller tip diameter is unchanged and the hub diameter at the impeller eye is fixed by the shaft diameter, the flow passage variations alter the efficiency as well as flow characteristics of the impeller.

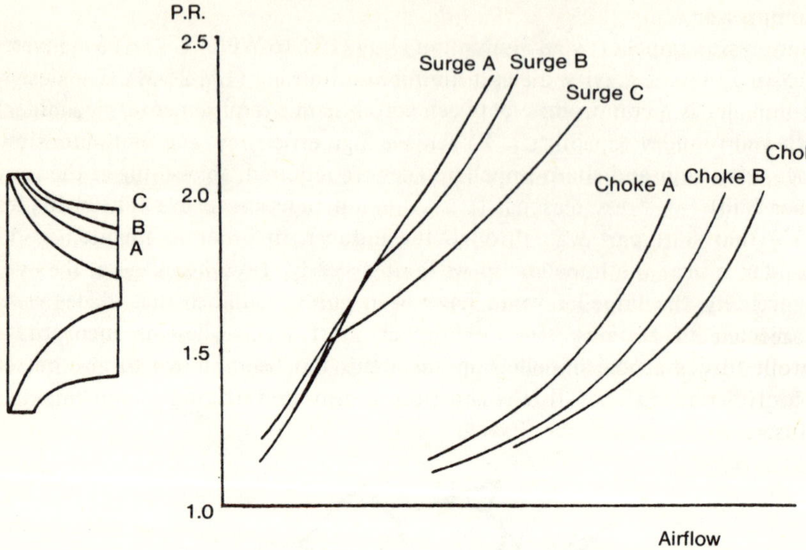

Fig. 33 – A range of compressor 'trims', machined from the same impeller casting.

The compressor can be a loose or slight interference fit on the shaft, clamped by the compressor end nut. Impellers of most turbochargers are balanced before assembly onto the shaft, so that components can be interchanged without re-balancing.

Vaneless diffusers (Fig. 31) are used on all except very high pressure ratio compressors. Relative to the alternative vaned designs, the vaneless diffuser is slightly less efficient owing to a longer gas flow path and poorer flow guidance, but has a substantially wider range of high operating efficiency. This is important in truck and passenger car applications where engine speed, and therefore mass flow range, is large. The volute acts not merely as a collection of air leaving the diffuser, but is usually designed to achieve a small amount of additional diffusion in its delivery duct. Generally the volute slightly overhangs the diffuser (Fig. 31) in order to reduce the overall diameter of the turbocharger. The volute and impeller casing are invariably formed as a single component.

7.2 Turbine

Radial inflow turbines are universally used, usually friction or electron beam welded to the shaft. The turbine wheel must sustain the same high rotational speed of the compressor and operate at gas temperatures up to 900 K. The turbines are investment cast in high temperature creep resistant steels, such as 713 C Inconel. Its properties exceed the requirements but it is a readily available material, from the gas turbine industry. The turbine housing must stand the same high gas temperature as the impeller but is not subject to rotating stress. However it should be strong enough to sustain a rotor burst. Sand or shell cast S.G. iron (spheroidal graphite nodular) is used in most applications, and is free of scaling at temperatures up to 900 K. Ni-resist is used for higher temperatures but is more expensive, and prone to cracking.

Vaneless stators are used except for a very few high pressure ratio applications. These are cheaper, and enable the gas angle at stator exit to vary to some extent with the mass flow, giving high efficiency over a transient flow range. On the rare occasions when stator nozzles are incorporated in turbocharger turbines of this size, a nozzle ring is cast in Ni-resist or the nozzle vanes are fabricated from nickel–chromium alloys. The gas angle at rotor inlet is controlled by the nozzle vane angle or the geometry of the vaneless housing. In the latter case (Fig. 34) the inlet scroll and vaneless stator are combined. The cross-sectional area defines the tangential gas velocity for a given mass

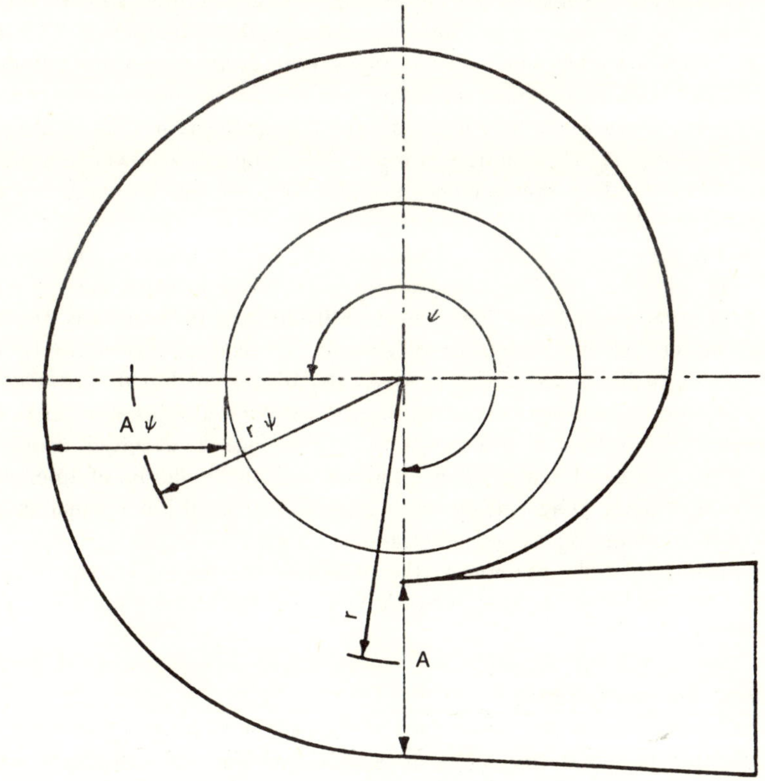

Fig. 34 – Volute-type turbine scroll.

flow rate and inlet density. The scroll is designed to spill the mass evenly around the circumference of the rotor. Again, for a given mass flow rate and density the radial component of velocity entering the impeller is fixed by the cross-sectional area at the rotor tip. This, and area A control the flow angle at rotor entry. Altering this flow angle alters the effective flow capacity of the turbine. Thus a range of casings with varying area A, or nozzle stator rings with varying blade angle, are available to match each rotor design to the requirements of a particular engine.

Gas flow through a turbine is predominantly accelerating, whereas in a compressor the gas diffuses. Fluid dynamic flow control is much easier to achieve in an accelerating flow, hence turbine design is less critical than compressor design. The turbine is more tolerant for mass flow variation, hence a single rotor may be used with a number of different area vaneless casings. Only a few turbine rotor trim variations can cover the requirements of a large range of engine sizes.

Most turbine casings have twin entries as shown in Fig. 31. The twin entries are used to separate the exhaust gas pulses coming from cylinders whose exhaust valves are open at the same time. Thus for example, the exhaust manifold of a typical six-cylinder engine will have two pipes, each connecting three cylinders to a twin entry turbine. Single entry casings are used on passenger car engines in order that a single waste-gate may be used to bypass some gas around the turbine at high engine speeds. This reduces turbine and compressor power and hence boost pressure, avoiding the tendency for the turbocharger to overboost the engine at full speed and load.

Since flow guidance is less critical than in the compressor, the turbine impeller has fewer blades. For stress and manufacturing reasons, the vanes are thicker than compressor vanes. They must withstand the pulsating gas pressure developed in the exhaust system of the engine. To reduce rotating inertia the back of the impeller is cut away between the blades. Although this slightly reduces efficiency, the benefit in terms of reduced stress and faster acceleration is substantial. Considerable rotor-to-housing clearance is essential owing to thermal effects, large thrust bearing clearance and the build-up of tolerances between components. Thus efficiencies are lower than those normal in gas turbine practice owing to leakage and clearance losses.

The turbine inlet flange mounts the turbine to the engine via the exhaust manifold. This avoids the need for expansion joints in the exhaust manifold.

7.3 Bearings and centre casing

For low cost, simplicity and ease of maintenance, the bearing system must be designed to use the lubricating oil of the engine. Automotive turbochargers use simple journal bearings, since ball bearings are more expensive, have a short life at very high speeds and are difficult to replace. The inboard (between compressor and turbine) bearing location imposes a short distance between bearings which, when combined with very light radial loads and a heavy overhung mass at one end only (the turbine), leads to complex bearing behaviour. In addition the very high rotational speed dictates that the rotor runs through both the first and second critical vibration frequencies of the rotating assembly. Thus bearing design is primarily concerned with the stability of the system.

Two fully floating sleeve bearings are used, rotating at a speed less than the rotor speed, and determined by bearing clearances. The outer oil film imparts an additional degree of damping into the bearing system, but both the shaft-to-sleeve and sleeve-to-housing clearances are large for improved stability. Typical clearances are 0.02 to 0.05 mm between shaft and sleeve and 0.07 to 0.1 mm between sleeve and housing. These large clearances mean that oil filtration down to about 20 μm only is required, which is easily met by full flow paper engine oil filters. A large oil flow rate is required owing to large clearances and the need to cool the turbine end bearing. Bearing material is typically leaded bronze with added tin flashing.

Axial thrust loads are relatively light, except when an exhaust brake is used, upsetting the compressor/turbine pressure balance. A simple thrust assembly is located at the cooler, compressor end, with flat or tapered lands, made from sintered leaded bronze.

The power required to overcome bearing losses is quite significant, typically being 5 to 10% of turbine power at full speed. More important is the fact that this percentage increases at lower engine speeds, which is a disadvantage in vehicle applications where high boost pressure is difficult to achieve at low engine speed. Lower viscosity oil would help but is incompatible with the requirements of the engine.

The bearing sleeves run directly in the bore of a high grade grey iron centre casting of the supercharger. This casting also acts as an oil drain and holds the compressor and turbine casings. For good oil drainage the casing must be installed with the oil inlet and drain at top and bottom, in the vertical position, with the rotor shaft horizontal. The oil seals at compressor and turbine ends are a difficult design problem owing to the need to keep frictional losses low, large excursions due to large bearing clearance, and to adverse pressure gradients under some conditions. The piston ring seals shown in Fig. 31 are typical, the rings being a press fit in the housing. They act as a labyrinth seal due to small clearance between the stationary ring and the sides and base of the groove in the rotor shaft. In practice, they leak unless care is taken to keep as much oil as possible away from them. Rotating flingers are designed into the shaft for this reason and an oil shield is frequently fitted at the compressor end since a depression can be generated in the compressor when the engine is idling.

The compressor and turbine housings can usually be rotated so that the compressor delivery and turbine inlet can be located in convenient positions. Set screws or stainless steel V-clamps are used to hold the components together.

8. TURBOCHARGER PERFORMANCE

The performance of turbochargers can be defined by the pressure ratio, mass flow rate and efficiency characteristics of the compressor and turbine, plus the mechanical efficiency of the bearing unit.

The isentropic efficiency of the compressor is defined as the power required for isentropic compression divided by that for actual compression for the same inlet conditions, mass flow rate and pressure ratio. Assuming that air is a perfect gas, this becomes:

$$\eta_{cTS} = \frac{\dot{m}\, C_p\, T_{01} \left[(P_2/P_{01})^{\frac{\gamma-1}{\gamma}} - 1 \right]}{\dot{W}_c} \tag{17}$$

or

$$\eta_{cTS} = \frac{(P_2/P_{01})^{\frac{\gamma-1}{\gamma}} - 1}{\dfrac{T_{02}}{T_{01}} - 1} \tag{18}$$

where P is pressure, T is temperature, $\dot{m}$ is mass flow rate, $\dot{W}$ is actual compressor power. C_P is specific heat (constant pressure) and γ is the ratio of specific heats. Subscript 0 denotes total or stagnation conditions; 1 is inlet and 2 is outlet. Thus, the above equations define efficiency from total inlet to static outlet. Exit velocity is assumed to be dumped, not converted to a further pressure rise in the inlet manifold, hence static exit pressure is appropriate for this application.

Similarly for the turbine

$$\eta_{tTS} = \frac{\dot{W}_T}{\dot{m}\, C_p\, T_{03} \left[1 - (P_4/P_{03})^{\frac{\gamma-1}{\gamma}} \right]} \tag{19}$$

or

$$\eta_{tTS} = \frac{1 - (T_{04}/T_{03})}{1 - (P_4/P_{03})^{\frac{\gamma-1}{\gamma}}} \tag{20}$$

where subscript 3 and 4 denote turbine inlet and exit, respectively.

8.1 Non-dimensional representation of compressor and turbine charactersitics

The mass flow rate, efficiency and temperature rise ($\Delta T = T_{02} - T_{01}$) of a compressor or turbine can be expressed as a function of all possible influencing parameters, as follows

$$\dot{m}, \eta_{is}, \Delta T = f(P_{01}, P_{02}, T_{01}, N, D, R, \gamma, \mu) \tag{21}$$

where N, D and μ are rotational speed, diameter and the kinematic viscosity of the gas, respectively, and $\gamma = C_p/C_v$. These can be reduced, using dimensional analysis, to the following non-dimensional groups.

$$\frac{\dot{m}\sqrt{(T_{01}R)}}{P_{01}D^2}, \eta_{is}, \frac{\Delta T}{T_{01}} = f\left(\frac{ND}{\sqrt{(RT_{01})}}, \frac{P_{02}}{P_{01}}, \frac{\dot{m}}{\mu D}, \gamma\right) \tag{22}$$

For the compressor, the value of γ remains constant, except for a very small variation with temperature, hence the last term may be ignored. γ does vary with air/fuel ratio, but its influence on turbine performance is small and is therefore also ignored. Fortunately the Reynolds number ($\dot{m}/\mu D$) also has only a small effect on performance and can be ignored. A relationship between η_{is}, $\Delta T/T_{01}$ and P_{02}/P_{01} has already been given, hence equation (22) may be reduced to

$$\frac{\dot{m}\sqrt{(T_{01}R)}}{P_{01}D^2}, \eta_{is} = f\left(\frac{ND}{\sqrt{(RT_{01})}}, \frac{P_{02}}{P_{01}}\right) \tag{23}$$

(for a turbine, subscripts 3 and 4 replace 1 and 2.)

For a particular turbocharger, the diameter remains constant and, for turbocharger applications the gas constant remains fixed, hence the variation in performance with running conditions is given by

$$\frac{\dot{m}\sqrt{(T_{01})}}{P_{01}}, \eta_{is} = f\left(\frac{N}{\sqrt{(T_{01})}}, \frac{P_{02}}{P_{01}}\right) \tag{24}$$

The complete performance map of the compressor (Fig. 35) can be shown by plotting a graph of $\dot{m}\sqrt{(T_{01})}/P_{01}$ against P_{02}/P_{01} showing lines of constant $N/\sqrt{(T_{01})}$ and efficiency (u_{is}). The advantage of this presentation is that it uniquely describes the performance of the turbomachine, regardless of inlet conditions (pressure and temperature). However the terms in equation (24) are no longer truly dimensionless.

8.2 Compressor performance

A typical performance map, from a turbocharger compressor designed for a vehicle engine, is shown in Fig. 35. The central area is the stable operating zone, bounded by the surge line on the left (low mass flow rates), and a regime of high rotational speed and low efficiency on the right (high mass flow rate). A detailed explanation of the causes of surge has yet to be fully accepted, but it is clear that when the mass flow rate through the compressor is reduced whilst maintaining a constant pressure ratio, a point arises at which local flow reversal occurs in the boundary layers. Complete flow reversal can develop in a 'stall cell' somewhere in the compressor, with normal flow elsewhere. Once several stall cells have developed the complete flow pattern can break down, hence mass flow and pressure fall. A stable flow pattern becomes re-established at a lower pressure ratio, allowing the mass flow to build up again to the initial value. The flow instability repeats in a surge cycle. The compressor must not be asked to work in this region of operation but it must be realized that surge is influenced by the complete intake, compressor and inlet manifold system. Thus the surge line drawn by the turbocharger manufacturer is only a guide, and in practice it varies from one engine installation to another.

The area of high rotational speed and low efficiency is a result of choking of the limiting flow area in the compressor. Extra mass flow can only be achieved by increasing rotational speed, which must be limited by stress constraints. If the diffuser chokes, rather than the rotor, then compressor speed eventually rises substantially with little increase in airflow.

Constant efficiency loops are shwon in Fig. 35. Note that these tend to be parallel to the surge line.

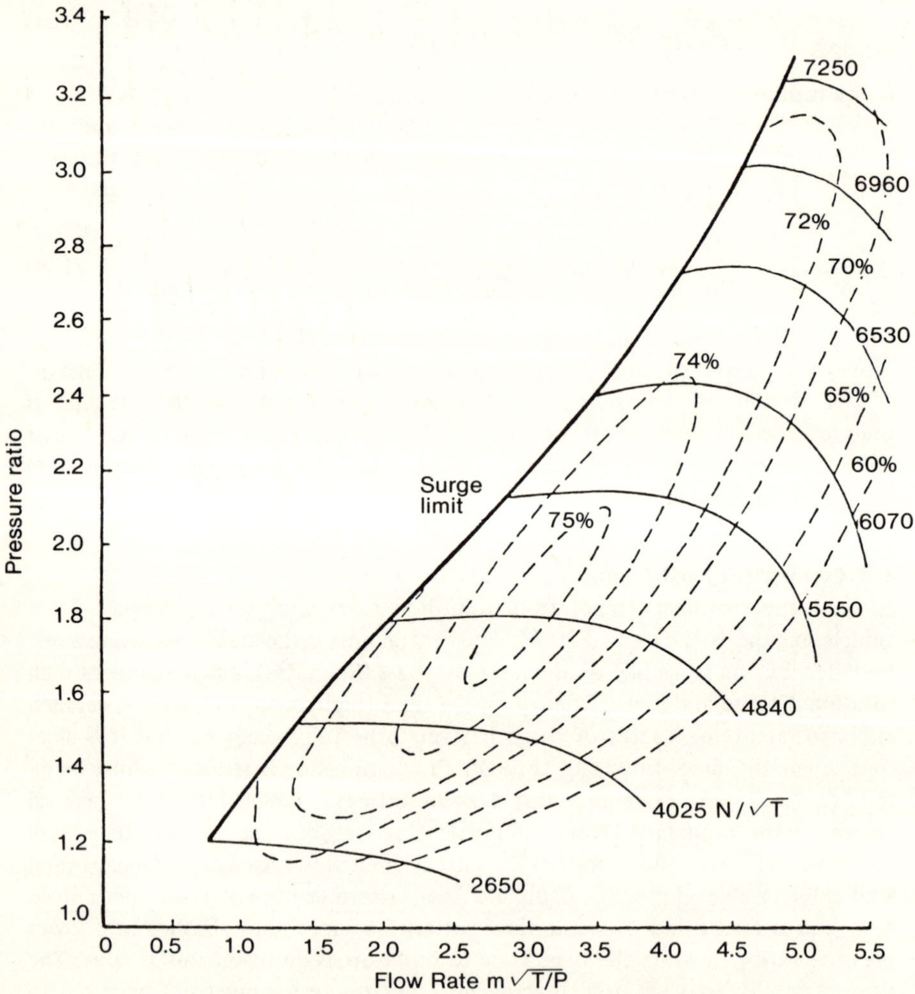

Fig. 35 – Turbocharger compressor performance map.

8.3 Turbine performance

In a radial flow turbine, rotational speed of the rotor influences the pressure at stator exit owing to centrifugal effects. Thus the overall pressure ratio (stator inlet to rotor exit) at which the stator chokes is dependent on rotor speed. Figure 36 shows pressure ratio versus mass flow rate curves for a radial flow

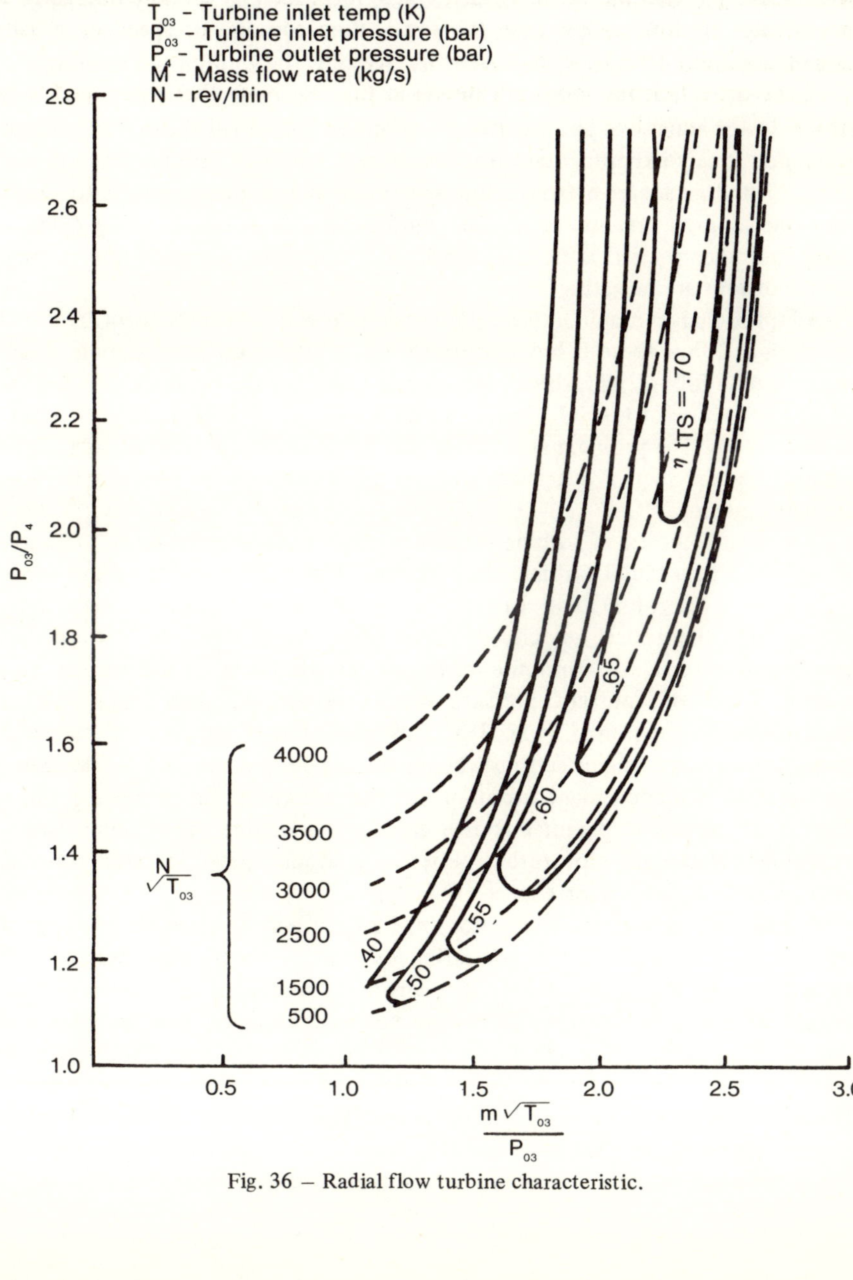

Fig. 36 – Radial flow turbine characteristic.

turbine, with stator nozzles, illustrating the variation of choked pressure ratio and mass flow rate with rotor speed.

Nozzleless radial tubines exhibit the widest variation of mass flow with rotor speed. However it is rare for the data obtained by the turbocharger manufacturer to cover the full pressure ratio versus mass flow range along each constant speed line, even for these turbines. It reduces test time and is rarely required by the user, for the following reason. For a given turbocharger, it is possible to link the compressor and turbine characteristics and plot the equilibrium running line on compressor and turbine maps. This comes from balancing the compressor and turbine power equations and imposing a common rotational speed, factors that are implicit in any experimental turbocharger test. This equilibrium line is shown in Fig. 37 and runs across from a low speed line at low pressure ratio to a high speed line at high pressure ratio. The equilibrium line is based on steady-state testing and theory, and is not necessarily the operating regime if the turbine inlet conditions are pulsating.

All turbocharger manufacturers present compressor data in the form described in section 8.2, but there is less uniformity of turbine data. For example, some manufacturers follow gas turbine practice and transpose the ordinate and abscissa of Fig. 36 to those of Fig. 37. Further variations occur in presentation of turbine efficiency data. The problem arises from the fact that the operational area of the turbine occupies such a restricted area on the pressure ratio versus mass flow parameter map.

Common is a plot of turbine efficiency against blade speed ratio (u/C). u is the tip speed of a radial turbine rotor (or the blade velocity at the mean blade height of an axial rotor). The rotor blade velocity (u) is non-dimensionalized by dividing by a theoretical velocity (C) that would be achieved by the gas if it expanded isentropically from the turbine inlet condition to the turbine exit pressure. Lines of constant pressure ratio or speed parameter $(N/\sqrt{T_0})$ are superimposed on the map. (Fig. 38). This presentation happens to be useful for the turbocharger manufacturer when matching compressor and turbine rotor diameters but is inconvenient for an engine manufacturer. He desires a graph is whose axes represent parameters relevent to this engine. Thus some turbocharger manufacturers plot turbine efficiency against pressure ratio, showing lines of constant turbocharger speed (Fig. 37).

In practice, diagrams such as those in Figs. 37 and 38 show the product of turbine and mechanical efficiency (bearing losses predominate) since it is difficult in practice to separate them. Indeed it is surprisingly difficult to achieve accurate turbine efficiency measurements, owing to heat transfer, non-uniform and unsteady flow effects, etc. As a result most turbocharger manufacturers have developed their own standard test techniques whose results can be reliably compared across their range of turbochargers. However these results cannot be directly compared with those of another manufacturer who, for example, tests with a different standard turbine inlet temperature and therefore measures a

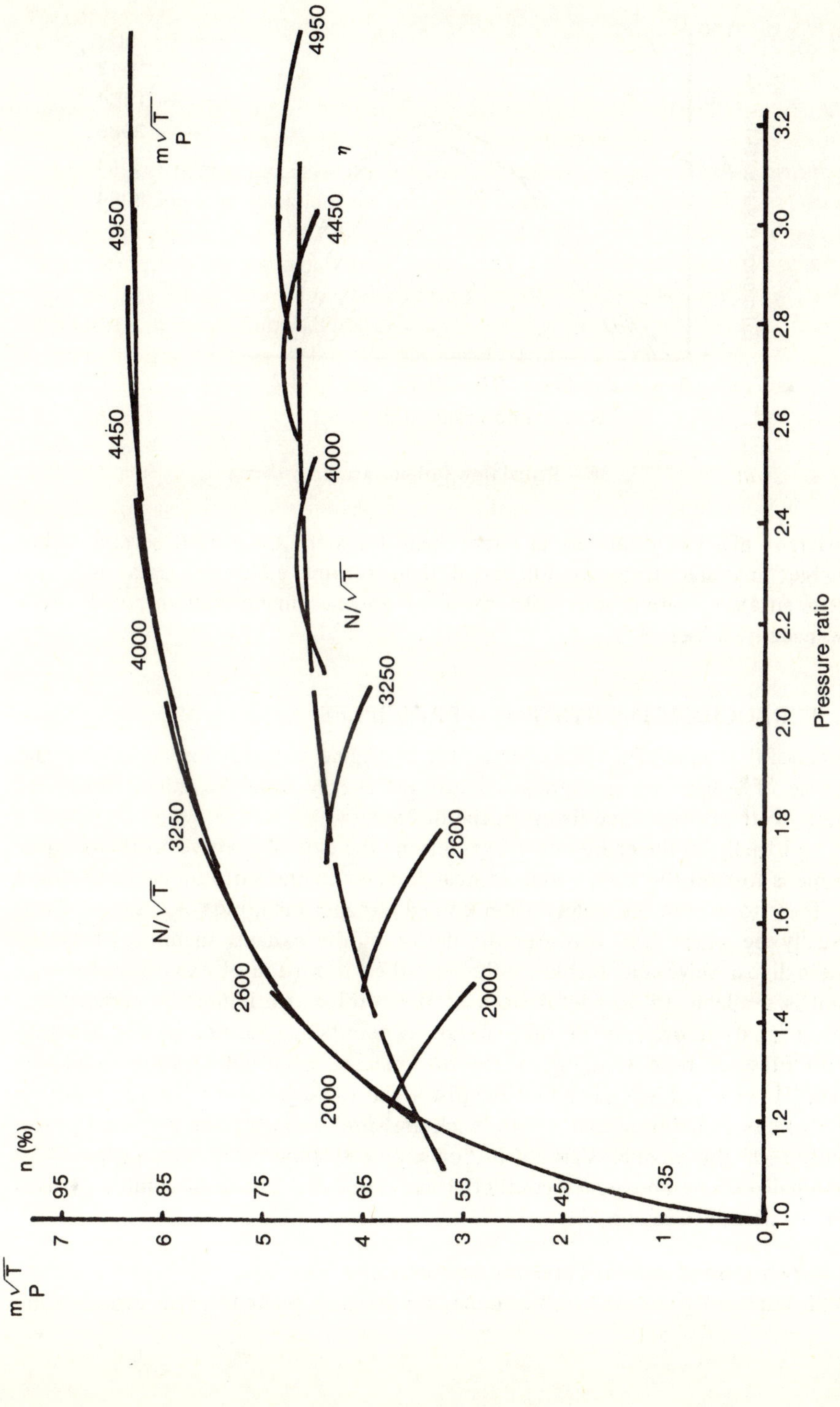

Fig. 37 – Turbine performance map, alternative presentation.

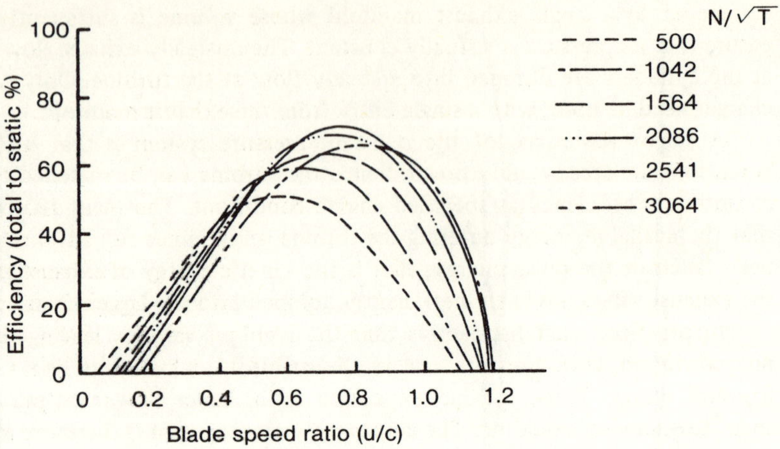

Fig. 38 – Radial flow turbine efficiency curves.

different efficiency due to different heat transfer. As a result several turbocharger manufacturers do not reveal their turbine efficiency data since unrepresentative comparisons with those of another manufacturer could easily be made by a user.

9. TURBOCHARGING SYSTEMS – PRINCIPLES

Successful design of a turbocharged diesel engine is highly dependent on the choice of system for delivering exhaust gas energy from the exhaust valves or ports, to the turbine, and its use in the turbine.

Virtually all the energy of the gas leaving the cylinders arrives at the turbine. Some is lost on the way, owing to heat transfer to the surroundings, but this is unlikely to exceed 5% unless water cooled exhaust manifolds are used, and will usually be much less. However the design of the exhaust manifolds between the exhaust valve and turbine influences the proportion of exhaust gas energy that is available to do useful work in the turbine. An important parameter is the stagnation pressure in the exhaust system, since turbine power increases with pressure ratio (P_{03}/P_4), hence the exhaust manifold pressure should be high. However, this implies that the piston has to push the combustion products out of the cylinder against a high 'back-pressure', reducing the potential power output of the engine. Various turbocharging systems have been proposed to rationalize these apparently conflicting requirements. The most commonly used will be described herein.

9.1 Principles of constant pressure turbocharging

With constant pressure turbocharging, the exhaust ports from all cylinders are connected to a single exhaust manifold whose volume is sufficiently large to ensure that its pressure is virtually constant. The unsteady exhaust flow processes at the cylinders are dumped into a steady flow at the turbine. Only one turbocharger need be used, with a single entry from the exhaust manifold.

A major advantage of the constant pressure system is that turbine inlet conditions are steady and known, hence the turbine can be matched to operate at optimum efficiency at specified engine conditions. The main disadvantage is that the available energy entering the turbine is low, since full advantage has not been taken of the pulse energy, that is the kinetic energy of exhaust gas leaving the exhaust valves has been inefficiently dumped into the large exhaust manifold.

In practice, what happens is that the high gas velocity leaving the port is not converted to a pressure rise in the manifold, which acts like a constant pressure dump. Thus, turbine expansion ratio, hence power output, is lower than it otherwise would be. The constant pressure system is therefore inefficient at transferring energy from cylinders to turbine.

From a purely practical point of view, the exhaust manifold is simple to construct although it may be rather bulky, particularly relative to small engines with few cylinders. However, for large engines with many cylinders, the convenience of being able to join all cylinders to a common exhaust manifold with a single turbocharger on top, or at either end, is useful. A major disadvantage of the constant pressure system arises from the use of an exhaust manifold having a large volume. When the engine load is suddenly increased or a rapid engine speed increase is required, the pressure in the large volume is slow to rise. Hence the energy available at the turbine increases only gradually. Turbocharger, and therefore engine response, will be poor. The poor response of the constant pressure turbocharging system restricts it from consideration for applications where frequent load (or speed) changes are required.

The turbocharging system will affect the engine through three parameters only, the boost pressure and temperature in the inlet manifold and the pressure in the exhaust manifold. Hence it is these factors that must be examined when considering a turbocharging system. The effect of the first two are obvious. The importance of the exhaust manifold pressure depends on whether the turbocharged engine is a four-stroke or a two-stroke.

It is useful to consider, when comparing turbocharging systems, that several arrangements might enable a certain boost pressure to be developed. The merits of the different systems can then be compared by considering the exhaust manifold pressure developed and its effect on engine performance.

By considering the energy balance for the turbocharger (equation (15)) when running with a constant exhaust pressure it is simple to derive a relationship between the exhaust manifold pressure (P_3) and the boost pressure (P_2). Compressor power must equal the product of turbine power and turbocharger mechanical efficiency.

Thus the relationship between the inlet manifold pressure (P_2) and exhaust manifold pressure (P_3) is a function of the overall turbocharger efficiency (η), the turbine inlet temperature (T_3) and, to a lesser extent, the air/fuel ratio (AFR).

The air/fuel ratio at full load will be governed by thermal loading or the onset of black smoke in the exhaust. The turbine inlet temperature will also be dependent on air/fuel ratio, the amount of cool scavenge air passing through the cylinders and heat loss from the exhaust manifold.

Equation (15) is plotted, for a compressor pressure ratio of 2, in Fig. 39, showing the dominant effects of turbocharger efficiency and turbine inlet temperature on the pressure ratio between inlet and exhaust manifolds. It is important to realize that if the turbine inlet temperature is held constant (for metallurgical reasons), then higher turbocharger efficiencies are required to maintain a favourable scavenge pressure ratio at higher boost pressures.

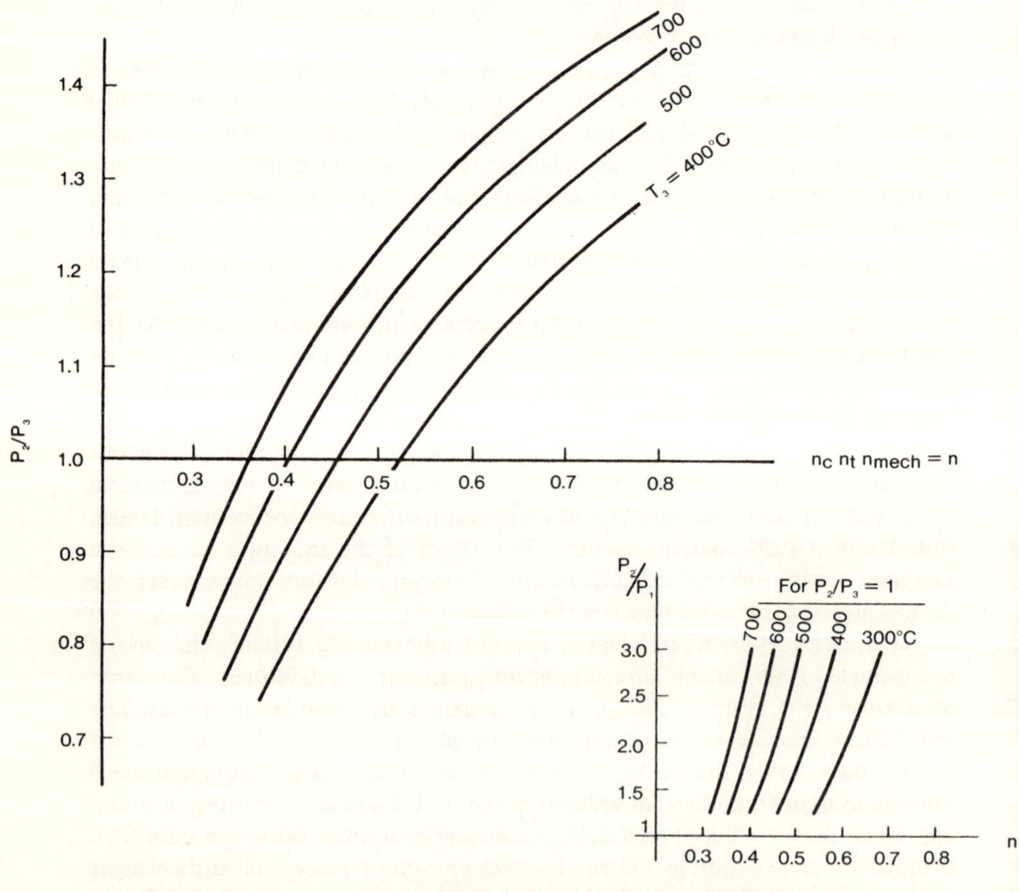

Fig. 39 – Effect of overall turbocharger efficiency on the pressure drop between inlet and exhaust manifolds (constant pressure system) for $P_2/P_3 = 2$.

Although the turbocharger efficiency (and air/fuel ratio, via the turbine inlet temperature) governs the relationship between exhaust and inlet pressures, it is the turbine area that controls the exhaust pressure itself, since it acts as a restricting orifice. By fitting smaller and smaller turbines to a four-stroke engine it is theoretically possible to develop very high exhaust pressures and correspondingly high levels of energy available at the turbine. This should allow very high boost pressures to be obtained (subject to the relationship described above). However the practical effects of turbine speed and inlet temperature limitations combined with decreasing turbocharger efficiency at high pressure ratios and poor scavenging prevent very high pressure ratio being developed in practice. Thus it is the turbine area that principally decides the exhaust manifold pressure and the air/fuel ratio and turbocharger efficiency that fix the relationship between inlet and exhaust pressure at the cylinders.

9.2 Principles of pulse turbocharging

The majority of turbocharged engines use the pulse, not the constant pressure, turbocharging system. The objective is to make the maximum use of the high pressure and temperature which exist in the cylinder when the exhaust valve opens, even at the expense of creating highly unsteady flow through the turbine. In most cases the benefit from increasing the available energy will more than offset the loss in turbine efficiency due to unsteady flow.

Consider an exhaust port connected to a turbine by a very small exhaust manifold, the turbine also having a small effective flow area. Initially, before the exhaust valve opens, the manifold may be at atmospheric pressure. As the exhaust valve begins to open exhaust gas flows from cylinder to manifold under the influence of the large pressure drop. Mass flow rate rapidly increases as the valve opens. Since the turbine acts as a flow restriction, the pressure in the manifolds builds up, governed by the difference in flow into the manifold and out through the turbine. Thus exhaust manifold (turbine inlet) pressure tends to rise towards cylinder pressure. Once the peak mass flow rate into the manifolds has passed, exhaust gas may be flowing out through the turbine faster than the flow rate from cylinder to manifold, hence manifold pressure gradually drops, reaching the atmospheric value shortly after the exhaust valve closes.

With this turbocharging system, energy is transmitted to the turbine in the form of a pressure pulse. Turbine inlet pressure is unsteady but rises above that of a constant pressure system at its maximum and less at its minimum. The important point is that most mass flows through the turbine when the pressure ratio is high, hence turbine power is generally greater than with the constant pressure system and is less sensitive to variations in overall engine mass flow rate and speed. This is mainly due to the non-linear pressure ratio versus mass flow turbine characteristic (Fig. 37), which gives a proportionally greater increase in pressure ratio at high instantaneous mass flow rate.

The pulse turbocharging system requires the exhaust manifold to be as small as reasonably possible in order that turbine inlet pressure should rapidly rise to almost equal cylinder pressure when the exhaust valve opens. It is also desirable to open the exhaust valve rapidly for the same reason. In practice, valve train inertia and cam stresses limit valve acceleration, and turbocharger location influences exhaust manifold volume. The turbocharger must be mounted as close to the cylinders as possible to reduce the length of the exhaust manifold. The cross-sectional area of the manifold cannot reasonably be significantly less than that at the exhaust valve when the latter is at full lift. These together define the minimum exhaust manifold volume. Virtually all turbocharged engines are multi-cylinder, hence several exhaust ports must be connected to the turbocharger turbine. To keep exhaust manifold volume small, the exhaust ports are connected to the turbine by short, narrow diameter pipes.

A typical example illustrating the gain in available exhaust gas energy when employing narrow pipes is shown in Fig. 40. These test data were obtained on a single-cylinder, loop-scavenge, high speed two-stroke engine with three types of exhaust manifold, a compact volume (pipe diameter = pipe length) and two narrow pipes (pipe area/cross-section port area ratio of 1.78 and 1.03). All the tests refer to the same engine speed, BMEP, boost pressure, airflow rate and turbine area. The main diagram shows how turbine available energy varies with exhaust pipe volume (non-dimensionalized over datum of available energy if manifold volume equals half of cylinder volume). Note that increasing exhaust manifold volume by a factor of 5 halves turbine available energy. It is also interesting to note that there is little difference between the three types of manifold (volume or long narrow pipe) provided that their volumes are equal. The top set of diagrams shows the pulsating exhaust manifold pressure, for the three types of manifold, with five gradually increasing total exhaust manifold volumes. In the first two cases (exhaust manifold volume = 0.5 or 0.8 times cylinder volume), the type of manifold, whether a long, narrow pipe or a plenum, has little effect. However, in the last two diagrams there are substantial differences. These differences are caused by pressure wave reflections in the very long, narrow pipes.

If the exhaust pipe is narrow, then the cross-sectional area of the pipe itself is sufficient for pressure to build up at the exhaust valve end as the exhaust valve opens. A pressure pulse is built up (hence the term pulse turbocharging). This pressure pulse or wave, travels at sonic velocity along the pipe to reach the turbine. Thus energy is being transmitted along the pipe at sonic velocity to the turbine. At the turbine the pressure wave is reflected with reduced amplitude since the turbine is a partial flow restriction. Thus a pressure wave is generated by the gas initially released at high pressure at EVO by the cylinder, which travels forwards and backwards along the pipe with gradually diminishing amplitude. This differentiates the performance of a long, narrow pipe from a plenum of the same total volume, although the available energy at the turbine is not greatly different (Fig. 40).

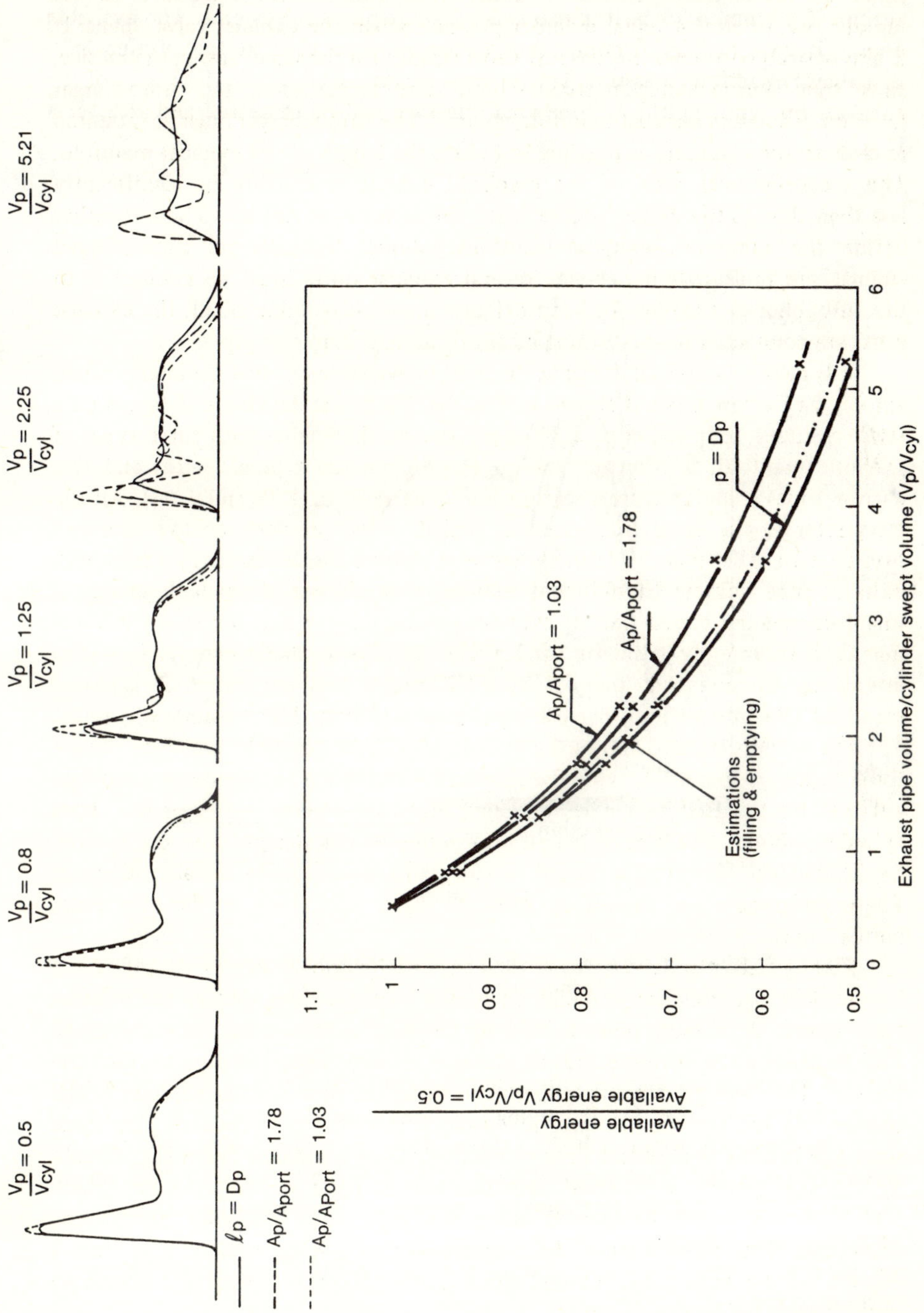

Fig. 40 – Effect of exhaust manifold shape on available energy.

The basic method of increasing (or reducing) the available energy at the turbine is common to both pulse and constant pressure systems. The available energy is governed by, among other factors, the pressure at the turbine inlet or exhaust manifold pressure. This in turn is controlled by the flow area of the turbine. By reducing the turbine area, pressure will be increased and vice versa (Fig. 41).

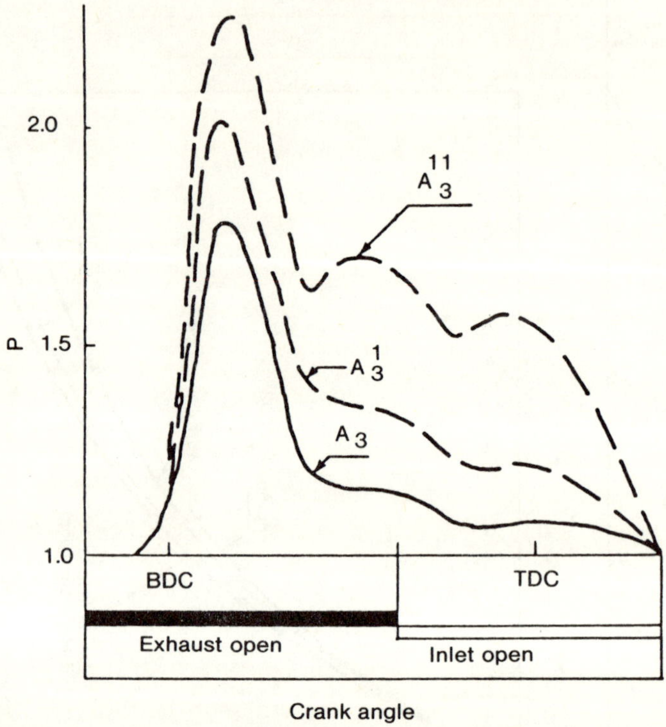

Fig. 41 – Effect of reducing turbine area on instantaneous pressure at turbine entry ($A''_3 > A'_3 > A_3$).

Up to this point, factors influencing the available energy at the turbine have been considered. Pulse turbocharging increases available energy but produces a highly unsteady flow through the turbine. How does the turbine behave under unsteady flow, and what is its efficiency? The energy is contained within the exhaust gas flow and can be found by applying a turbine power equation instantaneously and integrating over an engine cycle (Fig. 42). By this means the variation in turbine inlet temperature and pressure plus turbine efficiency are accounted for.

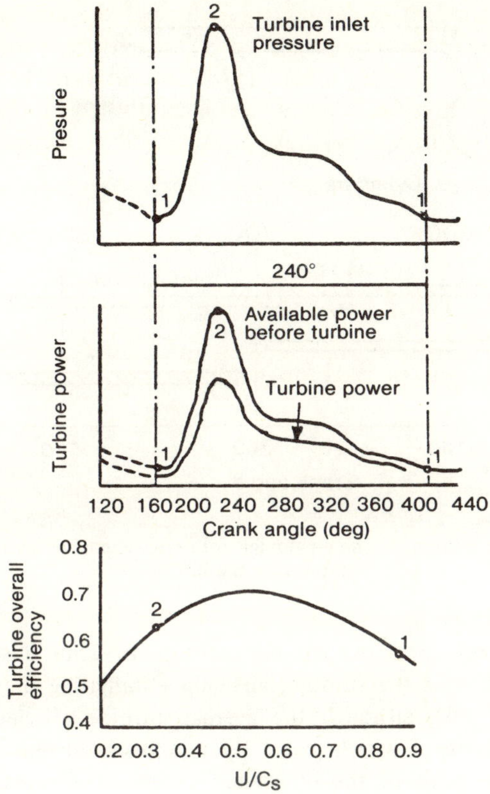

Fig. 42 – Turbine inlet pressure power and efficiency curves.

In Fig. 42 (lower) a typical turbine efficiency curve is shown. If quasi-steady flow is assumed, then the instantaneous turbine efficiency will be that corresponding to the appropriate instantaneous value of blade speed ratio (u/C_s). The values of efficiency coinciding with the peak and a low point of the pressure diagram are illustrated. For much of the remainder of the engine cycle the turbine will be windmilling and decelerating. When the next exhaust pulse arrives, part of its energy will be used to accelerate the turbocharger again, hence much energy is wasted. Thus energy is wasted owing to a low average turbine efficiency resulting from the very unsteady flow and windmilling no-flow conditions.

On a multi-cylinder engine, narrow pipes from several cylinders can be connected via a single manifold to one turbine. Consider the three-cylinder four-stroke, automotive engine shown in Fig. 43. Owing to the phase angle between cylinders, the opening periods of the exhaust valves follow successively

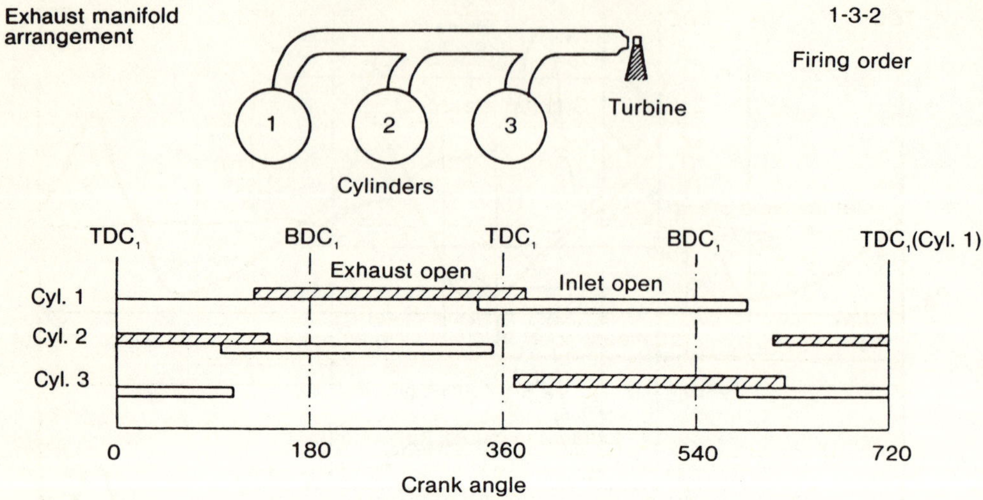

Fig. 43 – Pulse system – three-cylinder four-stroke engine with automotive type valve timing.

every 240° with very little overlap between them. Thus a steady 'train' of pressure pulses arrives at the turbine, virtually eliminating the long periods of pure windage (Fig. 44), although the average turbine efficiency will remain lower than that obtained with a correctly matched constant pressure system (operating near the peak of the efficiency curve). The remaining important point to consider is the exhaust pressure close to the valves, during the valve overlap (scavenging) period. As with the constant pressure system, a good pressure drop between inlet and exhaust manifolds during the period when both valves are open is important in the case of a four-stroke engine with significant valve overlap and vital for a two-stroke engine. In Fig. 44 pressure history in the inlet and exhaust manifolds (at a valve) is shown, the pressure drop during the period of valve overlap being cross-hatched. Clearly, at the running condition shown, the pressure drop is satisfactory.

The diagrams in Figs. 43 and 44 have been deliberately drawn for a three-cylinder four-stroke engine with short, narrow exhaust pipes and automotive type valve timing. Such an engine is rarely made, but is a convenient unit on which to present some important aims of pulse turbocharging, namely the way to increase average turbine efficiency by reducing windage periods, whilst avoiding interference with scavenging of one cylinder due to the effect of the pulse from another. The pressure pulse exhausting from a cylinder travels along the manifold until it reaches a junction. At the junction it divides into two pulses (each of smaller magnitude due to the effective area increase) travelling down each adjacent pipe. One pulse will travel towards the turbine, the other will

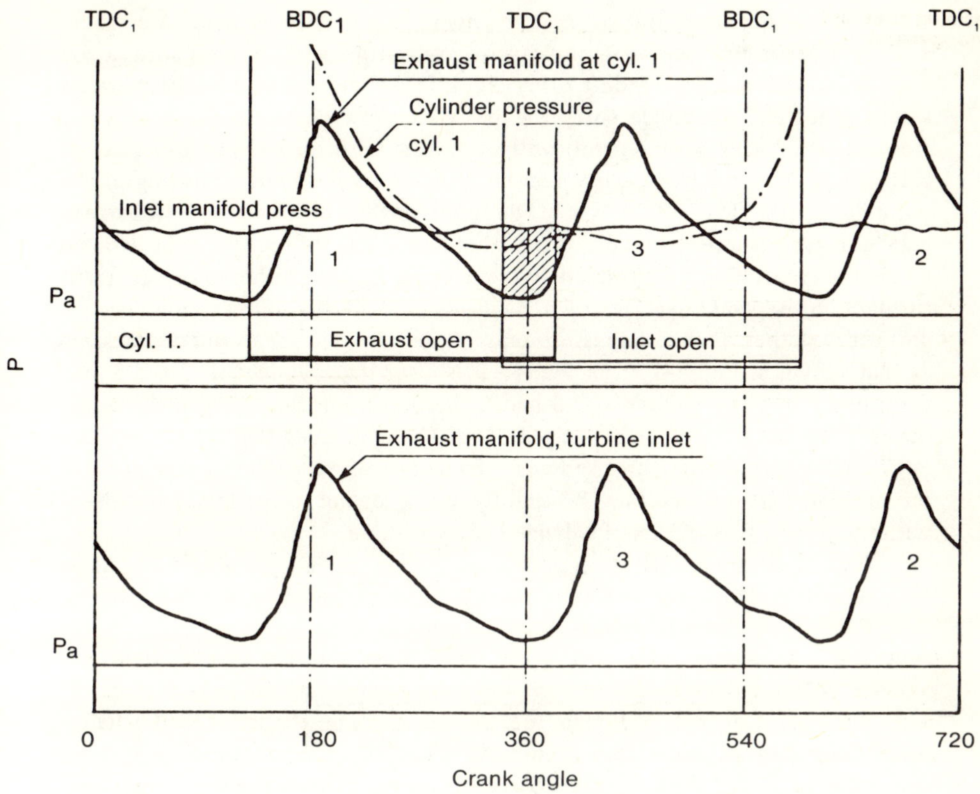

Fig. 44 – Exhaust pressure diagrams – three-cylinder four-stroke engine with automotive valve timing.

arrive at the exhaust valve of another cylinder. It is the later pulse, from cylinder number 3 (Fig. 43), that has arrived near cylinder 1 just at the end of the scavenge period of cylinder 1, that could be a problem. If it had arrived earlier (perhaps due to shorter exhaust pipes) it would have interfered with scavenging. This type of interference due to the *direct* action of a pressure wave from another cylinder is quite separate from the action of a pressure pulse *reflected* from the turbine, whether the latter started from cylinder 1, 2 or 3. The pipe lengths in Fig. 43 have deliberately been kept short so that the reflected pulses are almost superimposed on the initial pulse.

This will not always be the case, so the effect of both types of possible pulse interferences must be considered. It should be mentioned that if the direct pulse meets a closed exhaust valve at cylinder 1 (as it has in the figure) it will be reflected and will eventually arrive at the turbine some time after the first component of the pulse from cylinder 3.

Most engines have four or more cylinders, but it is convenient to consider a six-cylinder engine next. Figure 45 shows valve timing of a typical automotive six-cylinder diesel engine and its firing order. It is obvious that if all six cylinders were connected to a single entry turbine via narrow pipes, the pressure waves from each cylinder would significantly interfere with the exhaust processes of each other during valve overlap and the exhaust stroke, thus increasing piston pumping work. The effect would be poor engine efficiency. A two-stroke engine might not operate under these conditions. The difficulty can be avoided by simply connecting the cylinders in two groups of three, either to two different turbines, or separate sections of a single turbine. If the correct cylinders are grouped together, then the pressure pulse system in each group will be the same as that shown in Fig. 44. From Fig. 45 it is clear that cylinders 1, 2 and 3 may form one group and cylinders 4, 5 and 6 the other, but the arrangement would differ if the firing order were changed. It may be concluded that the six-cylinder engine is similar to the three cylinder, from the turbocharging point of view, but turbine performance may be slightly worse owing to the losses associated with the join of two sectors of a divided entry turbine.

It is disadvantageous to connect more than three cylinders to a single turbine entry. Thus, for the four-cylinder engine shown in Fig. 46 pairs of

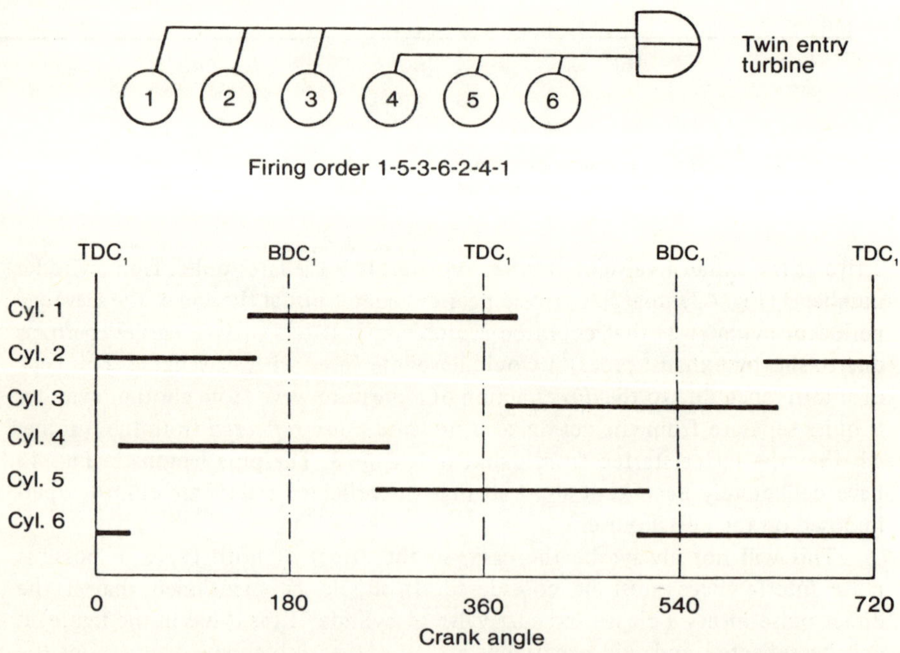

Fig. 45 – Exhaust valve timing for an automotive six-cylinder four-stroke engine.

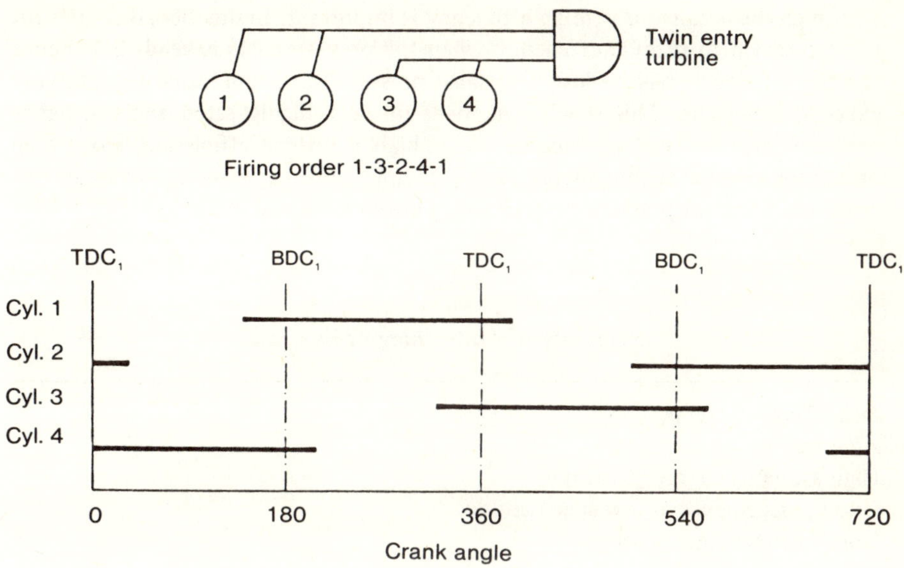

Fig. 46 – Exhaust valve timing for an automotive four-cylinder four-stroke engine.

cylinders (1–2 and 3–4) would be connected to a double entry turbine. On engines with other numbers of cylinders, the general rule will be to connect cylinders whose firing sequences are separated by 240° crank angle (in the case of four-stroke) and 120° (two-stroke) to a turbine inlet, and select those cylinders whose exhaust processes are evenly spaced out. However, this is not always possible. For example, on a vee-form engine, the vee angle will introduce an additional phase difference to the firing intervals between cylinders. In such cases, the more basic rule of avoiding direct pressure wave interference must be observed.

In the case of four-cylinder engines with very little valve overlap, turbocharger performance when all four cylinders are connected to one turbine entry is very little worse than with a two entry system. Even six-cylinder engines can operate reasonably with a single entry turbine if valve overlap is small, since no scavenge flow or reverse flow can occur. Thus, single entry turbines are used when an exhaust waste-gate is fitted, to avoid the need for two waste-gates.

The choice between pulse or constant pressure turbocharging systems is governed by engine duty, performance plus economic and maintenance considerations. In practice, these criteria mean that the constant pressure system is used exclusively on very large, highly rated two-stroke engines and some industrial medium-speed engines. On these engines the ratings are such that a very large pressure pulse would be generated with the pulse system. Since most of the exhaust pulse energy coincides with the peak of the pulse, matching this point with high instantaneous turbine efficiency is important. In practice, it is difficult to maintain high turbine efficiency when the pressure ratio exceeds 3:1, hence turbine efficiency will be low if exhaust pressure pulse amplitude substantially exceeds this value. This is what happens on very highly rated engines, hence constant pressure systems operate with higher turbine efficiency, more than offsetting lower available energy. The principal advantages and disadvantages of pulse and constant pressure turbocharging are summarized in Table 1.

Table 1

Summary of Turbocharging Systems

Pulse Turbocharging
Advantages
High available energy at turbine
Good performance at low speed and load
Good turbocharger acceleration

Disadvantages
Poor turbine efficiency with one or two cylinders per turbine entry
Poor turbine efficiency at very high ratings
Complex exhaust manifold with large numbers of cylinders
Possible pressure wave reflection problems (on some engines)

Applications
Automotive, truck, marine and industrial engines; two and four-stroke;
low and medium rating (e.g. up to 17–18 bar BMEP on four-stroke engines)

Constant Pressure Turbocharging
Advantages
High turbine efficiency, due to steady flow
Good performance at high load
Simple exhaust manifold

Disadvantages
Low available energy at turbine
Poor performance at low speed and load
Poor turbocharger acceleration

Applications
Large industrial and marine engines operating at steady speed and load, highly rated; two and four-stroke

Various designs of multiple entry turbines have been used. Today almost all turbocharger manufacturers use similar designs, dependent on whether the turbine has axial or radial flow. With large axial flow turbines the toroidal inlet area to the stator is equally divided into two, three or four separate sectors, each connected to separate pipes from groups of engine cylinders. Radial turbines with stator blades are available with single or twin entries, in the latter case (Fig. 47) with the periphery of the rotor divided into two 180° sectors. Radial turbines with nozzleless stators are also available in single or twin entry form. In this case the twin entry systems are divided meridionally (Fig. 47). This has the advantage of eliminating pure windage periods in any of the rotor passages and hence improving turbine efficiency when the flow in one entry is momentarily zero and the flow in the other is large.

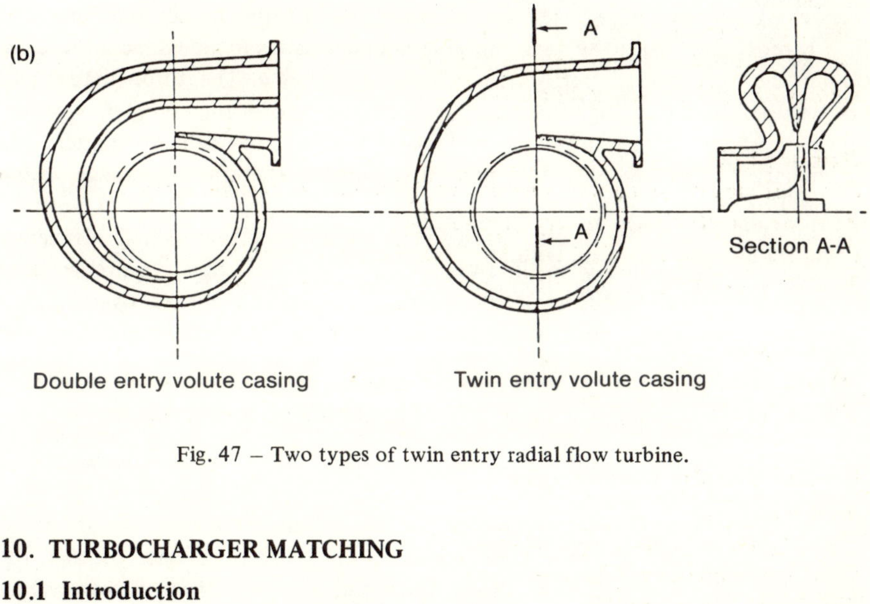

Fig. 47 – Two types of twin entry radial flow turbine.

10. TURBOCHARGER MATCHING

10.1 Introduction

Naturally aspirated diesel engines are capable of operating over wide speed ranges. The maximum useful speed will usually be limited by poor volumetric efficiency, the inertia of reciprocating parts or, in the case of some small, high speed engines, high frictional losses and poor combustion. An engine that is designed for variable speed operation will usually exhibit some deterioration in performance, both at extreme low and high speed. This is due to high gas friction losses in the inlet valves, the use of valve timing optimized in the mid-speed range and a gradual mismatch between fuel injector characteristics and swirl. However, the useful speed range can be wide, since reciprocating machinery is well suited to cater for a wide range of mass flowrate.

The performance of turbomachines is very dependent on the gas angles at entry to the impeller, diffuser and turbine rotor. The blade angles are set to match these gas angles, but a correct match will only be obtained when the mass flow rate is correct for a specified rotor speed. Away from this 'design point' the gas angle will not match the blade angle and an incidence loss occurs owing to separation and subsequent mixing of high and low velocity fluid.

These losses will increase with increasing incidence angle, hence turbomcahines are not well suited for operation over a wide flow range. Their use as superchargers is due to their high design point efficiency and their ability to pass high mass flow rates through small machines.

It is clear that a turbomachine is not ideally suited to operate in conjunction with a reciprocating machine, hence the combination of diesel engine and turbocharger must be planned with care. 'Matching' of the correct turbocharger to a diesel engine is of great importance and is vital for successful operation of a turbocharged diesel engine. The overall objective of turbocharger matching is to fit a turbocharger with the most suitable characteristics to an engine in order to obtain the best overall performance from that engine. The turbocharger will not be operating at its high efficiency flow condition over the complete working range of engine speed and load. It follows that it is possible to 'match' the turbocharger correctly only at a particular point in the operating range of the engine. For example, if an engine is required to run for most of its life at constant speed and full load, the turbocharger will be chosen such that its high efficiency operating area coincides with the pressure ratio and mass flow requirements of the engine at that condition. If the engine is required to operate over a broad speed and load range, then a compromise must be made when matching the turbocharger. This compromise will principally be governed by the duty for which the engine is required.

The basic size of the turbocharger will be determined by the quantity of air required by the engine. This will be a function of swept volume, speed, rating (or boost pressure), density of air in the inlet manifold, volumetric efficiency and scavenge flow. If these parameters are known an initial estimate of the air mass flow rate may be made from equation (1). With little valve overlap, the volumetric efficiency will be less than unity and may be estimated from values obtained under naturally aspirated operating conditions or from previous experience. With large valve overlap, the clearance volume will be scavenged and some excess air will pass into the exhaust.

The boost pressure will have been estimated for the engine to produce its target power output, subject to expected thermal and mechanical stresses. By assuming a value for the isentropic efficiency of the compressor (total to static) the boost temperature may then be estimated and inlet air density calculated.

The isentropic efficiency of the compressor may be taken from compressor maps. If the compressor efficiency is quoted on the total-to-total basis, a correction of a few percentage points down on the total-to-static value is required.

For an engine employing aftercooling the boost temperature has to be reduced to manifold temperature. From known values of ambient temperature and pressure, and required compressor pressure ratio, the mass flow rate at maximum power can be estimated.

By looking at the basic guidelines presented in the turbocharger manufacturers' literature, or the complete compressor characteristic curves (Fig. 35), a basic 'frame-size' of turbocharger may be selected. The final choice of compressor will be made bearing in mind the complete operating lines of the engine over its whole speed and load range super-imposed on the compressor characteristic. The compressor 'trim' or diffuser will be chosen to allow a sufficient margin from surge whilst ensuring that the operating lines pass through the high efficiency area. This will be discussed in detail later.

Once the basic frame-size and compressor have been established the turbine must be matched by altering its nozzle ring, or volute size (if it is a radial flow machine). The effective turbine area will change, raising or lowering the energy available at the turbine and hence adjusting the boost pressure from the compressor.

Turbocharger matching, particularly on a new engine, can be a lengthy process since many dependent parameters are involved. Although the basis is to match the turbocharger to the engine, it may well be necessary to improve the performance of the combination by engine design changes at the same time. Some such changes, adjustment of the fuel injection system, for example, are obviously essential. Others, that may be desirable, include alteration to the valve area and timing. It will not be possible to discuss all these factors in detail and the broad principles of matching fuel injection systems, swirl, optimization of valve timing, etc. are not substantially different from those pertaining to naturally aspirated engines.

10.2 Airflow characteristics of engine and turbocharger

10.2.1 *Four-stroke engines*

The airflow rate through a turbocharger (non-aftercooled) diesel engine will be a function of the engine speed, compressor delivery air density and the pressure differential between intake and exhaust manifolds during the period of valve overlap. If the engine is run at constant speed, but steadily increasing load, then the mass flow rate will increase approximately with the increasing charge density. The airflow through the engine may be superimposed on a turbocharger compressor characteristic, as shown in Fig. 48, the slope being governed by the density ratio. When matching a turbocharger to an engine with this operational requirement, the objective will be to choose a compressor such that the constant engine speed line falls through the middle of the high efficiency area of the compressor map. If an aftercooler is fitted then, as load increases the cooling effect will increase charge density more rapidly for a corresponding boost pressure, hence the slope of the constant engine speed airflow line on the compressor characteristic will be less steep than those shown in Fig. 48.

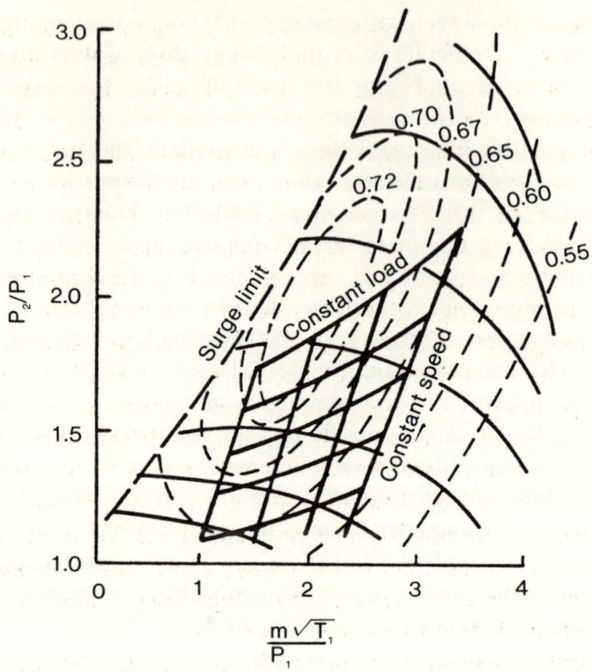

Fig. 48 – Compressor characteristic, with airflow requirements of a four-stroke truck engine superimposed.

Consider next an engine running at constant load but increasing speed. As the engine speed increases so will the volumetric flow rate of air. The effective flow area of the turbocharger turbine remains almost constant, hence turbine inlet pressure will rise. As discussed in section 9 the result is an increase in energy available for expansion through the turbine and hence increased boost pressure at the compressor. Thus the constant load line of the engine will not lie horizontally on the compressor characteristic, but will rise with engine speed (Fig. 48), the slope depending on whether the engine is aftercooled or not. If the engine is required to operate over a range of speeds and loads (e.g. an automotive unit), then a set of constant speed and constant load lines may be drawn on the compressor characteristic to represent the operating range (Fig. 48). The complete engine characteristic must lie between the compressor surge line and the limit imposed by low efficiency or possibly turbocharger overspeed at high mass flow rates.

The margin between surge and the nearest point of engine operation must be sufficient to allow for three factors. First, pulsations in the intake system may well induce surge when the mean flow lies clear of the nominal surge line.

Secondly, if the air filter becomes excessively blocked in service, the airflow rate through the engine will reduce, but the turbine work will be maintained by a hotter exhaust as the air/fuel ratio gets richer. Thus boost pressure may not fall and some movement of the engine operation line towards surge may occur. Thirdly, larger movement towards surge will result if the engine is operated at altitude. The effect of altitude operation on turbocharged engine performance is discussed in section 11. The combined effect of the three factors affecting surge margin will vary from one engine for one application, to another. In general, however, a margin of at least 10% (of mass flow rate) between surge and the nearest engine operating lines should be allowed. On engines with a small number of cylinders a 20% margin will sometimes be required.

Generally, the turbocharger turbine can operate efficiently over a wider mass flow range than its compressor. It follows that it is more important to examine the engine airflow plotted on the compressor map than on the turbine map. This is indeed fortunate since if the turbine is operating under the pulse system with highly unsteady flow, then it is not realistic to plot a 'mean' value on the turbine map. The result can be quite misleading. To accurately assess the operating area on a turbine map would require a plot of instantaneous gas flow and pressure ratio over a full range of engine operating conditions. The information is very difficult indeed to measure accurately.

The compressor may be matched initially by choosing the best combination of impeller and diffuser such that the engine operating characteristics lie within the guidelines given above. Final matching will depend on the type of power or torque curve required from the engine for a certain application, and will be discussed later. However, it has been pre-supposed that the turbine is able to provide sufficient power to drive the compressor and produce the airflow conditions discussed. Since the useful flow range of the turbine is wider than that of the compressor, the turbine supplied by the turbocharger manufacturer will inevitably be able to cope with the necessary mass flow. Whether it produces sufficient power depends on its efficiency and the turbine area (since this dominates the energy available for useful expansion). The turbine nozzle ring or volute controls its effective area. Thus the effective area at the turbine will be adjusted (by changing the components mentioned above) to achieve the desired boost level from the compressor. If turbine area is reduced, then the compressor boost pressure and the mass flow rate will go up, the former by a larger amount than the latter since charge temperature will also rise. The effect, in terms of the engine operating lines superimposed on the compressor characteristic, is shown in Fig. 49.

10.2.2 *Two-stroke engines*

The airflow characteristics of a two-stroke engine will depend on whether the engine is fitted with a turbocharger alone or has an auxiliary scavenge pump or blower. Consider first the situation when a turbocharger is used on its own.

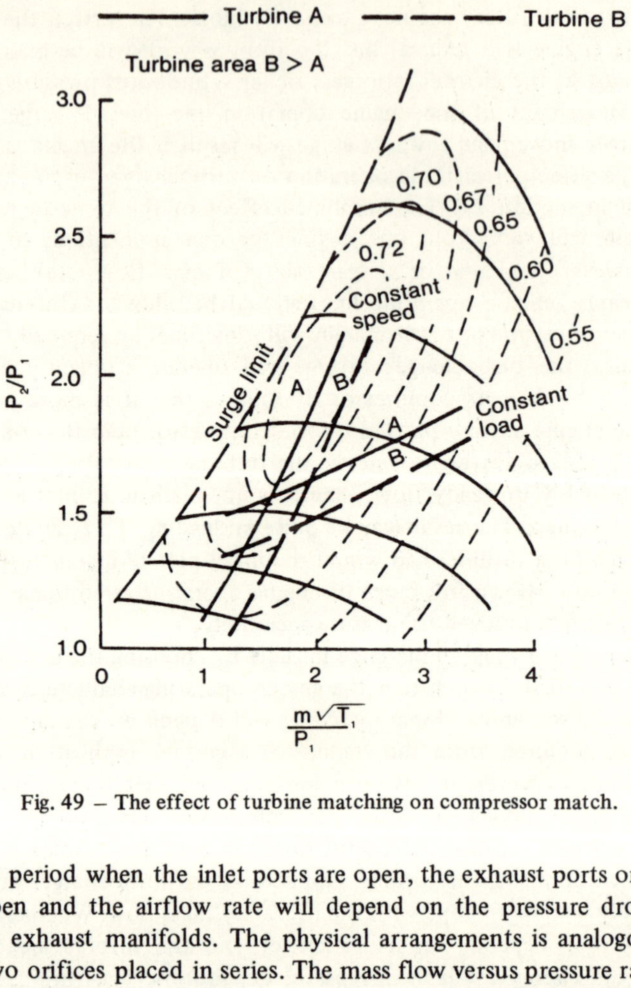

Fig. 49 – The effect of turbine matching on compressor match.

During the period when the inlet ports are open, the exhaust ports or valves will also be open and the airflow rate will depend on the pressure drop between intake and exhaust manifolds. The physical arrangements is analogous to flow through two orifices placed in series. The mass flow versus pressure ratio characteristic for steady flow through two orifices in series is a unique curve and it follows that the two-stroke engine will exhibit a similar characteristic. Thus the engine operating line, when superimposed on the compressor characteristic, will be a unique curve, almost regardless of engine load or speed (Fig. 50). Fortunately the compressor characteristics are well suited to this type of demand, and it is relatively easy to match the turbocharger. If an aftercooler is fitted (which it is on most engines of this type) the mass flow versus pressure ratio curve will have a different slope (smaller gradient) which suits the compressor characteristic even more.

If scavenge pumps or compressors are placed in series with the turbocharger compressor on a two-stroke engine, then the airflow characteristic will be dominated by the scavenging device. In general, the matching problem will not

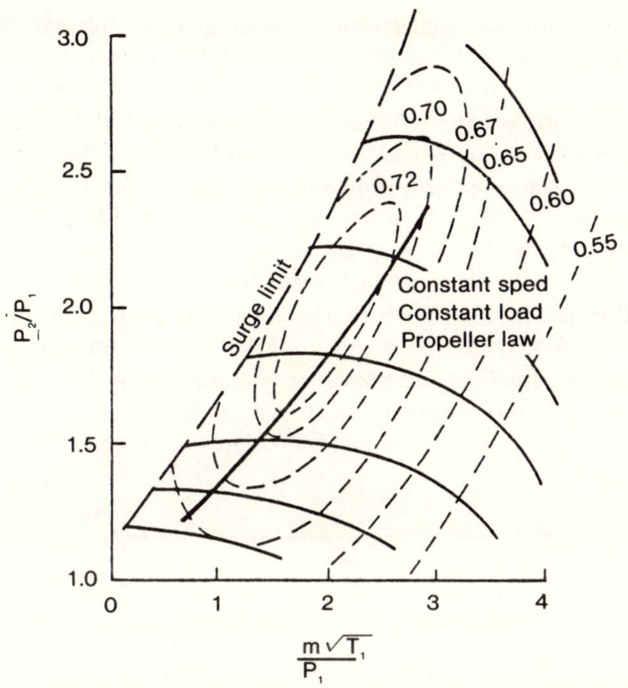

Fig. 50 – Compressor characteristic with airflow requirements of a two-stroke engine superimposed (with no additional compressor).

be significantly different from that of a four-stroke engine and the comments previously made apply. The additional difficulty that does arise is that the scavenge blower must be matched to the engine at the same time, as a compromise must be reached between scavenge blower and turbocharger work.

Generally the turbocharger will be expected to do as much of the compression work as possible. Depending on the quality of the scavenging system, turbocharger efficiency, use of pulse or constant pressure turbocharging, etc., the turbocharger may well be able to provide sufficient boost pressure at full power but not at low load. It follows that the work division between turbocharger and auxiliary compressor may well be governed by the requirement for adequate scavenging (and hence engine performance) for part-load operation. The final balance will vary from engine to engine since many different factors are involved.

10.3 Matching for constant speed operation

The most common application requiring a constant speed diesel power source is electricity generation. It is a relatively simple requirement from the point of view of turbocharger matching.

The basic compressor characteristics resulting from this varying load at constant speed application are shown by a constant speed line on Fig. 48. Small adjustments in turbine area will affect both the boost and exhaust pressures, hence pumping work and the resultant fuel consumption of the engine. Within the limits of acceptable mechanical and thermal loading of the engine, turbine matching will be used to achieve optimum performance and fuel consumption. Figure 51 shows the effect of turbine area changes on the specific fuel consumption with varying load at constant speed. Matching will be a compromise between performance at low and high load. If the diesel generator is required for base-load operation, then the turbocharger will be matched at the rated load (full line in Fig. 51). Otherwise a larger nozzle ring or volute will be fitted to improve part load efficiency. The small nozzle increases piston pumping work and hence increases fuel consumption, except at high load, when the engine benefits from the extra air which results from more turbine and compressor work.

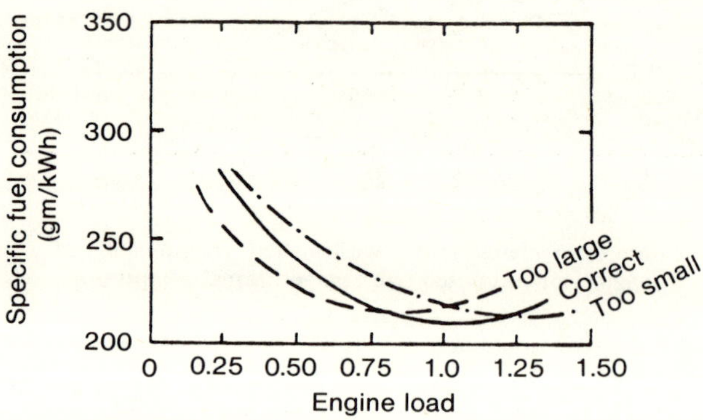

Fig. 51 – Optimum turbine matching for fixed load, constant speed.

In Fig. 52 the engine airflow rates are superimposed on the compressor map with three turbine trims. Clearly the choice of compressor is correct with the medium turbine, but a smaller compressor trim would be needed to avoid surge with the smallest turbine, and a slightly larger variant to achieve optimum compressor efficiency with the largest turbine.

10.4 Matching the four-stroke vehicle engine

10.4.1 *Torque curve*

Turbocharger matching for many industrial and marine duties is relatively straightforward owing to the limited speed and load ranges required. Matching the turbocharger to an automotive engine is considerably more difficult owing to the wide speed and load variations encountered. Although the power required

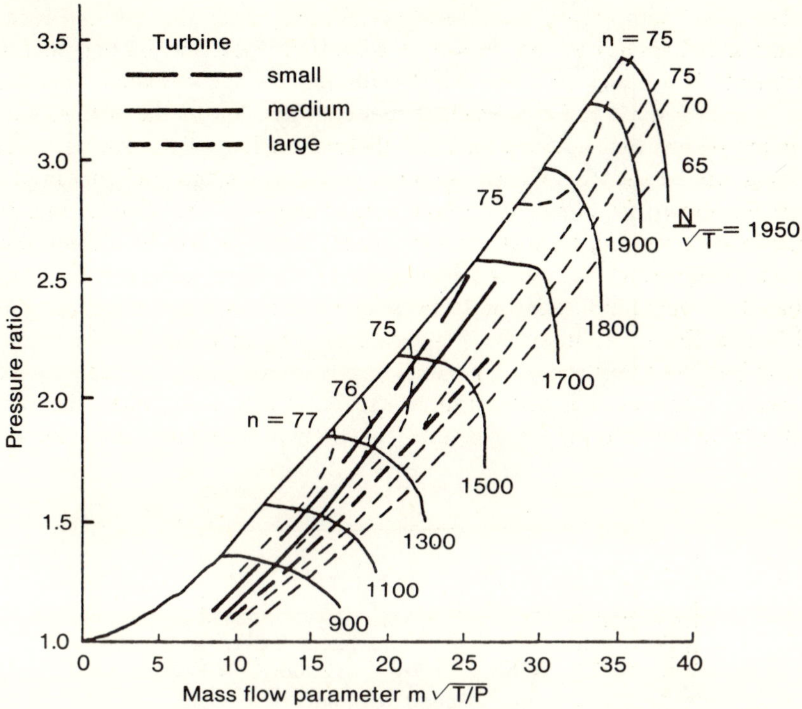

Fig. 52 – The effect of turbine matching on compressor requirement for constant engine speed operation.

to propel a vehicle increases rapidly with speed, a torque curve that rises as speed falls reduces the number of gearbox ratios and gear changes required.

Such a torque curve is said to have good 'torque back-up'. Pulse turbocharging is essential in obtaining good torque back-up at low engine speed. With good torque back-up, the vehicle will benefit from high torque at low speeds to provide a margin for acceleration and to allow the vehicle to lug up very steep hills.

Clearly the turbocharged automotive diesel engine should not be matched at full power and a compromise must be reached between power and a suitably shaped torque curve. Figure 53 shows a typical torque curve for a turbocharged automotive diesel engine. Torque back-up may be defined as:

$$\frac{\text{maximum torque} - \text{torque at maximum speed}}{\text{torque at maximum speed}}$$

Maximum torque of this engine occurs at 58% of its maximum speed, and the torque back-up is 29%. Since the engine will not normally be required to work

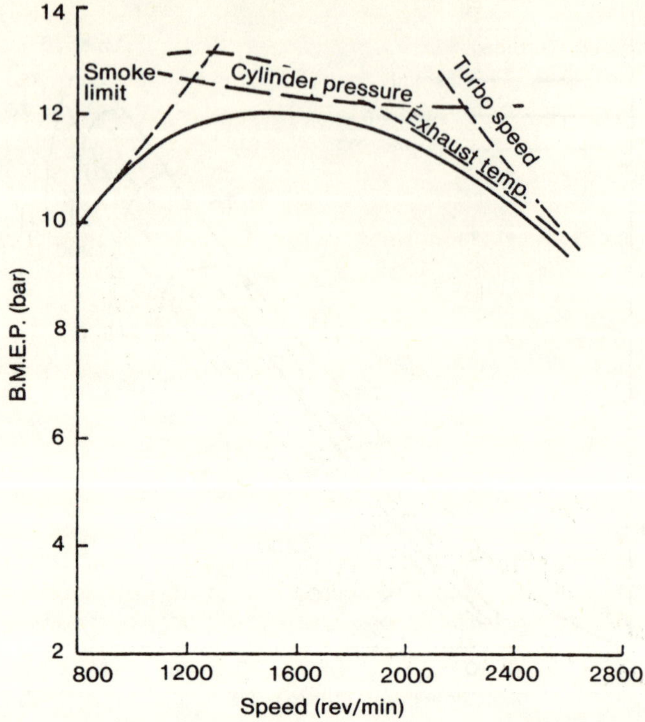

Fig. 53 – Turbocharged truck engine torque curve with limiting smoke, cylinder pressure, exhaust temperature and turbocharger speed lines (Goodlet [3]).

below 40% of the maximum speed, torque rises with reducing speed over half the useful speed range. This characteristic is, fortunately, a reasonable compromise between power and low speed torque, enabling trucks to use five or six-speed gearboxes.

Engine performance will typically be limited (for reliability) by a maximum value of cylinder pressure and possibly exhaust temperature (since the latter is an approximate guide to thermal loading of four-stroke engines, for example at the exhaust valve). In addition, a limit will be imposed by the acceptable (or legislated) exhaust smoke level, which will be determined largely by the quantity of air delivered by the turbocharger. The turbocharger will be limited by a maximum safe rotational speed and turbine inlet temperature (governed by the creep and scaling properties of the turbine wheel and housing). Depending on the engine rating, some or all of these factors will limit engine performance. This limits can be superimposed on the torque (or BMEP, since torque $\propto$ BMEP) curve as shown in Fig. 53 where the maximum possible torque curve is shown.

However, the position of most of these limiting lines will move if the turbocharger match is changed.

It is clear from Fig. 53 that the factor that is most restrictive when trying to achieve a desirable torque characteristic is the low speed smoke limit. This comes as no surprise since it was explained that it is normal for boost pressure to rise with engine speed, as a direct result of the flow characteristics of the turbocharger turbine. For example, the flow characteristics of three radial turbine trims (volutes) are shown in Fig. 54. The smallest area volute will generate the highest turbine inlet pressure and therefore the highest specific available energy at the turbine. Allowing for the variation of turbine efficiency over the pulsating and mean flow range, specific energy recovered by the turbine also increases substantially with engine speed. This characteristic is a consequence of the almost constant effective flow area of a fixed geometry turbine. Thus the natural characteristic of the turbocharged vehicle engine will be compressor work, and therefore boost pressure, rising with speed, The smoke limit is caused by insufficient boost pressure, and hence airflow, at low engine speeds.

In order to achieve an acceptable torque curve, the fuel delivery (per cycle) is held relatively constant over the speed range and efforts are made to raise boost pressure at low speed. Two techniques are available to achieve this. Either the turbocharger efficiency must be raised at this operating point or the thermodynamic availability of energy delivered to the turbine (i.e. specific available energy, Fig. 54) must be increased. Both techniques are usually adopted, and will be discussed under the headings of turbine and compressor matching.

10.4.2 *Turbine matching*

Figure 54 shows that by reducing turbine area, for example from match 1 to match 3, specific available energy at the turbine increases at all speeds. If the fuel delivery schedule is unchanged, then boost pressure increases as shown in Fig. 55, weakening the air/fuel ratio and reducing low speed smoke. Match 2 is based on a small turbine area reduction, hence the increase in available energy at low speed is small. However, turbine operation is more efficient with this build, hence the benefit in actual turbine work is more substantial.

The benefit of reduced smoke at low speeds does not come without accompanying disadvantages. Figure 54 shows that with the smallest turbine (match 3), the expansion ratio across the turbine will be very high at the maximum engine speed, when airflow is greatest. Thus the piston must pump the exhaust gases out against a high pressure, resulting in poor net power output and fuel consumption. This is seen as low BMEP and high BSFC at high engine speed (2000 rpm) with match 3.

In addition, the engine exceeds the allowable limits of maximum cylinder pressure and turbocharger speed. Thus match 2 is a reasonable compromise, except that maximum power (BMEP at maximum speed) is marginally less than that achieved with match 1.

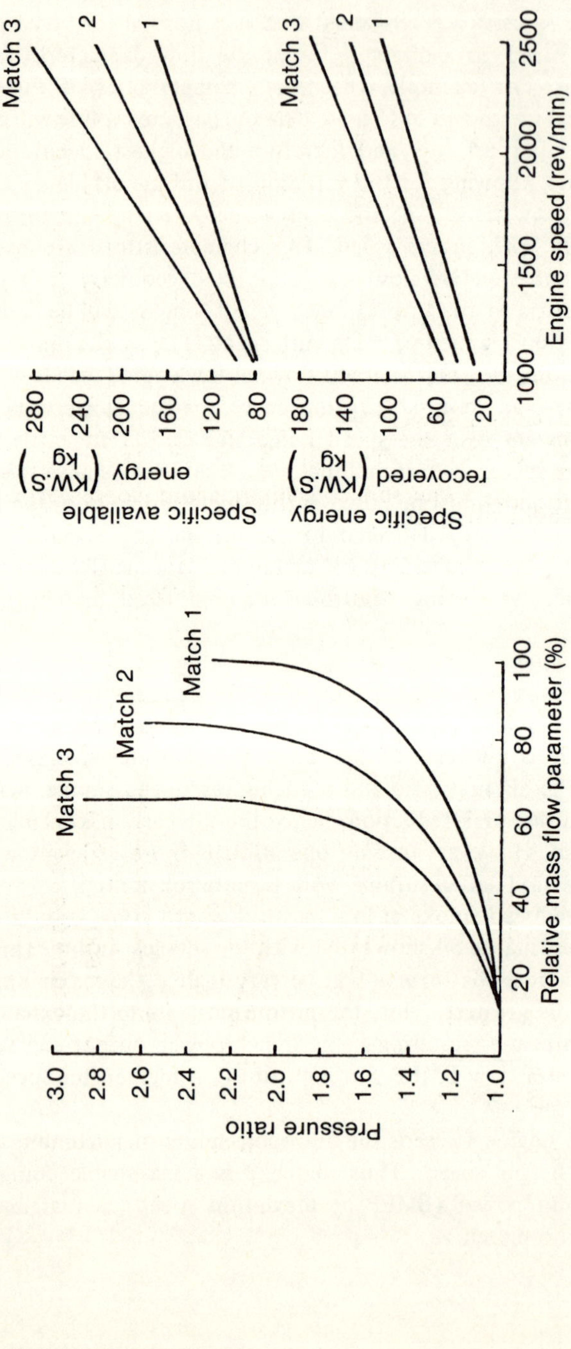

Fig. 54 – Characteristics of three truck turbocharger trims, and their effect on turbine energy on one particular engine.

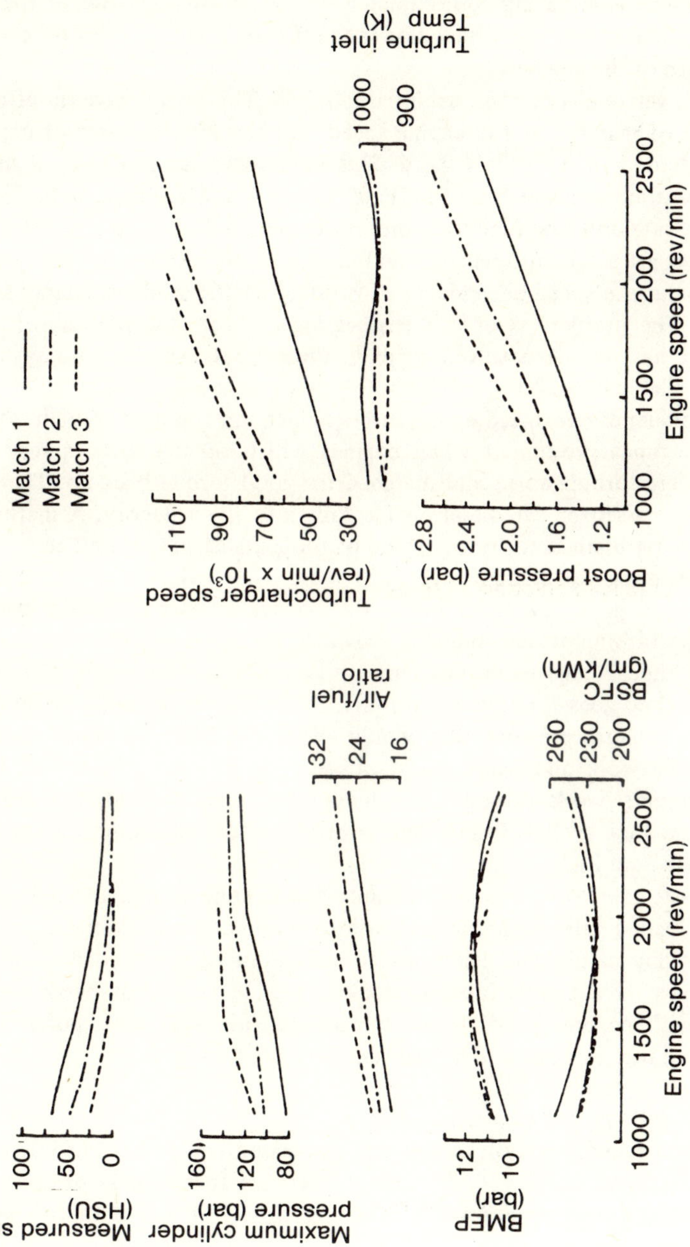

Fig. 55 – Rematching the turbine to improve low speed torque and smoke, with no change in fuel delivery.

Further 'fine-tuning' of the turbine may be achieved by selecting from different components having approximately the same effective area at the mid-engine speed condition, but whose flow and efficiency variations differ over the working range of the engine.

Ideally a variable geometry turbine is required. This would have the effective turbine area of match 1 at full engine speed, but the effective area of match 2 at mid-speed and match 3 at low speed. This would offer the low exhaust smoke, low fuel consumption and high BMEP of match 3 at 1000 rpm, with the low cylinder pressure and low fuel consumption of match 1 at 2500 rpm. This is an ideal that has attracted turbocharger and engine manufacturers for many years, since it reduces the speed dependence of turbine specific available energy shown in Fig. 54. The problem is one of engineering a cheap, reliable and effective system, and has not been solved to date. Prototypes exist, but no system is currently available in mass production.

The problem of overspeeding the turbocharger and coping with high cylinder pressures becomes prominent when engines which operate over a very wide speed range are turbocharged and matched for good torque back-up. The small turbocharged passenger car diesel engine falls into this category. A method of avoiding this problem is to bypass some of the exhaust gas around the turbine (through a waste-gate) at high speed and load. Thus, when a small turbine is fitted to achieve good low speed boost, the massive increase in specific available energy at the turbine at high speed is alleviated by increasing the effective flow area out of the exhaust manifold. This has two effects. First, only part of the exhaust gas flow goes through the turbine. Secondly, the increase in flow area reduces the exhaust pressure that would otherwise build up. Both measures reduce turbine work and hence boost pressure. In addition, the second factor reduces pumping work during the exhaust stroke and would, for example, moderate the loss in BMEP and deterioration in fuel consumption shown in Fig. 55 match 3, at high speeds.

The waste-gate valve may be built into the turbine casing, and will consist of a spring loaded valve acting in response to the inlet manifold pressure acting on a controlling diaphragm. Different combinations of spring load, diaphragm area and valve area can be used to achieve a wide variety of boost pressure variations with engine speed. Disadvantages are increased cost, potential unreliability and the restriction to a single entry turbine housing.

10.4.3 *Fuel delivery and engine speed range*

Development of fuel injection pumps and associated equipment has introduced additional freedom to vary fuel delivery over the speed range of an engine. The turbocharger matching process must be closely linked with fuel system matching even after optimum injection rates, pressures, nozzle sizes and swirl have been achieved.

Tailoring of the fuel delivery characteristic is a method of achieving good torque back-up within the framework of engine and turbocharger limitations (Fig. 53). For example, maximum fuelling can be restricted at high speeds in order to limit the maximum turbocharger speed with a small area turbine. Thus impressive torque back-up would be achieved, but at the expense of a low maximum power output.

At the other end of the speed range, excessive smoke can be reduced by restricting fuelling until sufficient boost is available to generate a reasonable air/fuel ratio. Thus a spring loaded diaphragm senses boost pressure and allows the maximum fuel stop to open as engine boost increases. Since fuelling is restricted only when the boost pressure is zero or low, torque is only reduced at very low speeds, that is below that at which maximum torque is achievable. The device is commonly called an 'aneroid' (fuel controller).

The difficulty of achieving a satisfactory match over a wide speed range has been explained. In certain circumstances it may be advantageous to reduce the rated speed of an engine whilst increaseing BMEP to achieve the same maximum power output.

By reducing the turbine area and increasing fuelling, high BMEP is obtained. If maximum rated speed is reduced from 2500 rpm to 2000 rpm (Fig. 55, but an extreme case), excessive turbocharger speed is avoided. Naturally the final drive gear ratio of the truck must be raised to compensate, hence the engine is working at a higher load than would normally be the case at the same vehicle speed and load. Since specific fuel consumtpion reduces with load, fuel savings are possible.

10.4.4 *Compressor matching*

Since the truck engine operates over a wide speed and load range, the airflow requirements cover large areas of the compressor map. A typical superimposition of engine airflow on the compressor map is given in Fig. 56, showing lines of constant engine speed (1000, 1500, 1900, 2400 and 2800 rpm) and load (3.85, 5.17 and 8.47 bar), and the maximum torque curve.

Selection of the correct compressor is largely a matter of ensuring a sufficient surge margin (*A*/*B*, Fig. 56) and that the operating points at maximum torque and power (points X and Y, Fig. 56) occur at reasonable compressor speed and efficiencies. Thus the compressor shown in Fig. 56 is a satisfactory choice, since the operational area is clear from the surge line and lies in an area of high efficiency. However, a slightly larger compressor might result in more of the operating regime experiencing higher compressor efficiency, with a less generous surge margin. A small improvement in low speed could also be obtained if maximum compressor efficiency occurred at a lower pressure ratio (e.g. 1.6 compared with 1.8 in Fig. 56).

Reducing turbine area will raise the boost pressure, reducing the surge margin (compare Figs. 56 and 57) and therefore a smaller compressor trim may be required in some cases. However, the surge margin shown in Fig. 57 is adequate, the compressor being well matched.

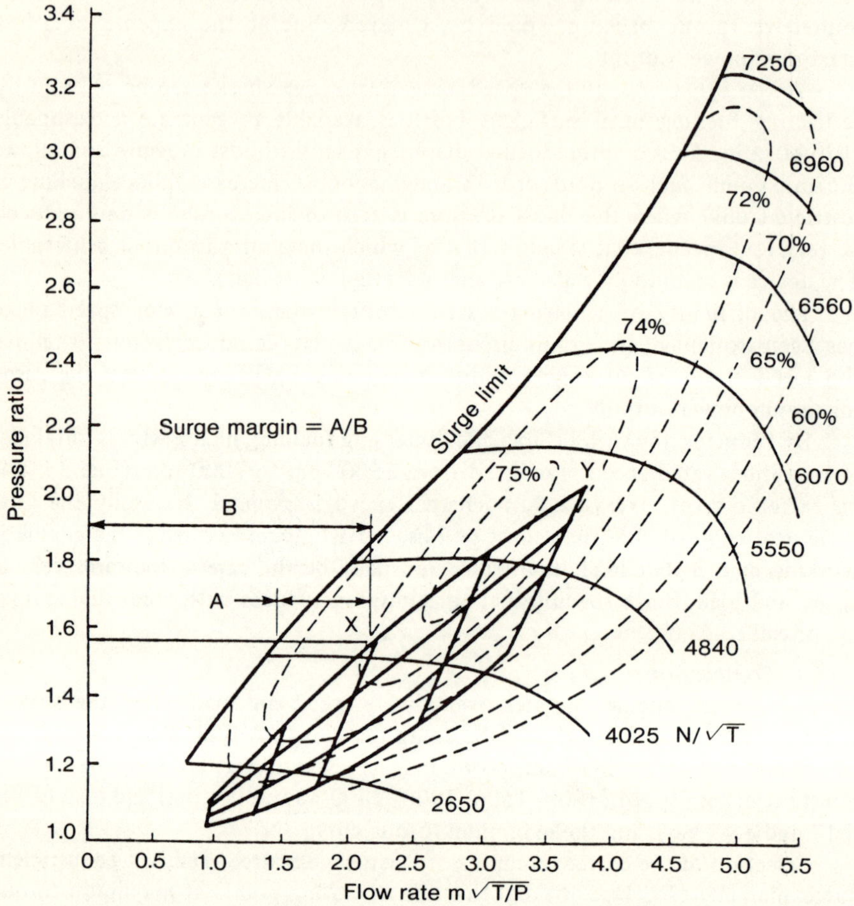

Fig. 56 – Engine operation area superimposed on compressor map, showing surge margin.

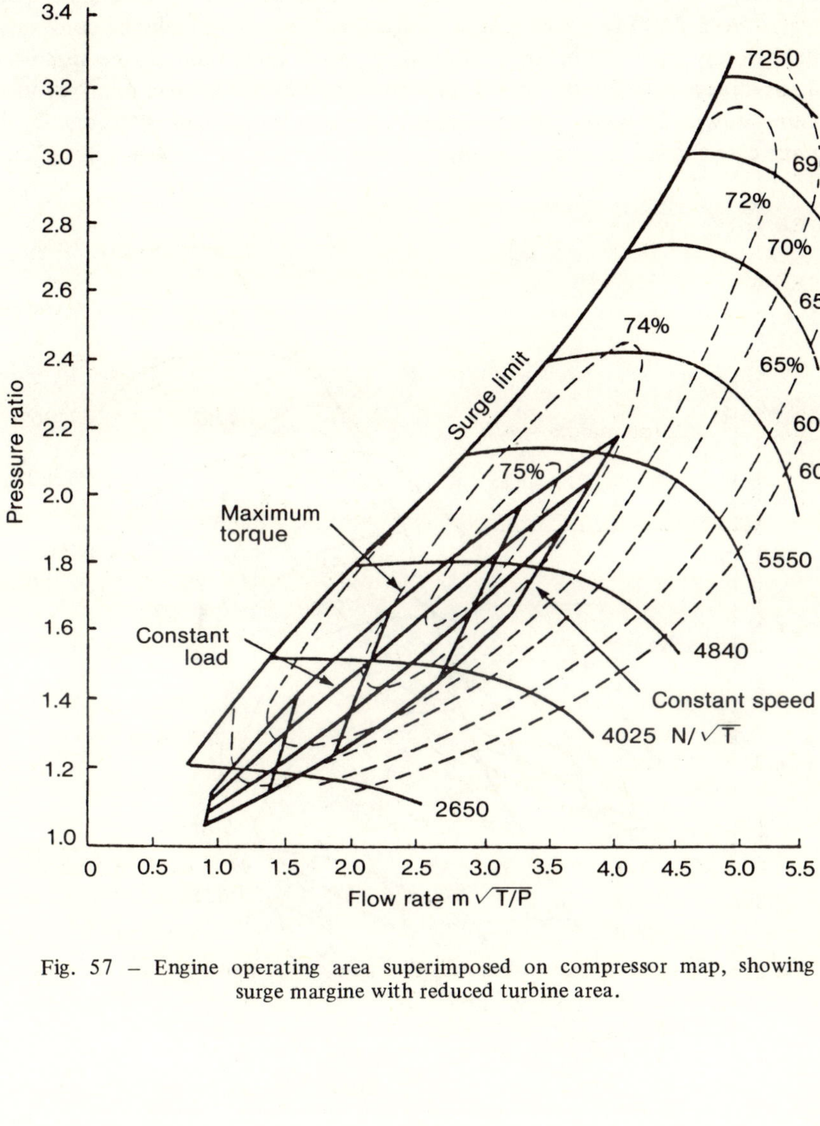

Fig. 57 – Engine operating area superimposed on compressor map, showing surge margine with reduced turbine area.

Figure 58 shows a poor compressor match, using too small a compressor trim on an intercooled engine. At maximum speed and load (point Y) the compressor efficiency is low. Boost pressure actually falls as the engine speed increases from 2400 to 2800 rpm as a result. It may occasionally be convenient to deliberately match into an inefficient region at maximum engine speed and load, in order to hold the boost pressure and therefore maximum cylinder pressure down. However an excessive reduction in efficiency occurs in the extreme case shown, and piston pumping work during gas exchange will suffer.

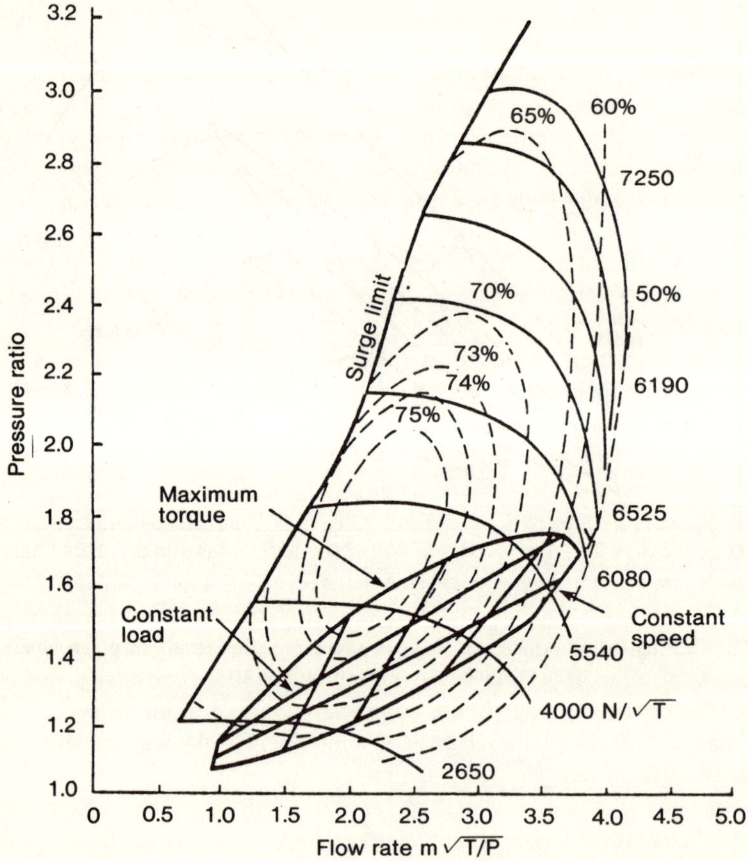

Fig. 58 – A very badly matched compressor.

11. CHANGES IN AMBIENT CONDITIONS

Changes in inlet air density may be caused by ambient temperature and pressure changes at sea level, or operation of the engine at altitude. The change of airflow may readily be predicted if the engine is naturally aspirated, but this is more difficult if the engine is turbocharged.

In mobile applications, such as a truck, the engine may be required to operate at sea level in a cold winter climate and perhaps at altitude during a hot summer. Thus the final compressor and turbine matches selected will be something of a compromise, particularly if the pressure ratio is high. The match selected will suit the normal operating environment of the engine, but with sufficient margins on surge, turbine inlet temperature and turbocharger speed to cover other conditions.

Although operation under changing ambient conditions introduces additional complications for the manufacturer of a turbocharged engine (such as reduction of the surge margin), the turbocharging system does offer partial compensation for reducing air inlet density at altitude. As air density and therefore air mass flow rate reduce, so the turbine inlet temperature will rise owing to the richer air/fuel ratio. This means that the ratio of compressor to turbine pressure ratios will be altered in favour of the compressor. Its pressure ratio will increase, partially offsetting the reduction in air inlet density. Also as ambient pressure falls, so the expansion ratio of the turbine increases, raising compressor pressure ratio, provided that the turbine inlet pressure does not fall at the same rate as ambient pressure.

An increase in ambient temperature, however, has an undesirable effect on the turbine-to-compressor energy balance, hence the turbocharger will tend to amplify the effect of such a change on airflow rate. Low ambient temperatures reduce the required compressor power, hence boost rises, sometimes causing compressor surge.

Large variations in ambient conditions can lead to problems due to compressor surge, excessive cylinder pressure, turbine inlet temperature, turbocharger speed or smoke emission. The actual performance of an engine under varying ambient conditions will depend on several factors that, for convenience of explanation, were assumed to be constant in the simplified analysis given above. For example, if air mass flow rate and compressor pressure ratio change, movement across the operating map of the compressor will be accompanied by an efficiency change. It follows that engines of similar performance at sea level will not necessarily perform comparatively at altitude.

The parameter that limits engine performance will depend on the design of individual engines. At high ambient temperatures the limits are likely to be smoke, due to reducing airflow, then turbine inlet or exhaust valve temperature or thermal loading of the engine. At low ambient temperature, compressor surge (due to high pressure ratio) or maximum cylinder pressure may be a limiting factor. It will be the limitations of a particular engine and turbocharger combination that will govern to what extent fuelling must be reduced, de-rating the engine for acceptable reliability or smoke emission.

Operation at altitude is usually, but not always, accompanied by a reduction in temperature. The turbocharging system offers partial compensation of the inlet air density reduction at altitude, thus an engine may have to be de-rated, but not by as much as a naturally aspirated engine.

The effect of altitude on a turbocharged truck engine at full power, with and without intercooling, is shown in Fig. 59. Although the absolute inlet manifold pressure reduces with altitude, the fall-off is slower than that of ambient pressure. Turbocharger speed increases owing to the increase in turbine

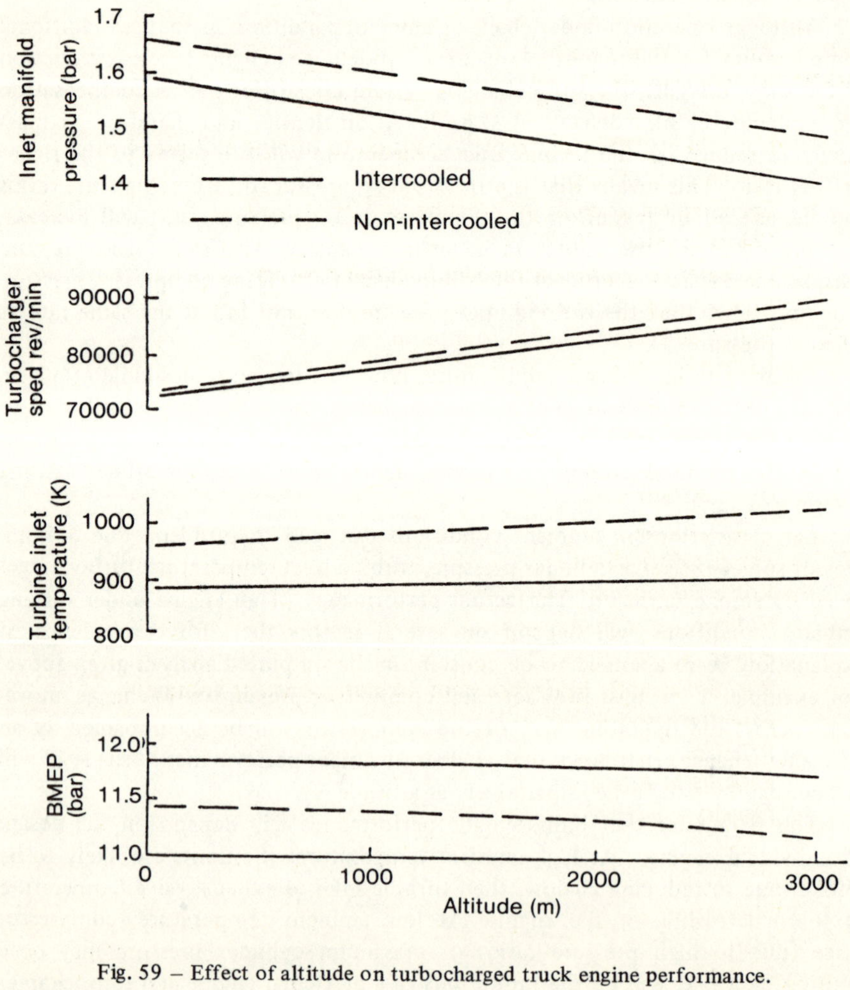

Fig. 59 – Effect of altitude on turbocharged truck engine performance.

inlet temperature and expansion ratio. It can be seen that thermal limits and the maximum permissible turbocharger speed will be the limiting factors, particularly the latter. Movement towards surge on the compressor map will be greatest for a non-intercooled engine.

ACKNOWLEDGEMENTS

The author wishes to acknowledge the help of Garrett-AiResearch and Holset Engineering Co. for the provision of diagrams and helpful discussions.

NOMENCLATURE

A	Area
AFR	Air/fuel ratio (by mass)
BMEP	Brake mean effective pressure
BSFC	Brake specific fuel consumption
C	Gas velocity after isentropic expansion to turbine exit pressure
C_d	Discharge coefficient
C_p	Specific heat at constant pressure
CV	Calorific value
D	Diameter (compressor impeller or turbine rotor)
F	Fuel/air equivalence ratio
ISFC	Indicated specific fuel consumption
l	Length
$\dot{m}$	Mass flow rate
N	Rotational speed
P	Pressure
R	Gas constant
T	Temperature
ΔT	Temperature change
u	Rotor tip tangential velocity
V	Volume
$\dot{W}$	Power
γ	Ratio of specific heats (constant pressure/constant volume)
ϵ	Effectiveness
η	Efficiency
μ	Kinematic viscosity
ρ	Density

Subscripts

$_0$	Stagnation
$_1$	Inlet
$_2$	Outlet

a	Air
c	Compressor
cool	Coolant
f	Fuel
i	Indicated
is	Isentropic
m	Manifold (inlet)
mech	Mechanical
st	Stoichiometric
sw	Swept (by pistons)
t	Turbine
TC	Turbocharger
TS	Total to static
vol	Volumetric

REFERENCES

[1] Zinner, K., *Supercharging the Internal Combustion Engine,* Springer-Verlag, 1978.

[2] Watson, N. and Janota, M. S., *Turbocharging the Internal Combustion Engine,* Macmillan Press, 1982 (U.S. edition, J. Wiley and Sons).

[3] Goodlet, I. W., 'Turbocharging of Small Engines', *Proc. I. Mech. E.,* **188,** No. 3/4, 1974.

6

Diesel exhaust emissions

R. W. WHEELER, Atlantic Research Associates

Clearly it will not be possible in this chapter to cover all aspects of this complex topic, therefore I shall concentrate on the technological aspects of the subject, leaving out legislation and the emission limits because much of this has little validity, being based on expediency.

Many diesel engineers have little sympathy with this subject and are quite rightly deeply suspicious of the motives of legislators. However there is no doubt that research into exhaust emissions has led to a considerable increase in knowledge of the diesel process, and, in the unlikely event of the legislators dropping the whole subject, it is certain that emissions' tests would still form part of diesel engine development.

1. GENERAL HISTORICAL INTRODUCTION

When the diesel engine was first invented it was intended for industrial purposes such as power generation, factory power, water pumping and ship propulsion. In these applications it did not intrude overmuch into the area of public amenity and its enormous economic advantages completely outweighed its problems of smoke, odour, noise, vibration and an evil smelling fuel.

During the 1920s and early 1930s when automotive versions of the engine were being pioneered, these problems came more to the fore since clearly the public would be intimately involved, especially as bus engines were an early objective. Naturally, smoke reduction followed the route of the control of optical opacity since, at that time, there was little suspicion of anything harmful in the soot. Interestingly, at about the same time the chemical structures of the polynuclear aromatic hydrocarbons (PNAH) were elucidated [1]. These materials are present in all soot and their control has recently posed a considerable challenge to the diesel industry.

Compared with the petrol engine the diesel burns its fuel much more efficiently. A modern engine emits only ~1% of its fuel as unburnt substances such as carbon monoxide (CO), soot and hydrocarbons (HC). Under equivalent conditions a petrol engine might waste ten times as much. For this reason the diesel engine acquired a reputation for low pollution which is justified if the masses of the pollutants only are considered and not their biological interactions.

One of the first groups of pollutants to gain notice comprised the aldehydes since these were recognized by their irritating odour in many diesel engines with poor combustion. Next the nitrogen oxides were observed as materials interfering with analyses for organic peroxides and aldehydes when these were investigated as combustion intermediates [2]. However, at that time there was not considered to be any threat to health from CO, aldehydes, or nitrogen oxides (NO_x) emitted from diesel engines running in the open air. In enclosed areas where ventilation may be less than adequate, such as in coal and other mines, the diesel engine needs to be a low pollution design because of stagnant pockets of low air movement where even the carbon dioxide in the exhaust may be a hazard. An early research at the U.S. Bureau of Mines determined the effect of engine load on CO, aldehydes and NO_x in the exhaust gas [3]. Even in 1979, however, the U.S. coal miners and their advisers remained unconvinced of the safety of diesel exhaust, a situation summed up fairly in a report produced for the American Mining Congress [4].

Other countries, notably Australia, South Africa and Great Britain, have carried out long-term epidemiology of the workers in dieselized mines and have not found any significant health effects, a result in line with that of a well-known research in London diesel bus garages by Professor P. J. Lawther [5]. Notwithstanding the lack of evidence of health effects, most national control bodies applied simple maximum levels for emissions of CO and in some cases NO_x. These varied with local needs and conditions and represented the first widespread acceptance of the need to control diesel emissions in special areas.

The vastly expanded use (and abuse) of fossil energy after 1945 led inevitably to disastrous situations such as that at Donora, Pa, in 1948 [6], in London in 1952 when many people of low resistance died [7, 8], and in Los Angeles where the infamous 'smog' makes life unpleasant for all and dangerous to the weak. Each of these events was caused by the overloading of the air by numerous pollutant sources. In the case of Los Angeles, clearly the gasoline car was the predominant source, but because this city like many others depended on the car for its life it was necessary to research the problem urgently on a nationwide basis. The Schenk Act of Congress of 1960 ordered the U.S. Surgeon General to report on the problem – which he did in 1962 [9]. This book – although the diesel engine is hardly mentioned – gives a good account of the overall situation and how inevitably regulation had to follow.

In 1966 the California Air Resources Board (CARB) applied the first regulations to cars powered by spark-ignition engines based on the non-dispersive infra-red (NDIR) analysis of carbon monoxide (CO) and volatile hydrocarbons (HC) in the exhaust gases emitted while the cars were driven over a simulated drive cycle on a chassis dynamometer. The U.S. Federal Government followed in 1970 with a similar law, but added nitrogen oxide (NO) analysis [10].

The acceptance of these laws by the U.S. public and eventually by the motor industry made it certain that the regulation of automotive diesel engines would follow if only for the sake of conformity.

The energy crises of the late seventies, and the fuel saving concept of Corporate Average Fuel Economy (CAFE) developed by the U.S. Federal Government led to a great upsurge in diesel development which in turn caused an environmentalist counter attack based on the alleged cancer implications of diesel soot, and introduced new biological assay technology into exhaust testing.

The battle is not yet over and what the diesel engine will look like at the end of it is not at all clear.

2. DIESEL POLLUTANTS

Gasoline engine exhaust emission technology was in large part adaptable to the diesel. The exhaust gas-volume of a diesel (being unthrottled) is far greater than that of an SI engine, presenting some handling problems. The NO_x concentration is of a similar order, but the CO level is much lower, entailing a more sensitive NDIR analysis. It was in HC analysis that difficulties arose because diesel hydrocarbons are in general too low in concentration for NDIR analysis, but also, more seriously because of the nature of the fuel, are much less volatile.

The answer to this problem was already available owing to General Motors researchers at (GM) on low-sac volume nozzles to cure diesel odour. Pearsall [11] showed that if the exhaust gas sample was maintained at c. 190°C it could be analysed by flame ionization (FIA) with minimal loss. FIA rapidly became the preferred HC analysis method, even for the gasoline engine (excepting in the EEC).

These, then, were the regulated pollutants, CO, HC and NO_x. Some explanation of the 'x' in NO_x is called for. NDIR analysis on an exhaust sample taken directly from an operating engine measures nitric oxide (NO) only, but NO in the presence of oxygen (O_2) is slowly oxidized to nitrogen dioxide (NO_2).

In any exhaust stream there is some oxygen and therefore some NO_2. This particularly applies to air diluted exhaust as in constant volume sampling (CVS) and of course the final fate of NO in the air is total conversion to NO_2.

Any practical exhaust sample therefore contains NO and NO_2, with an empirical formula NO_x, where 'x' lies between 1 and 2. For legislative purposes NO_x is calculated as NO_2 with a molecular weight of 46.

Actual NO_2 emissions vary strongly with engine conditions [12] and it has been argued that it might be considered to be a separate pollutant owing to its high toxicity than NO [13].

A more recent addition to the list of regulated pollutants is the particulate content of exhaust gas obtained by filtration under specified conditions. It cannot be said that this complex material has been satisfactorily defined as yet, neither has the analytical procedure been finalized to satisfy all critics. However, limits have been applied to light-duty diesel cars [14]. The problem has led to many papers; good collections are available [15, 16].

There are many more potential pollutants in diesel exhausts; a fairly complete list is as follows:

1. CO.
2. HC.
3. NO_x.
4. Particulates.
5. Carbon – the elemental C content of 4.
6. The HC content of 4 which consists of highly adsorbed high boiling point hydrocarbons.
7. Polynuclear aromatic hydrocarbons – PNAHs, carcinogens, typified by 3–4 benzpyrene, and derivatives such as epoxy, hydroxy and nitro PNAHs.
8. Aldehydes – irritants and odours typified by formaldehyde (H–CHO) and acrolein (CH_2 = CH–CHO).
9. Diesel odours – aromatic and oxygenate compounds typified by indans, naphthalenes, indanones, naphthaldehydes, phenols.
10. Sulphur compounds – pulmonary irritants, and acid-rain components such as sulphur dioxide (SO_2), sulphuric acid (H_2SO_4), and ammonium sulphate $(NH_4)_2SO_4$.
11. Blue–white smoke – an emission of hydrocarbons and oxygenates at cold-start and idle conditions – which is choking and highly offensive [17].

These substances have may interactions such that it is extremely difficult to assess their effects on man. Some may be artifacts of the sampling technology; for example NO_2 and PNAHs react to give highly mutagenic nitro-PNAHs [18]. Aldehydes and odour compounds have complex actions on the nasal receptors which are only separated with difficulty [19, 20].

The latest problem is the interaction of NO_2 and SO_2 in the atmosphere to give 'acid rain' or very dilute sulphuric acid. Diesel fuel contains from 0.1–1.0% of sulphur around the world, and there may be a call for its reduction and the introduction of a heavily desulphurized premium diesel fuel if the diesel car population grows, especially in California.

3. POLLUTANT SAMPLING AND MEASUREMENT

Without correct sampling procedures emissions' tests are pointless. There are many pitfalls – air leaks, back pulsing of air up exhaust pipes, probes too hot giving losses by oxidation, and too cold giving condensation – that it is vital to follow regulatory instructions exactly.

Raw, undiluted exhaust sampling as required for steady-state testing is best effected by using a heated pump–filter assembly (Fig. 1) situated near the exhaust system and feeding the analysis bench through a line heated to 190°C which maximizes the HC analysis response.

Soot seriously interferes with optical analysis procedures (NDIR) and clogs metering orifices; consequently exhaust filtration must be of high quality. Water condensation is seldom a problem, as in gasoline exhaust, owing to the high air/fuel ratios of diesels but it may be encountered with special fuels.

To convert raw exahust analysis into mass emissions the engine air consumption must be measured accurately. This is very difficult to effect for transient modes, so the constant volume sampling (CVS) system was invented.

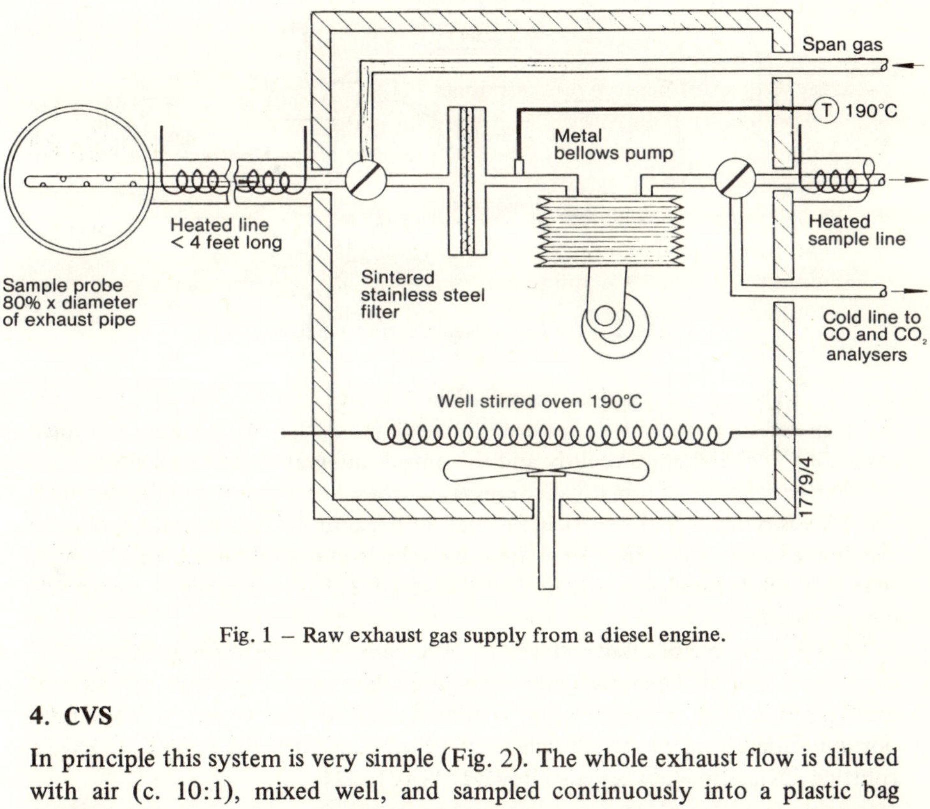

Fig. 1 – Raw exhaust gas supply from a diesel engine.

4. CVS

In principle this system is very simple (Fig. 2). The whole exhaust flow is diluted with air (c. 10:1), mixed well, and sampled continuously into a plastic bag (Tedlar). The total flow of diluted exhaust is handled by a positive displacement pump which runs at a known constant rpm for a known time. Hence the total flow is determined, while analysis of the bag sample gives the pollutant concentration as an average over the test duration. These two values combined with the pollutant gas density allows the emission to be calculated.

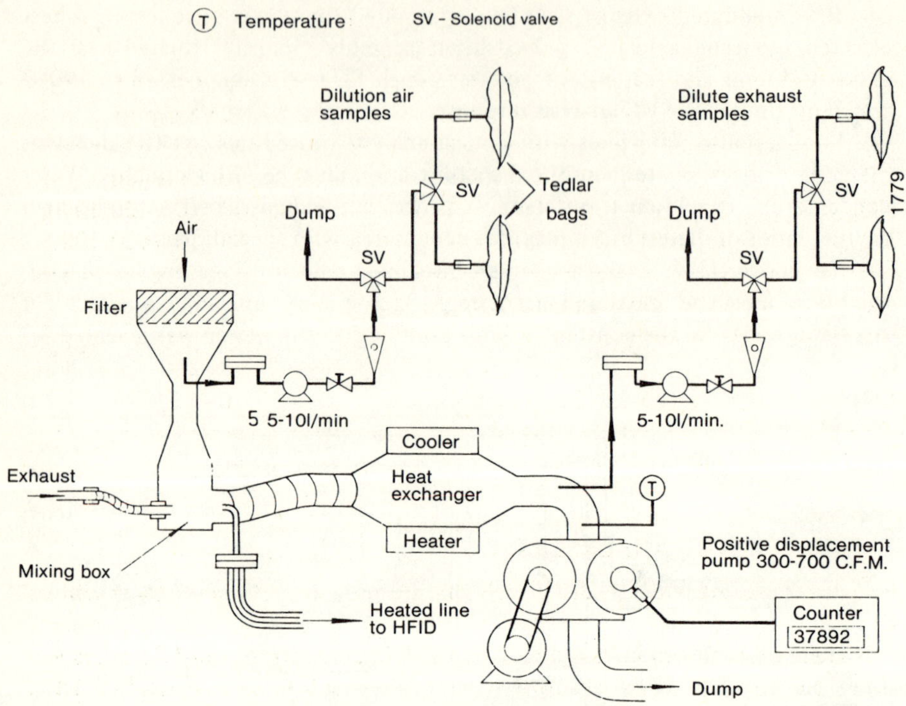

Fig. 2 – CVS equipment for diesel exhaust.

CO, CO_2 and NO_x can be measured in this way as they can be collected in the bag and integrated; HC, because it condenses in the bag, cannot, and must instead be analysed continuously and the output integrated electronically.

Particulates are automatically integrated when the trapping filter is weighed.

CVS is obligatory for particulate analysis because the dilution and cooling of the hot exhaust somewhat simulates the vehicle-in-use situation. EPA suggest that the diluted exhaust should never exceed 125°F for realistic particulate development.

Of the two major disadvantages of CVS, namely the need for analysers 10 × as sensitive and the large gas handling requirements of the apparatus, the latter is more serious. The application of transient cycle testing to heavy-duty transportation diesels in the future will need either huge CVS equipment or exhaust splitting devices to allow a scaled down CVS [21, 22].

5. POLLUTANT ANALYSIS EQUIPMENT

Before describing typical analysers the measurement ranges are listed, Table 1, as these to some extent dictate the analysis technology. These ranges are general experimental work; regulations will state specific requirements.

Table 1.
Analyser Ranges

Pollutant	Raw Exhaust (vpm)*	CVS (vpm)
CO	0–200/1000/5000	0–100/400
NO_x	0–1000/4000	0–100/400/1000
HC	0–100/1000/10000	0–100/400/1000
Gas	Raw exhaust % v/$_v$	CVS % v/$_v$
O_2	0–10/25	0–25
CO_2	0–5/15	0–1/2/5

*vpm – parts per million by volume.

6. PRINCIPLES OF ANALYSERS

Non-dispersive infra-red (NDIR) analysers are used for CO and CO_2 in almost all cases.

These gases absorb infra-red radiation at 4.7 microns (μm) and 4.3 μm, respectively. NO also has a weak absorption band, allowing it to be analysed by NDIR, but lack of sensitivity and interference by water vapour make it only suitable for crude work in its simple form.

An NDIR analyser measures the absorption at the required wavelength and compares the result to that given by calibration gases. It is a comparator not an absolute analyser and depends heavily on the quality of the pure gas blends.

As shown in a simplified instrument diagram (Fig. 3) a wideband infra-red source is split into two parallel beams which both pass through internally reflecting (gold plated) tubes and finally fall on either end of a differential gas (Luft) detector. The detector consists of a sealed container with two calcium fluoride I.R. transmitting windows. It is divided through the middle by a flexible diaphragm forming one plate of a condenser and is filled with the gas to be measured – CO, CO_2 or NO.

When the two beams enter the detector the components at the specified gas wavelength are absorbed by the detector gas which heats and expands, but if the beams are equal there is no unbalance of the separating diaphragm.

One beam, the reference, passes through a fixed path; the other passes through the analysis tube which contains the unknown gas. Here the beam loses strength by its absorption by the specific component, and on passing through into its half of the detector develops less heat of absorption, thus allowing the diaphragm to bulge towards it and change the condenser capacity. The beams are chopped at from 6–10 Hz, giving the system an A.C. output.

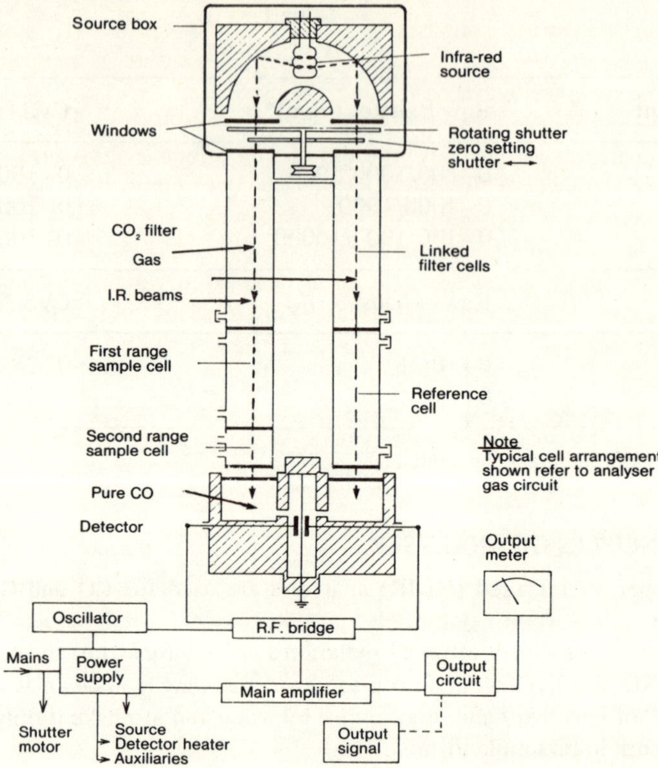

Fig. 3 – Schematic diagram of IR analyser.

Component selectivity given by the specific detector filling may be increased by narrow band IR Fabry-Perrot thin-film filters, or even by interfering component removal such as by the use of drying agents to remove H_2O from the sample stream in NO analysis.

7. CHEMILUMINESCENCE ANALYSERS (CLA)

When dilution sampling (CVS) entered the scene NDIR was plainly useless for NO_x analysis, but researchers at Ford came up with the CLA just in time [23].

This technique is based on the fact that NO reacts with ozone (O_3) to give NO_2 in an electronically excited state. This relaxes to the ground state and yields a quantum of red light.

$$NO + O_3 \rightarrow NO_2{}^* + O_2$$

$$NO_2{}^* \rightarrow NO_2 + h_v$$

The red light can readily be measured accurately using a photo-multiplier.

Any NO_2 in the sample stream can be thermo-catalytically converted to NO before entering the reaction chamber and analysed as an increased NO output when the sample is switched through the converter.

The instrument (Fig. 4) is basically very simple and reliable. A good design depends greatly on gas flow stabilities. The principle shown uses a reaction at c. 10 millibar where the quantum yield is high; however, the vacuum pump is heavy, expensive and if oil filled, needs an ozone removal scrubber. Another type uses a reaction at just sub-ambient pressure and is satisfactory in use.

CLA can be scaled down to a full range of 10 vpm or less and is extremely linear up to c. 2000 rpm. The Federal Register gives full instructions for calibration and test.

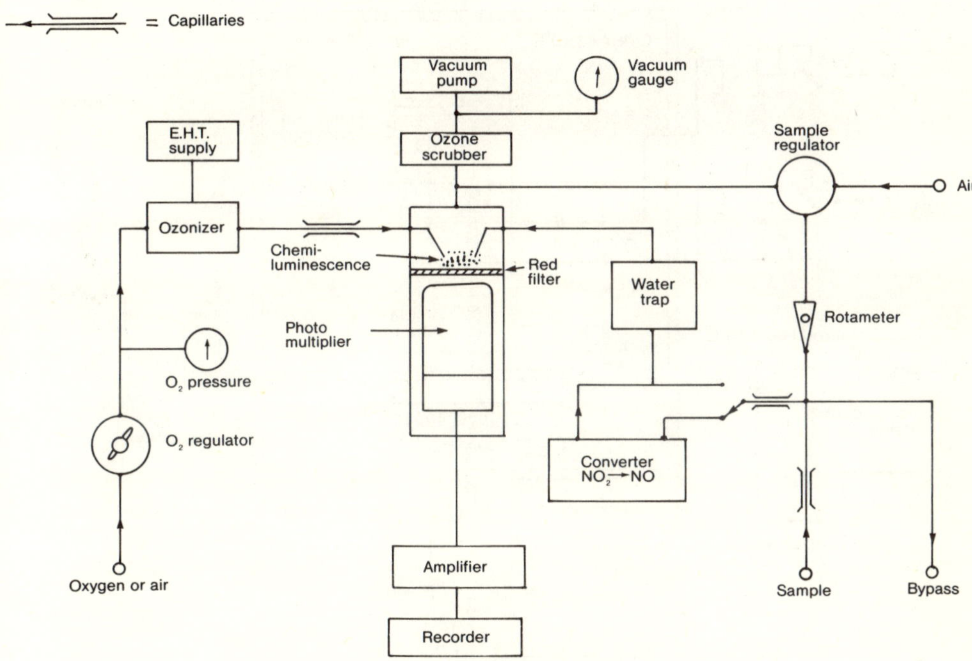

Fig. 4 – Chemiluminescence analyser.

8. HYDROCARBON ANALYSERS

Some hydrocarbons have an IR absorption at 3.4 microns but some others, notably aromatics, have almost none. Roughly 50% of exhaust HC is measured by NDIR which is therefore inadmissible as a technique.

It has long been known that a hydrogen flame is rendered conductive by gaseous HC. Flame ionization detection (FID), as it is called, was adapted for diesel exhaust by heating the whole instruement to 190°C [11] when it is then called HFID. A simplified instrument diagram is shown in Fig. 5.

FIDs have the unique property that the output of the detector is nearly proportional to both the HC concentration and to the number of carbon atoms in the HC molecule, i.e. 100 vpm propane (C_3H_8) gives half the response of 100 vpm hexane (C_6H_{14}). For this reason it is called a carbon counter and the output is expressed in terms of a single C atom HC as vpm C.

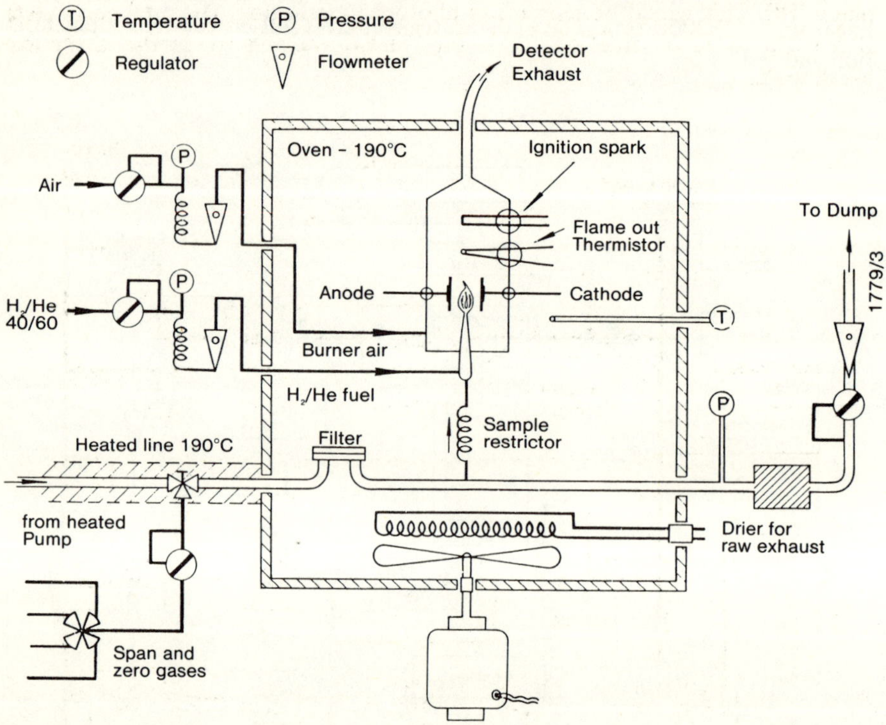

Fig. 5 – Heated flame ionization detector for hydrocarbons.

A cool diffuse hydrogen flame carbon counting to which end the flame fuel is 40% H_2 + 60% helium [24].

Total HC by HFID is an unsatisfactory measure to assess the human impact of diesel fumes in the atmosphere, being a combination of disparate elements such as:

1. Low molecular weight volatile HC.
2. Higher M.W. HC which condense to oily droplets in the air.

3. HC which is strongly bound to soot in the form of particulates [25].
4. Odorous HCs and oxygenates which cause the objectionable smell.
5. Polynuclear aromatic carcinogens and mutagens which are also associated with the particulates.
6. Irritant aldehydes such as acraldehyde.

With particulate legislation in force, the higher HCs are subject to two controls, both somewhat crude. It would be more sensible to abolish HFID and use ambient temperature FID as for gasoline engines to measure free atmospheric HC. This would be relevant to photochemical smog. The total particulate is relevant to air obscuration and soiling, and is available for further analytical breakdown.

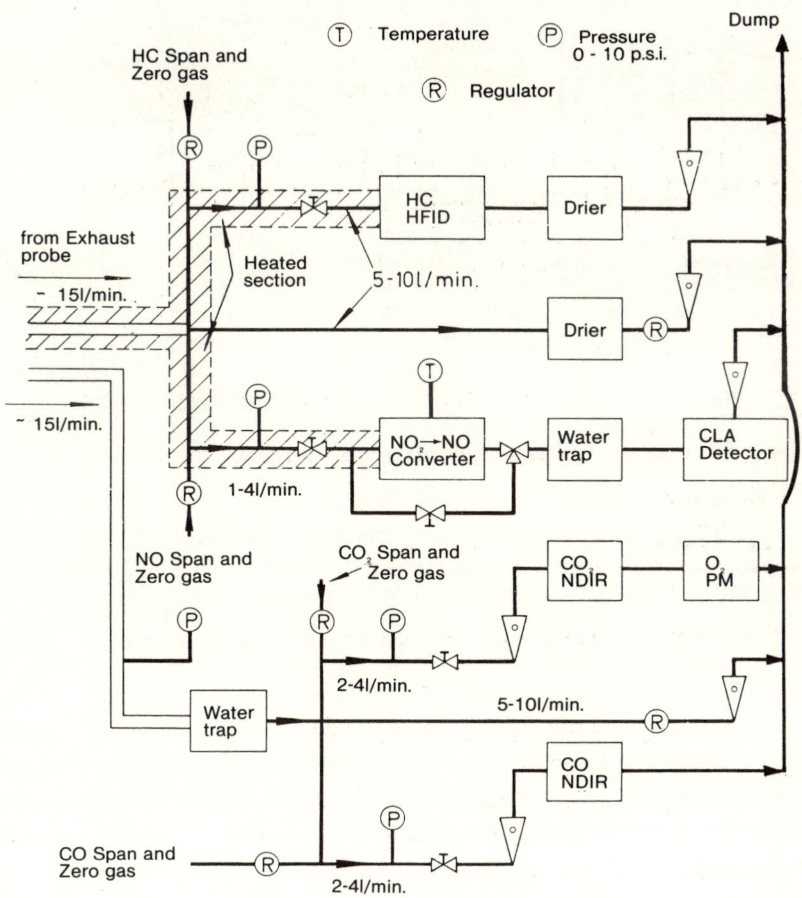

Fig. 6 – Diesel raw exhaust analysis train.

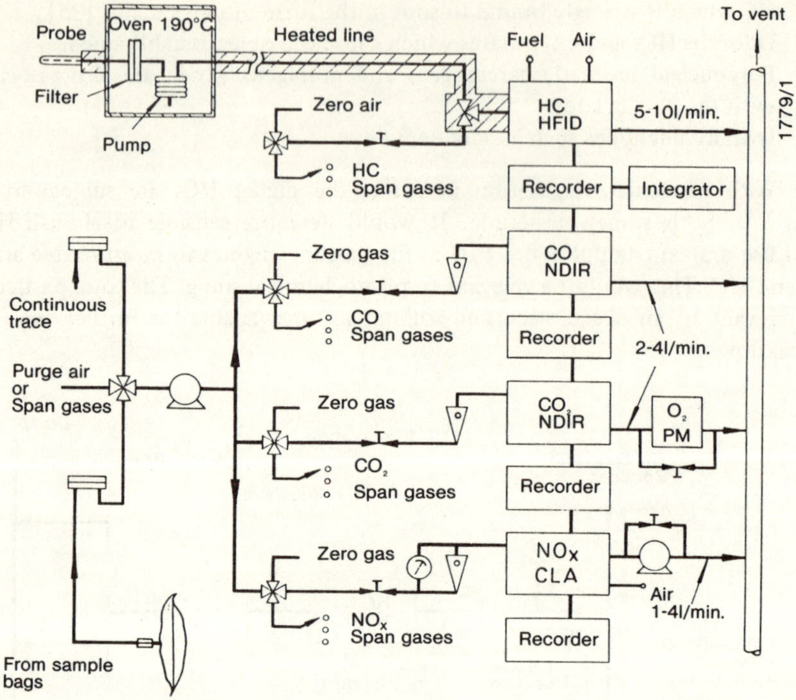

Fig. 7 – Analytical system for diluted diesel exhaust gases from a constant volume sampler.

9. ANALYSIS BENCHES FOR GASEOUS EMISSIONS

There are many ways to connect up analysers in order to obtain simultaneous analyses from an operating engine. Very much depends on the type of analyser used, particularly on its flow characteristics. An HFID needs a heated supply, as does a CLA, whereas NDIR analysers need a cool, and in the case of NO_x machines, dry sample.

Two suggested schemes with gas flows are shown in Figs. 6 and 7 for raw, and for CVS diluted exhaust samples, respectively.

Analyser manufacturers should be consulted before attempting to construct a bench. It is better, but much more costly, to purchase a professionally made system.

10. PARTICULATE ANALYSIS

The aim of particulate testing is to reproduce as nearly as possible the condensation and agglomeration of diesel fumes behind a vehicle. The dilution factor in

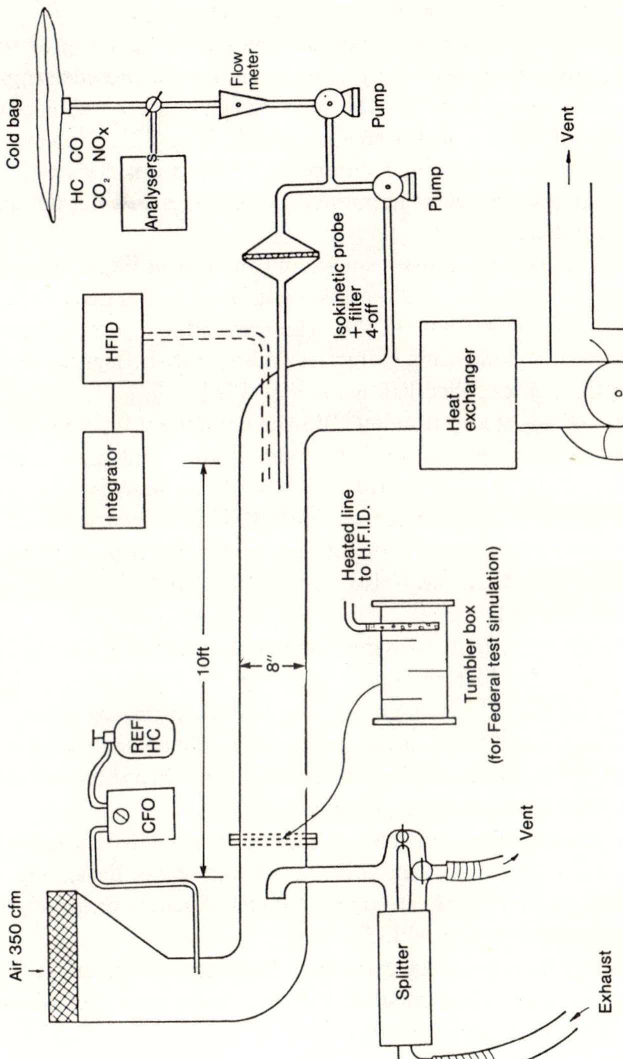

Fig. 8 – Simple CVS particulate tunnel.

real life is anything from 100 to 2000:1 and clearly this cannot be attained in dilution apparatus where the factor varies from c. 40:1 at idle to 2:1 at full speed. This restriction is for size and cost reasons. However, the particulate from a dilution tunnel is not too different from ambient material [25].

Optical opacity measurements of exhaust smoke are very useful to meet public complaints, but cannot assess oily droplet smokes or provide samples for further analysis. The classic work on the relationship between subjective and optical assessments of smoke was done at MIRA [26, 27]. However, with pollutants measured in grams per mile, smoke in terms of opacity would be out of place, whereas particulate weight harmonizes well and gives values of ambient air mass loading from diesel sources.

A CVS dilution tunnel apparatus is shown in principle in Fig. 8. The air flow for a 2-litre engine is from 250–350 cfm, being limited by the need to keep the mixed exhaust and air cooler than 125°F. The particulate can be filtered adequately for development work using fibreglass filters, but for legislative work a PTFE coated glass fibre filter called Pallflex is used [28].

If large truck engines need testing by CVS the equipment becomes huge and expensive. It is more economic to split the exhaust flow and use a 2-litre size CVS. Unfortunately, maintaining a constant split ratio, particularly over transient cycles, is difficult. A list of research particulate tunnel sizes, and some drawings, are given in [29]. Full instructions for light-duty engine particulate testing have been given by EPA [30]; they have also published regulations for heavy-duty engine gaseous emissions [31].

To separate the filtered particulates into their components it is usual to employ solvent extraction in a Soxhlet apparatus followed by liquid or gas chromatography on the extract [25]. Using high pressure liquid chromatography (HPLC) with fluorescence detection PNAH carcinogens can readily be analysed [29].

If the proportion of carbon to adsorbed HC is desired then thermogravimetry of the actual glass fibre filter gives a rapid answer [32].

To obtain transient response particulates presents difficulties. Optical opacity methods lack sensitivity and cannot present mass data, but a new sophistication called photoacoustic spectroscopy (PAS) seems promising to give useful modal data during drive cycles [33].

A rapid response direct weighing method has been described in which a filter becomes part of a vibrating element, the drop in frequency of which gives the mass of particulate which has deposited [34].

11. ODOUR ANALYSIS

It is possible to rate diesel exhaust odours using panels of trained observers who use dilutions of substances with diesel-like odour in water as bases of comparison. Amos Turk of the U.S. Public Health Service devised the best known example using 12 observers but it is very costly to run [35, 36].

To avoid panels, researchers at AD Little invented a diesel odour analysis system (DOAS) using liquid chromatography to separate samples of odorous material collected in traps filled with an absorbent, Chromasorb 102. The low resolution chromatogram separates the sampel into so-called 'aromatics' and 'oxygenates' and a total intensity of aroma (TIA rating) is calculated from their values [37].

The DOAS layout is shown in Fig. 9 as a relatively simple instrument with the two odour groups nicely separated.

A DOAS operating procedure as determined by a CRC group has been given in detail [29].

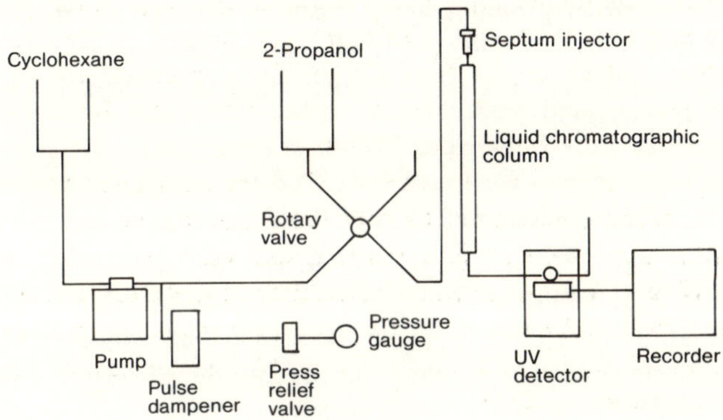

Fig. 9 – Liquid chromatograph.

12. ALDEHYDE ANALYSIS

The advent of alternative fuels (alcohols and oxygenates) will bring aldehydes to the fore but at present there is no practical direct instrumental procedure. The reactivity of the –CHO group in aldehydes is used to prepare derivatives such as oximes which can be separated by gas chromatography, or DNPHs which are separable by liquid chromatography [29].

Mass spectrometry may be applied but is too expensive for most occasions.

13. SULPHATE ANALYSIS

Most of the fuel sulphur emerges in the exhaust as SO_2; in fact, knowing the sulphur content and the air/fuel ratio, the SO_2 concentration can be accurately predicted. Some SO_2 is oxidized further to SO_3 which hydrates to sulphuric acid as it cools. The H_2SO_4 condenses to a particulate mist and is captured on a filter from which it is removed by extraction in aqueous isopropanol.

Four methods for sulphate analysis of the resulting extract are proposed:

1. Barium chloranilate (BCA) method using a liquid chromatograph and UV detection.
2. Barium perchlorate titration with Thorin end-point detection (BPT) method.
3. Combustion to SO_2 with NDIR detection Leco-S method.
4. Ion-chromatography (IC) method.

These methods are discussed and compared with a full bibliography [29].

14. OTHER MINORITY EMISSIONS

A vast and ever growing bibliography exists on this subject, much of it under the aegis of EPA [38] and CRC [29]. All that can be said here is that excepting the ill-defined mutagenic and carcinogenic compounds there is little cause for worry under this heading.

15. DIESEL COMBUSTION AND EMISSIONS

It is surprising that such a heterogeneous combustion with the seemingly unpredictable mixture of diffusion burning and pre-mixed flame that occurs in a diesel engine does not produce even worse emissions than it actually does. Fuel impinges on chamber walls and spray droplets run the gauntlet through hot flame regions where they may be pyrolized. The high compression ratio means a rapid expansion quenching the oxidation reactions in the hot gases and leaving potential emissions.

When, however, the details of the combustion and the 50-year-long optimization of the engine–fuel combination are considered it becomes less surprising. The chamber geometry, charge motion, compression ratio, fuel spray and timing, along with fuel viscosity, volatility and oxidizability (cetane rating) have all been matched to give the best power with minimum smoke.

It so happens that CO levels with acceptable smoke are far below SI engine levels, and because for the same reason diesels cannot operate on air/fuel ratios below c. 18:1, the NO_x level is also low. Figure 10 gives an illustration of the difference between petrol and diesel emissions as a function of AFR. It must be remembered that some diesel fuel burns at the rich limit (c. 8:1), giving soot and CO, so the AFRs are average values. Also remember that despite the low NO_x concentration, the engine's mass flow is high, giving a higher mass emission figure.

A fairly complete list of factors controlling diesel emissions follows:

Engine factors
Injection timing and control
Injection quality, secondaries, nozzle, pilot
Injection rate

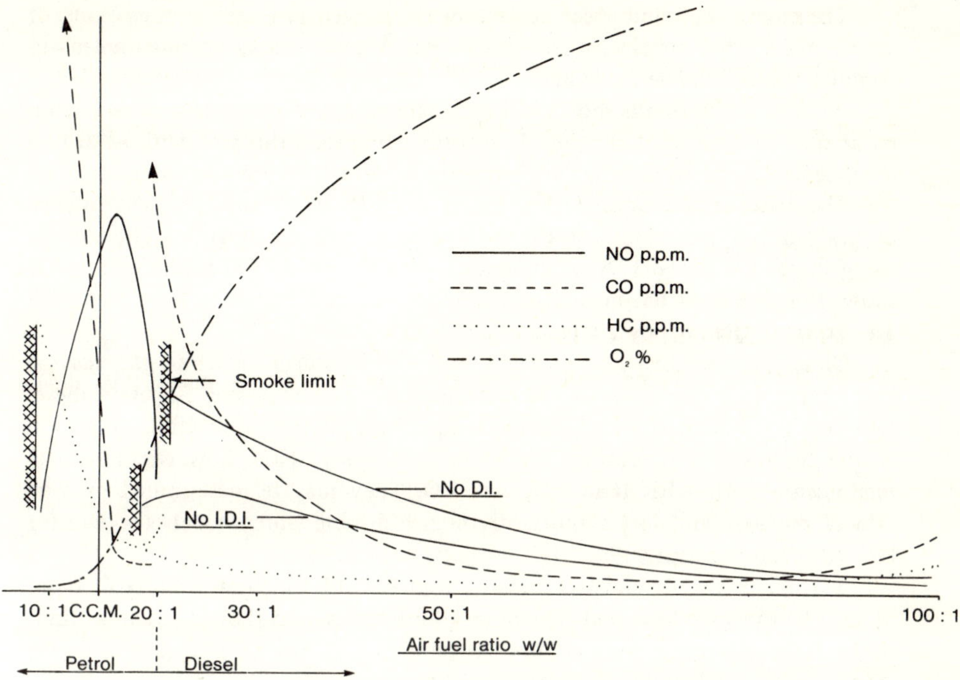

Fig. 10 – Emissions' trends v. air/fuel ratio. SI engine compared with a diesel.

Cylinder size, stroke/bore ratio
Compression ratio
Chamber shape and proportions (pre/main)
Pre-chamber throat area
Valve timings
Lub oil quality and consumption
Engine temperature

Outside factors
Turbocharger
Inlet air temperature
Exhaust back pressure
Exhaust gas recirculation, EGR
Catalyst, water injection, exhaust scrubbing
Governor type
Cold start devices
Fuel quality, cetane, boiling range

The engine load and speed requirements, as well as transient conditions, of course make a big impact on emissions. Only a few remarks on this immensely complex subject can be made here.

As long as the trends shown in Fig. 10 are accepted as averages derived from an unknown mixture of rich and lean combustion (and this also applies but to a much lesser degree to petrol) then some important points are illustrated.

The diesel smoke limits (DI and IDI) at AFRs of 21 and 18:1, with plenty of available oxygen, demonstrate the degree of rich combustion. The steeply rising diesel CO at 18:1 AFR compared with the petrol CO at the same AFR shows that this soot has insufficient time to oxidize through to CO_2, even with 3% excess oxygen, so some quenches out as CO.

Diesel HC remains low over the load (AFR) range owing to the ease of low temperature oxidation of the molecular species comprising 50 cetane diesel fuel combined with the high charge temperature. A rise around idle condition is due to lower temperatures and poor fuel spray. HCs, however, have less relationship with AFR than NO_x and CO. They may be determined by wall effects, crevices, and very importantly nozzle dribble which occurs too late for HC oxidation to occur [39].

NO curves for DI and IDI are shown in Fig. 11 which are similar to those of Fig. 10 but based on load. The point of interest here is that at its smoke limit of 18:1 the IDI only gives about 750 ppm of NO despite its nearness to the maximum (premixed charge) NO of 4–5000 ppm at 16:1.

The DI gives almost 2000 ppm so it must be assumed that a considerable premixed burn must occur near the four sprays. The IDI manages to avoid this to some extent by its two-stage combustion and the high turbulence in the prechamber which allows a retarded injection and a fast combustion – all good for low NO.

The best visualizations of combustion in these systems are undoubtedly the high speed films of W. M. Scott at Ricardo [40] which show permitted zones by green radiation from a copper tracer. Sooting areas are also prominent in rich spray centres.

Fuel mixing in these engines has been largely concerned with power, economy, smoke and noise. Little is known of the origin of the pollutant emissions within the chamber but concerted attempts are being made to trace the reactions by sampling [41] (which is limited by surface effects) and by total cylinder dumping at any point in the cycle [42].

16. EMISSIONS MODELLING

Much of the basic kinetic data concerning the production of the major pollutants and their subsequent decay reactions is known. Even soot production and oxidation can be treated reasonably adequately.

NO modelling has received very considerable attention on the basis of the

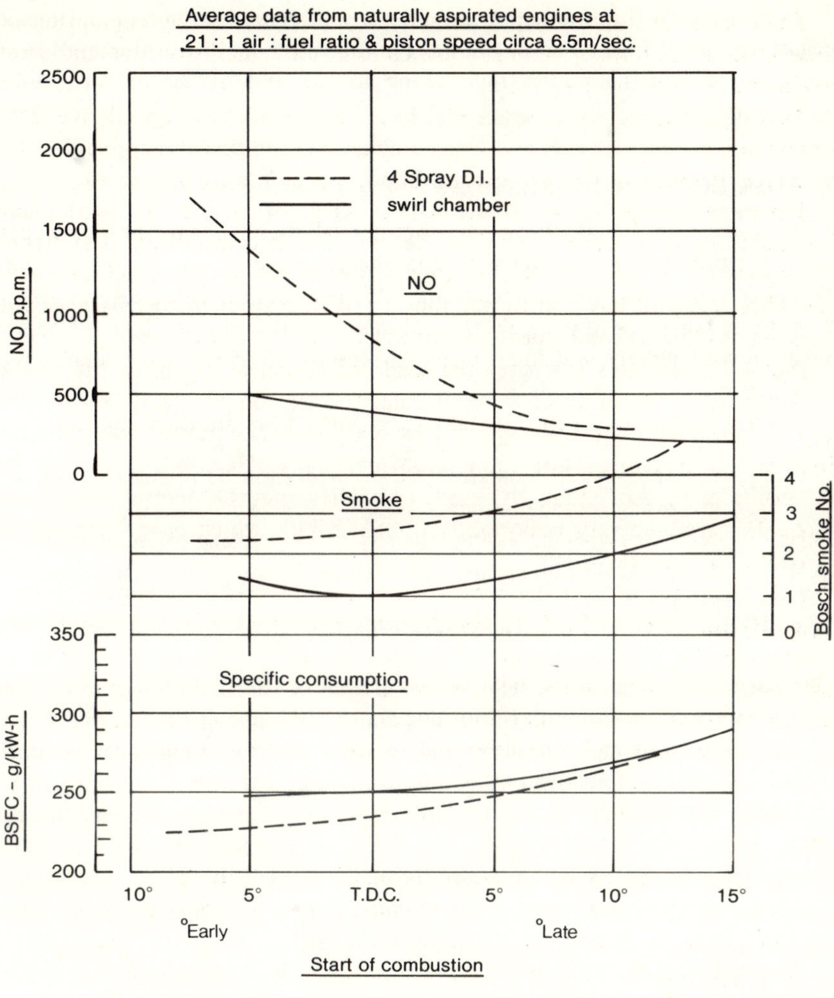

Fig. 11 – NO emission and performance characteristics of DI and swirl chambers.

famous Zeldovich mechanism and can be used to predict NO emissions within 10% of those actually observed. However, the chemical kinetics are only one side of the problem; the other, which is the fuel/air mixing process, presents considerable difficulties. It is necessary to know the history of the pattern of fuel/air ratios throughout the combustion and early expansion cycles.

A description of recent work at M.I.T. has been presented by Heywood [43]; he uses a novel mixing model. A good bibliography is appended. He obtains good NO predictions and interesting information on the soot production in a swirl-chamber engine.

It is unlikely that modelling can give any information on true particulates because these are formed from the basic combustion soot by agglomeration and adsorption of volatile bodies right along to the end of the exhaust system and beyond.

17. FUEL EFFECTS ON EMISSIONS

As has been mentioned above the commercial diesel engine has been painstakingly optimized to run on a fuel in the range of 45–55 cetane numbers. Now with economic refining technology this restricts the hydrocarbon composition to a fairly narrow band of qualities. For this reason, large effects on emissions are unlikely and indeed are not found. However, even small effects are important.

Some simple rules taken from an Exxon paper [44] are worth quoting.

B.S.CO decreases with increasing 90% distillation temperature and with increasing cetane number

B.S.HC decreases with increasing viscosity and increasing cetane number

B.S.NO_x decreases with increasing 50% distillation temperature and decreasing gravity

Smoke increases with the amount of 'heavy ends' boiling above 340°C.

These rules are logical in the light of the effects of the fuel properties on spray patterns, proportion of premixed burning and droplet pyrolysis.

Sulphur dioxide and trioxide emissions are of course proportional to sulphur content and may be controlled by fuel hydrodesulphurization.

Smoke opacity may be controlled by the use of barium and calcium containing anti-smoke additives [45] but these have undesirable side effects of toxicity, engine deposits and increased particulate weight [21].

Particulates are affected by fuel quality but in a complex manner which also depends on the engine combustion system [46].

At present, then, there is little pressure for an improvement in fuel quality; this will only emerge later with increasing diesel car populations bringing competitive pressures for a premium light low sulphur fuel of good startability and low filter plugging.

18. REDUCING EMISSIONS

Heterogeneous diesel combustion provides many sources of pollutants which escape in a relatively cool exhaust gas containing excess air. This precludes the use of the gasoline three-way-catalyst system which needs a hot exhaust gas from a stoichiometric combustion in order to reduce NO_x with simultaneous oxidation of both HC and CO. However, most other options are open, but must be used with care to prevent side effects on thermal efficiency, HC particulates, etc.

It is desirable to depollute within the combustion process as far is possible. The air/fuel mixing and the chamber geometry offer vast scope for experiment and every new engine needs its own treatment. Fast mixing and a high burn rate with a late injection as used in the IDI engine give low emissions, particularly of NO_x, but with a fuel penalty of up to 10% compared to the DI which has higher emissions all around. Intermediate systems such as the re-entrant bowl DI (squish tip) are possible [47].

Changes in injection rate and air swirl affect the proportions of premixed and droplet burning as well as the wall impingement of fuel with a direct effect on all emissions [48, 49].

Supercharging may also, by leaning AFRs and changing temperatures, reduce particulates and smoke, and raise or lower NO_x depending on the precise application because lean AFRs give lowest NO_x but any temperature increase causes an exponential rise in NO_x. Intercooling may be a vital factor in this equation.

To reduce emissions over a drive cycle needs much compromise, particularly between bsfc, NO_x and particulates. An elaborate FIE programme is called for to optimize the spray over the sensitive parts of the cycle. To help in this process it is clearly desirable to have external devices that can be called into play as required.

A number of these aids exist for various applications:

19. EXHAUST GAS RECIRCULATION (EGR)

EGR has proved very useful in NO_x reduction is SI engines and is believed to work by virtue of the higher heat capacity of the charge reducing the peak combustion temperature [50]. In the diesel case, unfortunately, serious side effects occur with the NO_x reduction. Thus the recycling of soot causes problems in intake systems, but more seriously the lower oxygen availability in the soot removal zones causes smoke, particulates and HC to rise, especially at higher engine loadings.

If uncooled EGR is used, the higher charge temperature at light (25%) loads may cause lower HC and particulates, as the lack of oxygen is less significant [51]. EGR's thermal recycle is good for better combustion with low noise at start-up [52].

Clearly, the application of EGR has many complexities, but many rewards. With a low particulate engine (turbocharged) and a careful control system it offers a solution to the NO_x problem.

20. WATER INJECTION OR EMULSIONS

This is a controversial mode of NO_x reduction, mainly for mining and stationary engines. Water admitted by carburation or spraying does reduce NO_x, but may also cause ring wear and oil sludging, whereas the use of either stabilized or unstabilized emulsions lengthens ignition delay which if corrected by more advance may even result in increased NO_x. Also HC and CO may increase, but soot decrease because of cooler soot pyrolysis zones in the flame.

21. CATALYSTS AND SOOT FILTERS

Catalysts have been used with limited effect in mining diesels in the form of monolithic matrices, or pellet containers sensitized by platinum or base metals (copper, chrome). The low exhaust temperatures of these de-rated engines prevent efficient catalysis. The automobile diesel engine offers better prospects with its higher rating and the hot exhaust of an IDI.

To obtain the best HC and CO reduction, a platinized matrix in the engine manifold is highly successful [53] and because the HC is removed the particulate consists mostly of soot and so is much reduced in mass.

Some soot may also be oxidized in these systems. A possible ideal configuration is shown (Fig. 12) using a port liner matrix for the highest temperatures and a cascaded wire element for multiple soot impactions. The final solution will use a total engine/catalyst integrated design.

Another solution is to filter the particulates using an aluminized wire wool matrix which is periodically burned off by a fuel 'torch' operated *in situ* [54]. A special monolithic ceramic filter also gives good soot filtration, the soot being removed by 'torching' or by the use of a 'soot burning' metal in the fuel, e.g. cooper, lead, manganese, etc. [54].

A side issue in these catalytic systems and filters is the problem of sulphur conversion to sulphuric acid.

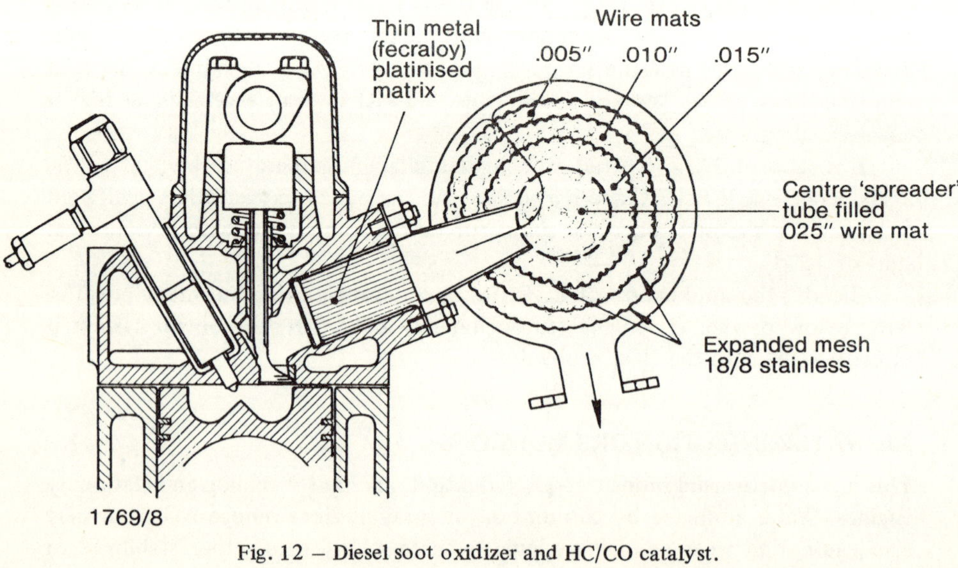

Fig. 12 – Diesel soot oxidizer and HC/CO catalyst.

22. WATER SCRUBBERS AND CONDITIONERS

Diesels in gassy mines must pass their exhaust through water to cool it and eliminate sparks; some soluble components are also removed such as aldehydes and sulphur oxides. A modern version is described in a 1978 Patent [55].

These devices must be cleaned frequently to remove sludge, although they do not trap much particulate as a percentage. They are not popular in humid mines owing to their water vapour emissions.

Reactive salts may be added to the water to optimize solvency: sodium sulphite solution removes aldehydes [56], and sodium carbonate/lime removes sulphur oxides.

23. FUMIGATION

The introduction of a voltaile fuel by carburettor into the intake of a diesel will reduce smoke; also, if 15% of the main fuel is atomized into the intake, smoke is reduced [57]. The reason is the greater degree of premixed combustion; NO_x will be expected to increase.

This system is little used but may occur in the future in some form of bifuel engine with hybrid combustion, perhaps using methanol.

These systems may be used in conjunction to obtain the ultimate clean diesel. The U.S. Bureau of Mines is researching water injection, coupled with a catalyst and water scrubber, to produce an exhaust gas acceptable to U.S. miners [4]. In principle a clean system might look like Fig. 13.

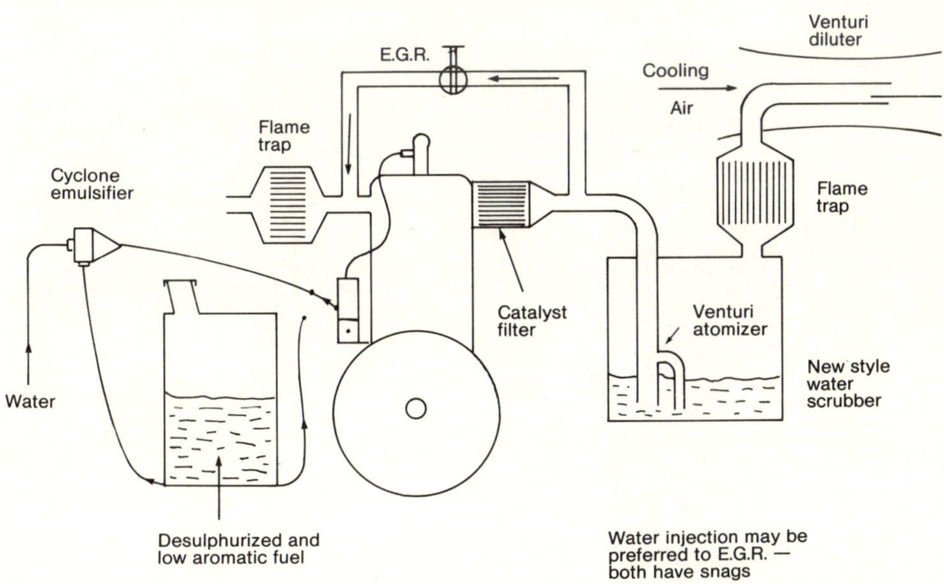

Fig. 13 – Clean diesel system for mines.

24. HEALTH EFFECTS

'If the human nose is to be trusted then diesel exhaust must be dangerous' and 'if cigarettes are bad why not diesel exhaust?' are two typical reactions from lay observers. However, the nose has no evolutionary wisdom on diesels – it's only been sniffing them for 50 years and cigarette smoke is quite unlike diesel exhaust smoke in chemical composition.

Cigarettes emit higher CO and NO_2 concentrations than diesels and more varied volatile organics which, unlike their diesel counterparts, are not tightly adsorbed on a soot substrate.

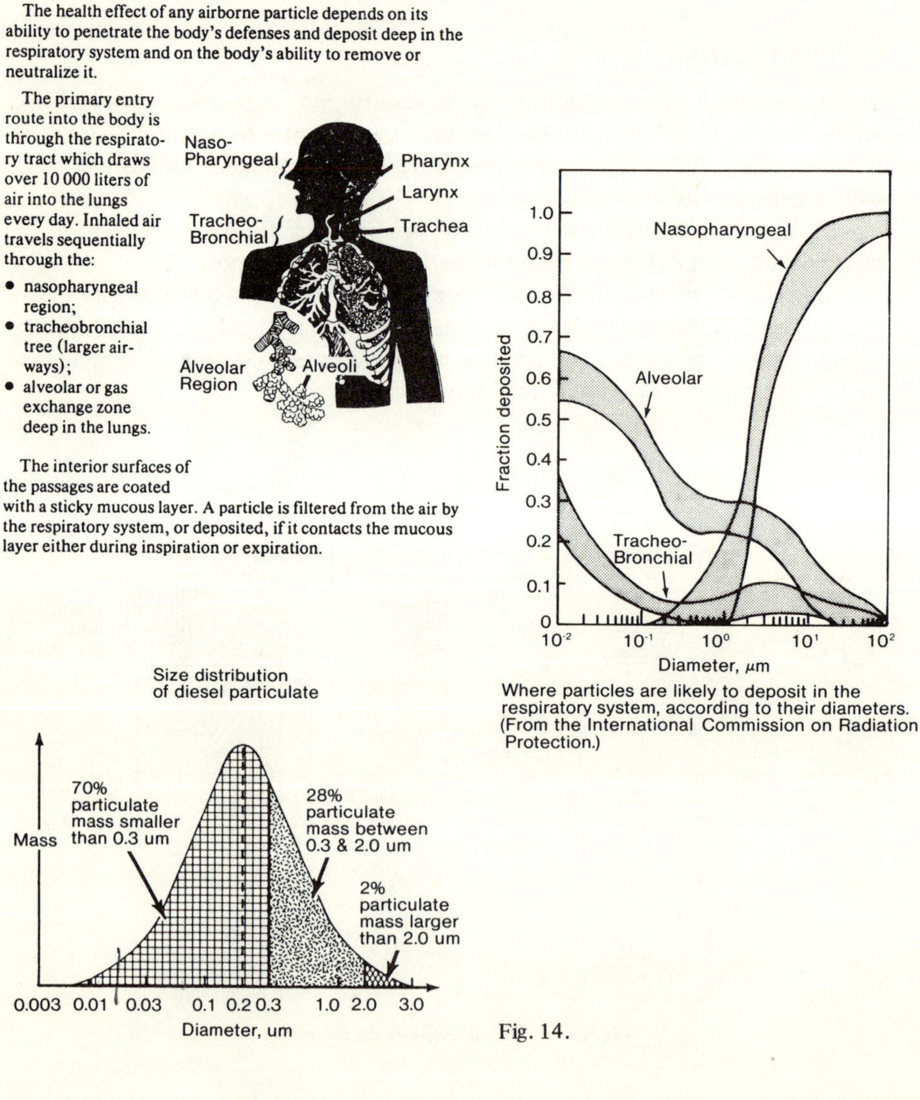

Fig. 14.

Potentially dangerous PNAHs can only be removed from diesel soot by high temperature evaporation or by very powerful solvents [32] which do not exist *in vivo*.

The diesel particulate and perhaps the NO_2 hold the key to any diesel exhaust toxicity. The former has been studied intensely [58, 59] and its toxicity examined in bacteria and in animals, and also by human epidemiology.

The size spectrum and lung penetration of particles are shown in Fig. 14. This GM publication gives the results of their animal tests as negative at much above ambient exposure levels. This coupled with the famous London bus garage study [5], they believe, goes a long way towards clearing the diesel.

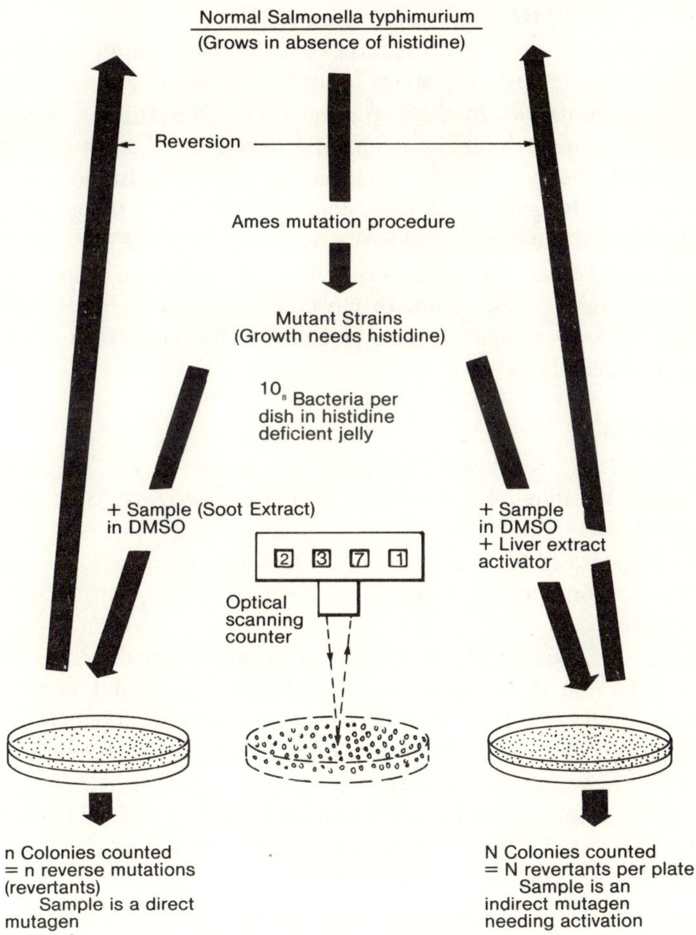

Fig. 15 – Ames test for mutagens.

It is indisputable that particulate extract contains mutagens to the uncomplicated salmonella bacteria in the Ames test [60]. A simple scheme showing how a special mutant strain is reverted to the original strain is given in Fig. 15; by a clever trick the mutagen is given a numerical value.

Many human carcinogens are mutagens to salmonella in the Ames test which is the reason it caused so much worry. The active agents are now believed to be nitro derivatives of PNAHs formed by direct action of NO_2 on the hydrocarbons [61]. An engine running on a nitrogen free atmosphere gave little mutagenicity in its particulates owing to no NO_x formation [62].

The amount of carcinogenic material in diesel exhaust is of the same order as in non-catalysed gasoline exhaust despite the latter fuel containing 10 x as much benzo-a-pyrene (BAP) – a highly carcinogenic PNAH. Maybe the diesel soot protects the PNAH from the hot gases [63].

There seems to be little justification for the 'witch hunt' against the diesel engine. Like all combustion systems it contributes some toxic emissions to the air, and will play its part in clean-up programmes. Nevertheless, where it fails is in the visibility of its fumes – an amenity destroyer, if not a toxic hazard. In this it has been badly served by crude smoke opacity legislation, but since the introduction of the particulate mass concept in c. 1975 things are improving fast design-wise, if not as yet production-wise.

At the present time the light-duty IDI engine's exhaust, which has benefited most from modern ideas, compares most favourably with that from the over-powered European gasoline engine bloated by its 'anabolic steroids' of rich mixtures and lead.

REFERENCES

[1] Cooke, J. W., Kennaway, E. L. and Mayneard, W. V., 'The Production of Cancer by Pure Hydrocarbons, Part 1', *Proc. Roy. Soc. B* 1932, **III**, pp. 455–484.

[2] Drinkwater, J. W. and Egerton, A. C., 'The Combustion Process in the Compression Ignition Engine', *I. Mech. E. Proc.*, April 1938, **138**, pp. 415–493.

[3] Holtz, J. C., Berger, L. B., Elliott, M. A. and Schrenk, H., 'Diesel Engines Underground 1 Composition of Exhaust Gas from Engines in Proper Mechanical Condition', *Report of Investigation R13508*, 1940.

[4] 'Health Effects of Diesel Exhaust Emissions – A Comprehensive Literature Review, Evaluation and Research Gaps Analysis', Environmental Health Associates Inc., 2150 Shattuck Avenue, Berkely, California 94704.

[5] Commins, B. T., Waller, R. E. and Lawther, P. J., 'Air Pollution in Diesel Bus Garages', *Brit. J. Industr. Med.*, 1957, **14**, 232–239.

[6] Schrenk, H. H., 'Air Pollution in Donora, Pa. Epidemiology of an Unusual Smog Episode of October 1948', *Public Health Bulletin No. 306,* Federal Security Agency, Washington D.C., 1949.

[7] Ministry of Health, Mortality and Morbidity During the London Fog of 1952, *Reports on Public Health and Medical Subjects,* HMSO No. 95, 1954.

[8] Logan, W. P. D., *Brit. M. J.,* **1**: 722–725, 31 March 1956.

[9] Motor Vehicles Air Pollution and Health. A report of Surgeon General to the 87th US Congress 2nd Session. Document No. 489, June 1962.

[10] *U.S. Federal Register,* **33**, No. 108, 4 June 1968.

[11] Pearsall, H. W., 'Measuring the Total Hydrocarbons in Diesel Exhaust', *SAE Paper No. 670089.*

[12] Hilliard, J. C. and Wheeler, R. W., 'Nitrogen Dioxide in Engine Exhaust', *SAE Paper No. 790791.*

[13] Wheeler, R. W., 'Vehicle Emissions and Urban Amenity', I. Mech. E. Conference: *Combustion in Engines,* 7–9 July 1975, Paper C100/75, p. 185.

[14] *U.S. Federal Register,* **45**, No. 34, 14496, 1980.

[15] *SAE Publication PT17,* 1979.

[16] *SAE Publication PT25,* 1982.

[17] Lyn, W. T., 'Some Observations on the Nature of Blue Smoke in Engine Exhaust', *Lucas Eng. Rev.,* July 1964, **1**, No. 1, pp. 10–13.

[18] Bradow, R. L., *et al.,* 'Sampling Diesel Engine Particle and Artefacts from Nitrogen Oxide Interactions', *SAE Paper No. 820182.*

[19] Degobert, P., 'Diesel Exhaust Offensive Effect – True Odour or Irritancy?', *SAE Paper No. 800423.*

[20] Iwai *et al.,* 'Measurement of Irritating Odour from D. I. Diesel Engines and Their Reduction Methods', FISITA Tokyo, May 1976, *Paper 2–11.*

[21] Collins, D., Cuthbertson, R. D., Gawen, R. W. and Wheeler, R. W., 'The Use of the CVS and Dilution Tunnel to Compare the Total Particulates from a Range of Automotive Engines', *SAE Paper No. 750904.*

[22] Dietzmann, H. E., Bradow, R. L. and Parness, M. A., 'Emissions from Trucks of Chassis Version of 1983 Transient Procedure', *SAE Paper,* October 1980.

[23] Niki, H., Warnick, A. and Lord, R. R., 'An Ozone-NO Chemiluminescense Method for NO Analysis in Piston and Turbine Engines', *SAE Paper No. 710072.*

[24] Black, F. M., *et al.,* (EPA), 'The Application of Total Hydrocarbon Flame Ionization Detectors to the Analysis of HC Mixtures from Motor Vehicles with or without Catalytic Emission Control', *Water Air and Soil Pollution,* **5**, (1975), pp. 52–62.

[25] Black, F. M. and High, L. E., 'Methodology for Determining Particulate and Gaseous HC Emissions', *SAE Paper No. 790422.*

[26] Dodd, A. E. and Reed, L. E., 'The Relationship between Subjective Assessment and Objective Measurement of Exhaust Smoke from Diesel Engine Road Vehicles', *MIRA Report 1964/12,* April 1964.

[27] Dodd, A. E. and Holubecki, Z., 'The Measurement of Diesel Exhaust Smoke', *MIRA 1965/10,* April 1965.

[28] A. D. Little Inc., 'Evaluation of Filter Media for Quantitative Collection of Particulate Matter from Engine Exhaust', *Final Report EPA Contract No. 68–02–1211,* March 1977.

[29] Co-ordinating Research Council, 'Informational Report on the Measurement and Characterization of Diesel Exhaust Emissions', *CRC No. 517,* December 1980.

[30] Particulate Regulations for Light Duty Diesel Vehicles, *Federal Register,* Feb. 1, 1979.

[31] Gaseous Emissions Regulations for 1984 and Later Model Year Heavy-Duty Engines, *Federal Register,* Jan. 1, 1980.

[32] Cuthbertson, R. D., Stinton, H. C. and Wheeler, R. W., 'The Use of the Thermogravimetric Analyser for the Investigation of Particulates and Hydrocarbons in Diesel Engine Exhaust', *SAE Paper No. 790814.*

[33] Osada *et al.,* 'Real-time Measurement of Diesel Particulate Emissions by the PAS Method Using a CO_2 Laser', *SAE Paper No. 820461.*

[34] Whitby, R., *et al.,* 'Real Time Diesel Particulate Measurement Using a Tapered Element Microbalance', *SAE Paper No. 820463.*

[35] Hare, C. T., Springer, K. J., Somers, J. H. and Huls, T. A., 'Public Opinion of Diesel Odour', *SAE Paper No. 740214.*

[36] Turk, A., 'Selection of Training Judges for Sensory Evaluation of the Intenstity and Character of Diesel Exhaust Odours', *PHS Publication 999–AP–32,* 1967.

[37] Levins, P. L., 'Analysis of Odorous Compounds in Diesel Exhaust', *Final Report A.D. Little–CRC NTIS Accession No. PB228–756–AS,* September 1973.

[38] Dietzmann, H. E. and Black, F. M., 'Unregulated Emissions Measurement Methodology', *SAE Paper No. 790816.*

[39] Ford, H. S., Merrion, D. F. and Hames, R. J., 'Reducing Hydrocarbons and Odour in Diesel Exhaust by Fuel Injector Design', *SAE Paper No. 700734.*

[40] Scott, W. M., 'Looking in on Diesel Combustion', *SAE Trans.,* **78,** pp. 33–51, 1969.

[41] Nightingale, D. R., 'A Fundamental Investigation into the Problem of NO_x Formation in Diesel Engines', *SAE Paper No. 750848.*

[42] Liu, X. and Kittelson, B., 'Total Cylinder Sampling from a Diesel Engine'. Part I', *SAE Paper No. 820360.*

[43] Mansouri, S. H., Heywood, J. B. and Radakrishnan, K., 'Divided-chamber Diesel Engine. Part 1: A Cycle-simulation which Predicts Performance and Emissions', *SAE Paper No. 820273.*

[44] Gross, G. P. and Murphy, K. E., 'The Effects of Diesel Fuel Properties on Performance, Smoke and Emissions', *ASME Paper No. 78–DGP–26,* July 1978.

[45] Miller, C. O., 'Diesel Smoke Suppresion by Fuel Additive Treatment', *SAE Paper No. 670093.*

[46] Hare, C. T. and Bradow, R. L., 'Characterization of Heavy Duty Diesel Gaseous and Particulate Emissions and Effects of Fuel Composition', *SAE Paper No. 790490.*

[47] Bertodo, R., *et al.,* 'A Method of Direct-injection Diesel Emissions Control', *I. Mech. E. Conf.,* Cranfield, July 1975, Paper C86/75.

[48] Khan, I. M., *et al.,* 'Effect of Air Swirl on Smoke and Gaseous Emissions from Direct-injection Diesel Engines', *SAE Paper No. 720102.*

[49] Greeves, G., 'Response of Diesel Combustion Systems to Increase of Fuel-injection Rate', *SAE Paper No. 790037.*

[50] Quader, A. A., 'Why Intake Charge Dilution Decreases NO Emissions from S. I. Engines', *SAE Paper No. 710009.*

[51] Kageyama, K. and Kinehara, N., 'Characterization of Particulate Emissions from Swirl-chamber Type Light-duty Diesel Engines as a Function of Engine Parameters', *SAE Paper No. 820181.*

[52] Watanabe, E. and Fukutani, I., 'Reduction of Diesel Idle Knock by EGR', *SAE Paper No. 810296.*

[53] Enga, B. E., *et al.,* 'Catalytic Control of Diesel Particulate', *SAE Paper No. 820184.*

[54] *SAE Papers Nos. 810112, 810114, 811236 and 810118.*

[55] 'Exhaust Gas Water Conditioner for an I.C. Engine', *Brit. Pat. No. 1517115.*

[56] Davis, R. F. and Elliott, M. A., 'Removal of Aldehydes from Diesel Exhaust Gas', *Trans. ASME,* **70**, pp. 745–750, August 1948.

[57] Alperstein, M., *et al.,* 'Fumigation Kills Smoke – Improves Diesel Performance', *SAE Trans.,* **66**, 1958, p. 575.

[58] Johnson, J. H., 'The Physical and Chemical Character of Diesel Particulate Emission – Measurement Techniques and Fundamental Considerations', *SAE Paper No. 789108.*

[59] Amann, C. A., *et al.,* 'Some Rudiments of Diesel Particulate Emissions', *SAE Paper No. 800251.*

[60] Ames, B. N., *et al.,* 'Methods for Detecting Carcinogens and Mutagens with the Salmonella Mammalian-microsome Mutagenicity Test', *Mutat Res.,* **31**, 1975, pp. 347–364.

[61] Pitts, J. N., *et al.,* 'Atmospheric Reactions of Polycyclic Aromatic Hydrocarbons – Facile Formation of Mutagenic Nitro Derivatives', *Science,* **202**, Nov. 1978, pp. 515–519.

[62] Lestz, S. S., *et al.,* 'The Role of Nitrogen in the Observed Direct Microbial Mutagenic Activity for Diesel Engine Combustion in a Single-cylinder D. I. Engine', *SAE Paper No. 820467.*

[63] Williams, R. L. and Swarin, S. J., 'Benzo-a-Pyrene Emissions from Gasoline and Diesel Automobiles', *SAE Paper No. 790419.*

7

Lubrication in diesel engines

LEN W. KILBOURNE, Automotive Products, Silkolene Oil Refinery.

1. INTRODUCTION

The lubrication of diesel engines is no different from the lubrication of any other piece of machinery and there are no mysteries.

The fundamental reasons for lubrication are to keep moving surfaces apart so that the friction involved is low and the heat generated will not damage these surfaces. Secondly, when oil is the lubricant, it is used as a transfer fluid to dissipate heat to the surrounding atmosphere.

Clearly, as in any engineering design, materials have to be used which are capable of withstanding the environment in which they work.

Continuing the idea of simplification, it is suggested that it is incorrect to consider diesel and petrol engines as separate entities. They are both hot air engines with air as the working fluid and, from a lubrication point of view, the differences are only of degree in terms of the environment.

The temperatures are different because of the higher compression ratios used in diesel units and products of combustion are different because of the fuels used, but essentially this chapter is concerned with lubricating sliding surfaces, gears, bearings and the cooling of parts.

2. BRIEF LOOK AT LUBRICATION THEORY

Lubrication is about friction reduction and this is accomplished by keeping a film of lubricant between the moving surfaces so that they do not come into contact and cause mutual damage.

The provision of this lubricant film is usually done mechanically and this includes the pumping action generated with a bearing between the rotating shaft and the stationary bearing – the hydroynamic effect. The journal is supported on an oil film, the thickness of which is affected by:

1. The load squeezing out the oil.
2. The oil viscosity since it affects the leakage rate from the bearing.
3. The oil temperature since this affects the viscosity.
4. The journal speed since this affects the pumping action between the moving surfaces.

It is not proposed to go deeply in lubrication theory but Fig. 1 shows the relationship between viscosity, speed and load on the coefficient of friction in a plain bearing.

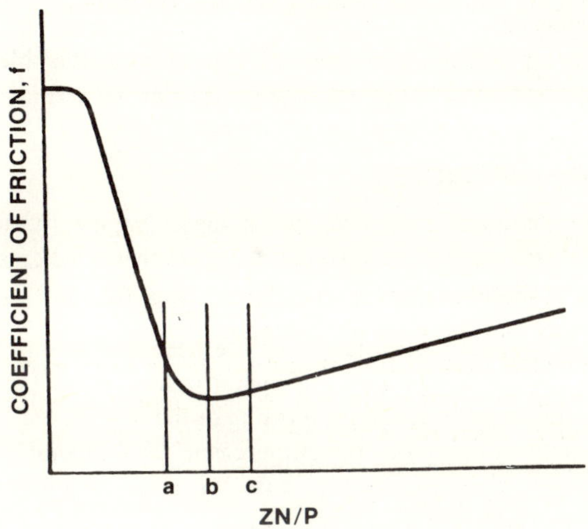

Fig. 1 – Effect of viscosity, speed and load on bearing friction. For each hydrodynamic film bearing, there is a characteristic relation, such as that shown, between the coefficient of friction and the factor combining viscosity (Z), speed (N) and load (P).

There is another condition which occurs called boundary lubrication and this applies under circumstances when the hydrodynamic situation described above breaks down such as under shock load, slow speed, very heavy load, high temperature or critically low viscosity.

In these situations there is intermittent contact between the surfaces, the oil film is not able to protect them adequately and some support in the form of additives must be provided.

In practice, most lubrication situations are a mixture of hydrodynamic and boundary so provision has to be made for both regimes.

3. CONSTRUCTION OF A DIESEL ENGINE LUBRICANT

It will be apparent already that the viscosity of an oil is important and arguably is the most important physical characteristic.

This focuses attention on the base material or base oil which forms the foundation for the finished product.

Generally, the base is a petroleum oil but synthetic lubricants and blends of mineral and synthetic lubricants are beginning to be used and will be considered separately later in this chapter.

Petroleum based lubricant is produced by the fractional distillation of crude oil, the lubricating oils having, typically, boiling ranges above 343°C.

The production of a lubricant base stock from crude involves subtractive processes to remove unwanted components so that finally an oil with the desired properties is achieved.

Base oils are produced in a range of viscosities so that blending of two or more may be carried out in order to produce the desired result.

4. NECESSARY PROPERTIES

Looking at the required properties for a diesel engine lubricant, several are base-oil related and cannot be influenced by additives whereas others can be so influenced; for example:

1. Volatility is an inherent property of the base and is important as it determines the rate of loss which the operator sees as oil consumption. Such losses are additional to leakages, burning in the engine, etc.
2. Flow characteristics in the operating range of temperature. These can be influenced by additives but the base oil plays an important part as the additive contribution is the lesser.
3. Ability to maintain the required characteristics for period of usage.
4. Must not degrade or be degraded by other materials with which it comes into contact in the user machine.

The word 'additive' has crept into the discussion so, in the context of lubricants, a definition is a material added to the base which enhances a desirable property or suppresses an undesirable one. For example: load carrying is a desirable property and the ability of a base oil can be enhanced by the incorporation of a suitable additive; on the other hand, oxidation is undesirable and this can be suppressed or delayed by an appropriate addition to the base oil.

It is not intended to dwell on the chemistry of additives but more on their functions, so these will be reviewed in turn.

5. DETERGENT/DISPERSANT

These properties are coupled together because they are not separately effective.

The need to keep the internals of an engine clean has already been noted in connection with the dissipation of the heat generated and to prevent deposit build-up which may reduce critical clearances – such as piston to cylinder – to a dangerous figure. The combustion process gives rise to the formation of carbonaceous solids; so inevitably these will be formed; hence, if the parts are to be kept clean provision has to be made for the solids to be dispersed in the lubricant in such a manner that safe operation of the engine is not impaired.

It is the operation of the detergent/dispersant additives that gives a used oil the characteristic blackness, even after a comparatively short period of use. The finely divided particles, too small to cause problems and indeed fine enough to pass through filters, are prevented from coalescing and forming sludges.

Two main sub-divisions of these additives are:

1. Metallic.
2. Ashless.

The metallic dispersants are sulphonates, either natural from petroleum, or synthetic, and thiophosphonates or phenates.

The ashless dispersants have better ability to handle low temperature deposits and sludges, so they tend to appear in oils designed for petrol engines rather more than in diesel engine formulations, but not exclusively so.

A simple, mechanical engineers' way of visualizing the action is to imagine each tiny particle of carbon being enveloped by a buffer at the time of formation; hence if no two particles can get together, then sludge cannot form.

In the mention of undesirable properties of an oil was its tendency to become oxidized, particularly under high temperature conditions which introduces the next class of additive:

6. OXIDATION INHIBITORS

Oxygen combination with the base oil causes thickening and acid generation within the oil which can lead to corrosion attack on bearing materials, so reduction in the rate of oxidation is important to provide an oil with an acceptable life in the engine.

Phenols are used and their action is believed to destroy free radicals, but probably the most used type in automotive engine oils is zinc dithiophosphate of which there are two derivatives, the alkyl and aryl. The former are better as anti-wear materials but the latter have better themal stability. In practice, the alkyl types are most common in diesel engine lubricants.

Mention of wear brings the consideration of anti-wear additives.

7. ANTI-WEAR ADDITIVES

Wear is caused by metal to metal contact with local seizure and transfer of metal, corrosion attack or abrasion.

The zinc dithiophosphates referred to in section 6 also have film forming abilities and are effective in controlling the first type of wear mentioned.

Corrosive wear is offset by providing an alkaline environment which must offset the acidity generated from combustion products, and phenates and sulphonates are used in a combined role with detergent/dispersant attributes.

Abrasive wear control relies on filtration of the oil while circulating in the engine and, perhaps more importantly, filtration of the air because this is the fluid passing through the engine in the greatest volume.

8. VISCOSITY INDEX IMPROVERS

The ideal oil would have a constant viscosity regardless of temperature, but few liquids reach this ideal. Water comes very close to this between the limits of 0°C when it freezes and 100°C when it boils, but oils are not so accommodating.

All oils get thinner as the temperature increases and the best that can be done up to now is to reduce the rate at which thinning occurs – it cannot be eliminated.

The improvers used increase the oil viscosity at all temperatures but their structure is such that their thickening powers increase with temperature and this offsets to some degree the thinning of the base oil.

The improvers are typically polymers soluble in oil and the molecule swells in volume as the temperature rises thereby modifying the oilflow characteristics.

These materials are subject to mechanical shear and this causes loss of viscosity of the oil.

If the loss is permanent, the oil gradually thins and approaches the viscosity of the base oil. On the other hand, modern developments have made the polymers much more stable and it is possible to introduce temporary shear which is a viscosity reduction in the bearing which gives lower friction losses yet the bulk oil remains at the nominal viscosity. Fuel economy oils frequently use this method of achieving the desired result.

9. POUR POINT DEPRESSANTS

Most oils contain waxes and indeed these waxes contribute to the 'oiliness', 'lubricity' or whatever you care to call it.

Unfortunately, these waxes congeal at low temperatures owing to crystallization of the paraffin wax.

It is not practical to de-wax the oil completely because it would seriously reduce the yield and, as already stated, it would reduce the lubricating abilities.

A compromise is effected whereby an additive is used to lower the pour point of the oil by dispersion of the wax crystals so that the volume is reduced and the flow is less hindered.

10. OTHER ADDITIVES

There remain some further commonly used additives such as anti-foam to prevent loss by frothing, anti-rust which is particularly important in two-stroke oils and storage oils, dyes for identification and film strength improvers.

11. SPECIAL ASPECTS OF DIESEL ENGINE LUBRICATION

Because of the higher compression ratios employed in diesel units, compared with petrol engines, the structure of diesel engines tends to be heavy and the combustion temperatures high.

Development is, therefore, towards increasing power output from the engine without increasing the weight and this means increasing the engine speed and the introduction of turbocharging.

What are the lubrication inferences of this?

As far as the rotating shafts are concerned, there is no problem – hydrodynamic lubrication occurs with occasional boundary conditions but it is in the piston area that the effects are most pronounced.

Piston temperatures will be increased so provision must be made for the oil to withstand these without degradation which, in turn, will lead to lacquering or other deposits on the piston skirt. If these deposits are not checked, they can build up to the extent that piston seizure can result.

The most critical area of the engine is round the piston rings for these form the gas seal for the working fluid. If the rings are not free to move in the piston grooves, compression pressures may be lowered, thus impairing the combustion efficiency, and the translation of the pressure form the heated air after combustion into work will be reduced.

The oil must have this ability to prevent deposit forming which will interfere with the free movement of the piston rings.

In general terms, therefore, increased output from the engine means an increased detergency/dispersancy need in the oil.

This is, however, not the only development route. Diesel engines, which in the U.K. have dominated the commercial vehicle power unit market, are slowly creeping into lighter vehicles such as passenger cars.

The emphasis here is on good cold-starting and fuel economy pointing to thinner oils of the SAE.10W/XX or even lower categories. Base oils in this range must be of high stability otherwise consumption is unacceptably high.

Although it might be thought from oil companies' advertising that each one had its own criteria and standards, their products are largely fitted into a pattern of internationally recognized standards as discussed below.

12. OIL SPECIFICATIONS

Probably the most widely used specifications are American Military ones and those current are:

MIL–L–2104 series
MIL–L–46152

MIL–L–2104 is at suffix 'C' at which level it gives an oil suitable for turbocharged diesel engines. The 'D' suffix is waiting in the wings until certain test problems are sorted out.

MIL–L–46152 is a rather broader based specification in that it is suitable for petrol engines as well as diesels.

Within these specifications are stipulated tests to check specific aspects of lubrication such as bearing wear measured by weight loss, piston deposits, valve gear wear and so on.

The British Military Authorities also have an approval system and are shortly to uprate OMD 75 which is an SAE.10W/30 high performance oil.

In addition to this military background, determined, of course, by predicted equipment needs, are tests called up by engine manufacturers such as Daimler Benz, Volkswagen, Volvo, Leyland, Ford, General Motors and so on.

These can be tests on specific items of the engine such as camshafts and tappets (Volvo) or piston deposits (Caterpillar).

The American Petroleum Institute (A.P.I.) attempted to simplify the performance ratings of oils using a letter code system.

For diesel engines the designation letter is 'C' and the performance letter ranges from 'A' to 'D' currently.

For example 'CA', which is obsolete, is equated to the American MIL–L–2104A specification and 'CD' covers MIL–L–2104C and the old series 3 lubricants. Some idea of the picture is given in Figs. 2 and 3.

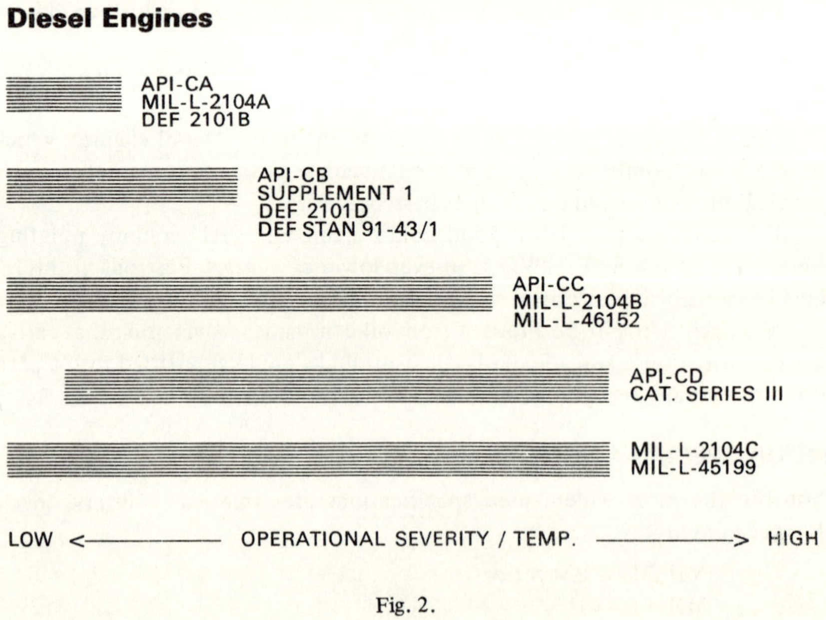

Fig. 2.

DIESEL ENGINES

Performance Level: LOW ↑ … ↓ HIGH

A.P.I. Service Designation	British Military Spec'n.	American Military Spec'n.	Other Recognised Spec'ns.	Typical Applications
CA	DEF 2101C	MIL-L-2104A		Obsolete
CB	DEF 2101C + 1% S endorsement DEF 2101D			
CC		MIL-L-2104B		Normally aspirated diesel engines
CD		MIL-L-45199B MIL-L-2104C	Caterpillar Series 3	High performance and turbo charged diesel engines

Fig. 3.

A good example of the gradual increase in severity of oil test conditions is given by the Caterpillar single-cylinder engine test which forms part of several specifications.

Remembering that two engines of differing bore sizes are involved, the BMEP and Horse Power per litre of oil charge in the sump figures are of interest – see Table 1.

Table 1.

	Caterpillar 1A	Caterpillar 1H2	Caterpillar ID	Caterpillar IG2
BMEP (lb/□″)	76	110	131	140
HP per litre of sump capacity	60.6	101.4	127.2	130.2

Figure 4 shows photographs of pistons after three of the Caterpillar tests, where satisfactory and unsatisfactory results have been obtained. After test, each piston is rated using an accepted system of demerit for such features as crown deposits, crown land cutting, due to hard carbon cutting grooves down the crown land, deposits on the ring lands and grooves and on the piston skirt. The condition underneath the crown is also taken into consideration.

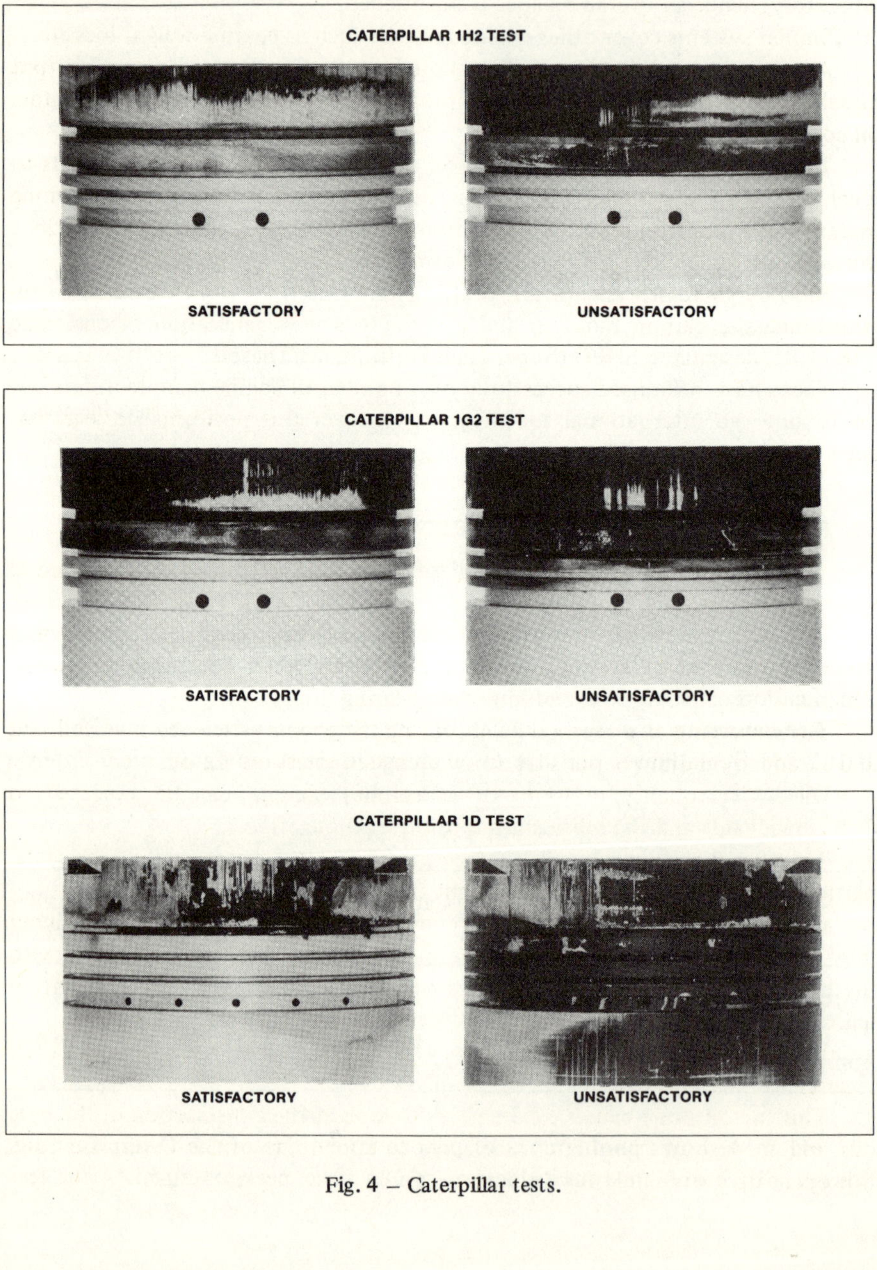

Fig. 4 – Caterpillar tests.

In order that pass or fail results shall be meaningful, great trouble is taken to standardize the rating system so that it can be truly comparable between engine test laboratories throughout the world.

By and large, this is successful and, in the case of British Ministry of Defence approvals, a panel reviews the engine parts after rating by the particular test laboratory; hence an overall balance is maintained.

Similar systems cover other engine features such as bearing weight loss etc.

A phenomenon which was first noted in certain European high output diesel engines, i.e. bore polishing is symptomized by a large increase in lubricating oil consumption.

The problem is related to carbon on the piston crown land which leads to local wear on the cylinder bores. This wear removes all the controlled honing marks which contribute to oil retention, and continued running results in a mirror finish which loses ring sealing control.

Blow-by increases and piston scuffing can result, so work was centred on controlling the carbon build-up and eventually a new generation of oils came into being, designated SHPD (Super High Performance Diesel).

These have been used successfully by a number of engine manufacturers but up to now no international specification to cover this performance level has been established.

13. FUTURE DIESEL ENGINE LUBRICANTS

The increase in popularity of small diesel engines, particularly in Continental Europe, is putting pressure on certain aspects of lubrication.

Power outputs from this class of engine have increased considerably and some engine makers have indicated that they need better control of ring wear, and hence oil consumption, and improved bearing life.

Turbocharging as a method of improving the engine efficiency is established and extended oil drain periods all throw increased stress on the oil.

Diesel fuel quality in terms of its sulphur content can be expected to drift downwards and the lubricant has to offset these effects.

There is, therefore, already the need for improved diesel engine lubricants, particularly in the passenger car and light truck sizes.

Fuel economy is an ever-present requirement and this leads towards thinner oils. By nature, thinner oils tend to be more volatile than thicker ones so with the higher temperature resulting from increased engine output, special attention has to be paid to oil volatility.

It is significant that Volkswagen have included a volatility restraint in their recent specifications for SAE.10W/XX oils.

This factor alone causes oil designers to look at the construction of the base oils and the use of 'synthetic' lubricants to support the mineral oils used and this opens up a wide field for discussion.

It is unfortunate that the word 'synthetics' has already been too widely applied and, in effect, it means a lubricant base not of mineral oil form.

Some synthetics are oil related which negates any idea of them coming to the rescue 'when the oil runs out', but as is so often the case, things are not as simple as this.

The synthetics most commonly encountered are:

1. Poly Alpha Olefins (PAO).
2. Alkylated Aromatics.
3. Esters.

1. PAOs are attractive at first sight because they behave like petroleum derived oils and can be described as synthesized hydrocarbons, or 'oil made backwards'. This means that the hydrogen and carbon components are brought together without all the unwanted substances that come along with crude petroleum.
2. Alkylated Aromatics. Again, only hydrogen and carbon are present, so they are near-relatives to mineral oils.
3. Esters. Polyol and Dibasic esters find application in lubricants and some are derived from vegetable oils or animal fats. There is some evidence that the properties of esters are very appropriate to internal combustion engine lubricants.

The main properties which mark synthetics out as improvements over mineral oils are:

1. They have high natural viscosity indices, so viscosity index improvers are not needed.
2. They exhibit high thermal stability, hence their use in aircraft gas turbine engine lubricants.
3. They have low volatility, even in the low viscosities, therefore, thinner oils can be made without necessarily increasing lubricant consumption.
4. Some synthetics, particularly certain esters, have excellent load-carrying abilities.

With future emphasis on fuel economy, there is no doubt that low viscosity oils are going to be required and these will give better cold starting performance which could lead to smaller and cheaper starter motors, for example.

The high viscosity indices mean that high temperature load-carrying will not be sacrificed, so it is reasonable to predict that synthetics will be used blended with mineral oils to enahnce their properties, then gradually a movement to wholly synthetic lubricants will take place.

It has to be noted that the synthetic lubricant is only the base carrier; other additives such as detergents/dispersants still have to be added to complete the product package.

So, when shall we see synthetic lubricants in the market place?

Because of the increased cost compared with conventional mineral oil based lubricants, synthetics will only become used when either or both the following situations arise:

1. Shortage of mineral oil.
2. Technical requirements which cannot be achieved with mineral oil.

This will happen, first in the racing and competition world, then in high performance passenger cars, followed by commercial vehicles – we shall see.

ACKNOWLEDGEMENT

The assistance of Lubrizol (GB) Limited, with diagrams, is gratefully acknowledged.

8

Condition monitoring and fault diagnosis in diesel engines

S. D. HADDAD, Dept. Engineering Technology, Western Michigan University

1. INTRODUCTION

Subjective assessment of the correct functioning of the diesel engine is a technique that has been employed by engineers from the early stages of the engine's development. This technique would rely on experience and close observation of various engine operating parameters such as temperatures, pressures, flows, noise and vibration. Any slight deviation from the correct type of operation could be readily noticed and by learning from the various faults verified by dismantling the engine. The discrepancies related to particular faults could be recognized. However, with increased engine automation coupled with inherent complexities and owing to greater manpower movement, objective techniques of preventive maintenance are being developed and adapted within the context of 'condition monitoring and fault diagnosis applied to engines'.

2. CONDITION MONITORING VERSUS OTHER MAINTENANCE TECHNIQUES

Before looking at condition monitoring of diesel engines it is appropriate to present a brief definition of all maintenance techniques as follows:

1. Operate to Failure (O.T.F.) where replacement or repair takes place after failure.
2. Planned Maintenance (P.M.) or Fixed Time Maintenance (F.T.M.) where replacement or repair or adjustment takes place at fixed periods, depending on accumulated knowledge.
3. Opportunity Maintenance (O.M.) where inspection and repair are based on some other item's maintenance schedule or inspection period.

4. Design-out Maintenance (D.O.M.): that is to design engine components with the view to minimize or eliminate maintenance requirements as justified in nuclear machinery.

and finally,

5. Condition-based Maintenance (C.B.M.) or Condition Monitoring (C.M.) in two ways:
   (a) Fixed time monitoring – inspection via equal or variable inspection periods; then adjust, repair or replace on condition.
   (b) Continuous monitoring – inspection on continuous basis; then adjust, repair or replace on condition.

Figure 1 shows their approximate positions on the failure rate–time, known as 'the bathtub'.

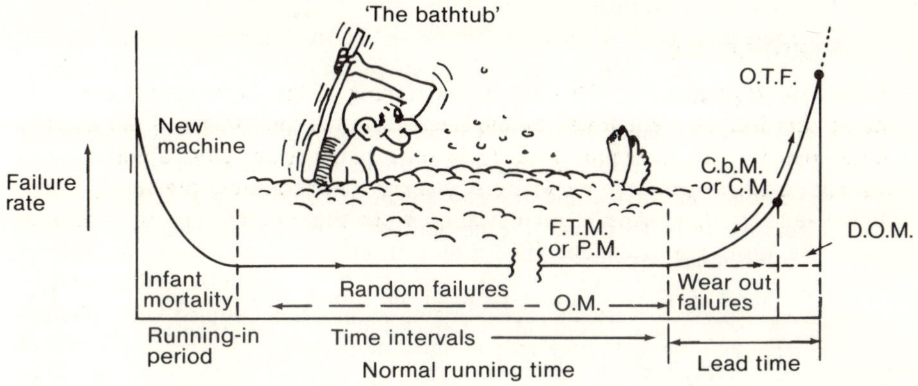

Fig. 1 – Typical failure rate variation for diesel engine parts.

Condition monitoring is now considered the most reliable, cost effective and efficient technique for maintaining the majority of engines in today's complex and sophisticated industrial set-ups. This technique relies on objective assessment or quantitative measurements of the machine in the following steps:

1. Trend Monitoring: that is the taking of regular measurements and their recording and analysis which gives a lead time on machine problems.
2. Condition Checking: that is the taking of numerate values to allow comparison with other machines to give knowledge of machine condition.
3. Finally, measurements may be analysed in considerable detail to provide guidance on possible causes of the problem.

3. ADVANTAGES OBTAINED BY THE USE OF C.M.

Condition monitoring leads to improved safety and increased output as follows:

1. Safety: that is resulting in reduced injuries and fatal accidents caused by machinery to personnel. Basically, this is due to the fact that C.M. enables the engine to be stopped safely when instant shutdown is not permissible, and it provides better engine knowledge if instant shutdown is permitted.
2. Output: that is increased engine availability, and increased rate of net output or service:
   - More running time as it enables shutdown for maintenance to be related to required production or service, and various consequential losses from unexpected shutdown to be avoided.
   - Less maintenance time as it enables shutdown without destruction or major damage requiring a long repair time. Also it enables the maintenance team to be ready with spare parts to start work as soon as the engine is shutdown.
   - Increased rate of net output as it permits increased load and/or speed; also it can detect reductions in efficiency or increased energy consumption.
   - Improved quality of product or service as it allows advanced planning to reduce the effect of impending breakdowns on the customer for the product or service and thereby enhances company reputation. Also it can be used to reduce the amount of product or service produced at sub-standard quality.

Application of C.M. in industry has been shown to give huge savings. For example, in 1978 it was estimated that the total conceivable level of net saving if C.M. could be applied 100% throughout British industry would be over £million 600.

4. FACTORS INFLUENCING THE CHOICE OF CONDITION MONITORING METHODS

The choice of monitoring instruments is dictated by the nature of the monitored engine and failures normally developed. These factors may be classified as follows:

1. Type of failure development (whether rapid, slow, intermittent. . .).
2. Possibilities of detection.
3. Frequency of this failure.
4. Use of failure detection to detect further development.
5. Cost effectiveness – especially with regard to savings on maintenance and improved performance.

The Ideal Condition Monitoring Technique

1. Simplicity in use with pass–fail indication or a simple scale, i.e. no range control and automatic calibration.

2. Insensitivity to sensor location so that repeatability is not lost by careless inspectors.
3. Insensitivity to the load condition of the engine so that tests are not omitted because of inconvenience.
4. Robust equipment to withstand the type of maltreatment common in industrial situations. For example, instruments should withstand dropping and cables withstand water, oil and heavy boots.
5. Intrinsically safe techniques in both the technical sense and manner of use.

Very few techniques meet all the above specifications.

5. TECHNIQUES FOR CONDITION MONITORING AND FAULT DIAGNOSIS IN DIESEL ENGINES

This broad area can be conveniently divided into two categories of techniques: those that can be carried out and the results presented almost immediately (which usually reveals how well the engine is performing at that instant given a particular mechanical condition); and secondly those techniques that require detailed or trend analysis. Into this category fall oil analysis techniques and methods used to establish wear rates. For convenience, they can be termed *condition checking* and *trend monitoring* techniques, respectively.

6. CONDITION CHECKING TECHNIQUES

Methods will be discussed for evaluation and diagnosis of each of the separate areas and sub-systems. These areas and techniques are:

- Overall engine power and friction.
- Cylinder power balance.
- Cylinder compression balance.
- Cylinder unit monitoring.
- Fuel injection system.
- Air charging system (including turbochargers).
- Exhaust system.
- Engine blow-by.
- Pressure pulsation analysis.
- Electrical system.
- Lubrication system.
- Cooling system.
- Bearings.

In this context the discussion would concentrate on surveying the above techniques in relation to what has been made available and possible developments.

6.1 Engine Power Estimation

Traditionally, this parameter is determined under steady state conditions using a dynamometer, for example see [2, 4, 5]. However, it is necessary that diagnosis takes place with the engine *in situ* in the vehicle.

Typically, many of the problems associated with engine sub-systems are reflected in the power output of the engine so this measurement is often a good indication of the overall condition of the engine.

The most common method of diesel engine power evaluation employs accelerating the engine against its own inertia whilst under maximum fuel delivery conditions [14, 37, 22, 23]. By knowing the rotational moment of inertia of the engine, brake torque could be determined. In addition, the frictional torque could be estimated by allowing the engine to decelerate from full speed by shutting off the fuel [14, 37]. Hence the brake power, indicated power and frictional power could all be calculated.

However, according to [38], under free acceleration conditions considerable distortion of the combustion process could occur, resulting in an overestimation of the power. This was claimed to be due to the amount of injected fuel and timing change during the course of engine acceleration.

The most common method of measuring engine load on marine diesel engines has been to fit strain gauges to the drive shaft [25, 39] and monitor the torque in the shaft. Then either a slip ring collector system was used, or even radio telemetry [40].

6.2 Cylinder Power Balance

In the systems developed by Hamilton Test Systems (Dieselsense) [22, 23], Siemens [15] and P.R.D. Electronics [21], all used the fluctuations in flywheel speed to determine power balance, as measured using an inductive probe actuated by the passage of flywheel teeth. The use of a microprocessor allowed data to be collected very quickly such that power balance could be calculated during a free acceleration test.

6.3 Cylinder Compression Balance

The STE/ICE [14], Siemens [15], VW [16], Autosense [18] and Dieselsense [22, 23] systems all used starter current or voltage as a measure of the relative compression of each cylinder. A large amount of current is required to drive each piston through the compression stroke. By comparing the peak values of current required for each cylinder, the compression balance between the cylinders could be obtained. Battery voltage could also be used as a measurement for this test.

The P.R.D. system [21] chose to use engine speed variations for monitoring engine compression as it was stated that a speed sensor was mandatory during engine testing whereas a current sensor was not, and also that some engines were fitted with air starters.

6.4 Cylinder Unit Monitoring

This included monitoring thermal load, cylinder liner temperatures, piston ring condition, combustion process and liner wear.

6.4.1 *Thermal Load*

It was found that combustion anomalies were mainly the cause of failures or heat loaded components of propulsion diesel engines. Combustion anomalies were found to cause an increased thermal load [27].

Thermal load monitoring was useful to prevent cylinder unit damage caused by high metal temperature, and to detect poor combustion and faults in the scavenging process [42].

The most representative method of determining the thermal load was to measure the temperature in the cylinder cover (head) [26, 27, 42]. In this way, local effects such as blow-by could be excluded.

6.4.2 *Surface Temperature of Cylinder Liners*

By monitoring the temperatures in the surface of the cylinder wall, it was found possible to detect the transient temperature caused by microseizure between rings and liner (scuffing) [27, 32, 42]. This was found to occur when the oil film between the piston rings and cylinder wall was broken down. Special sensors had to be developed for this measurement [35, 43]. This was essentially a special thermocouple with a very fast response capable of withstanding 3 to 4 mm of wear without its characteristics being altered. Figure 2 shows scuffing flashes recorded during tests on a medium-speed diesel engine when the piston cooling was intentionally restricted.

For monitoring purposes, the number of flashes detected during a certain period may be displayed as a condition parameter.

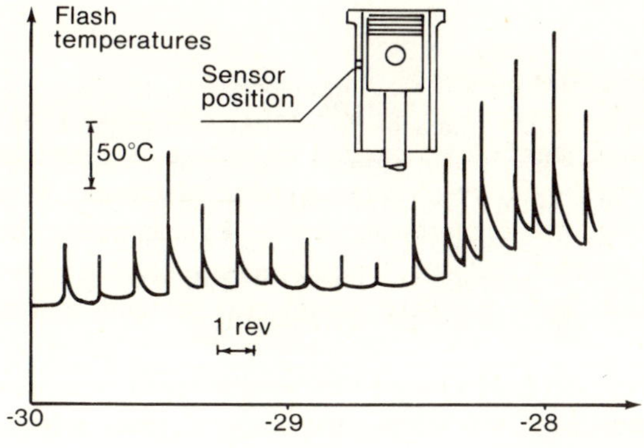

Fig. 2 – Scuffing activity as recorded from a test with a medium-speed engine.

6.4.3 *Piston Ring Condition*

Current functioning of the piston rings is vital to the reliability of the cylinder unit. A worn or burnt ring reduces the efficiency of the diesel engine. In addition, gas leakage (blow-by) past such a ring may cause secondary damage.

Proximity sensors in the cylinder wall could measure the distance between the liner wall and piston ring surface [27, 32, 42]. These sensors gave an output which was inversely proportional to the distance from the rings. Figure 3 illustrates the types of signal obtained. A certain positive signal meant that the ring was working correctly; a low positive signal meant that the ring was sticking in its groove or had collapsed; a negative signal indicated that the ring was missing or broken.

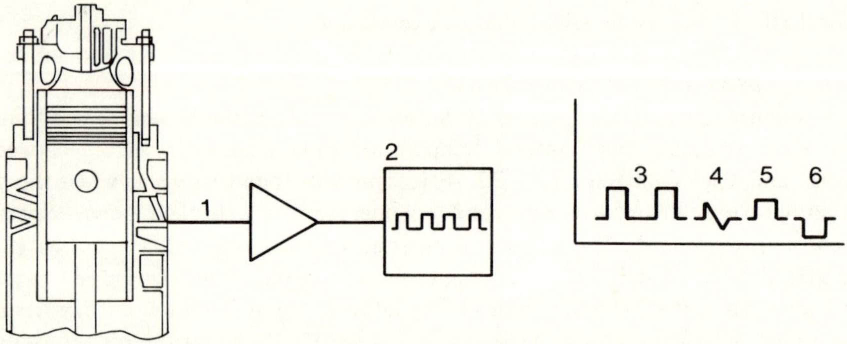

Fig. 3 – Location of piston ring sensor and examples of different signals: 1. Piston ring sensor; 2. Oscilloscope; 3. Normal rings; 4. Ring stuck in groove; 5. Burnt ring; 6. Part of ring missing.

6.4.4 *Combustion Monitoring*

By measuring and monitoring the cylinder pressure, it was possible to check if the combustion was taking place in an optimum manner [32, 42] which was essential for fuel economy, prevention of fouling and keeping wear of cylinder components to a minimum. The combustion process could easily be disturbed by several factors such as fuel, oil quality, fuel oil preheating temperature and condition of injection equipment. These distrubances could be reflected as changes in the following combustion characteristics:

(a) Ignition delay.
(b) Rate of combustion.
(c) Combustion intensity.
(d) Incomplete combustion.

From a diagram of cylinder pressure, Fig. 4, the following parameters could be obtained [42]:

- maximum cylinder pressure (related to ignition delay period and initial rate of heat release)
- cylinder pressure at a fixed crank angle on the expansion line (giving information about after-burning and delayed combustion)
- ignition timing (ignition delay and delayed combustion associated with this parameter)
- compression pressure
- mean indicated pressure

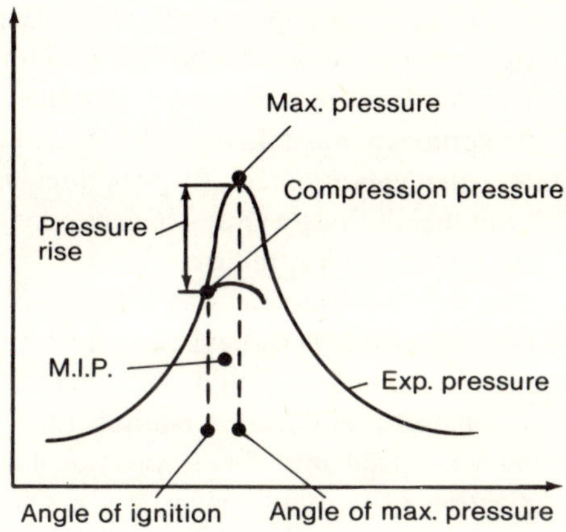

Fig. 4 – Cylinder pressure analysis.

6.4.5 *Liner Wear*

Cylinder liner wear would be very moderate under ideal conditions, yet in practical service some combustion anomalies and other effects could lead to rapid wear. Although methods of oil analysis had been known (and are discussed later), to measure the wear rate directly and continuously was a distinct advantage. Such measurements were thought to be valuable, since they gave instantaneous information about abnormal operating conditions not detected by other instrumentation.

Special wear sensors were developed for this task [35, 43]. The sensor was inserted through the cylinder liner, flush with the liner surface. As the liner wore down the resistance would change and the sensor would function, even if the liner had worn down several millimetres.

By using the sensor, monitoring of the accumulated wear as well as the wear rate was possible. The most suitable position in the cylinder liner for fitting such sensors was found by experience where wear was statistically high and the piston speed high [31].

6.5 Fuel Injection Systems

Probably the earliest examples of an analyser for fuel injection systems were developed by AVL in conjunction with Mobil Oil [19]. This electronic device used pressure transducers to measure the instantaneous pressure in each line. The pressure signal from cylinder 1 was used to syncrhonize a motor with crankshaft rotation and the pressure trace from each injection line could be displayed simultaneously on the oscilloscope.

From an analysis of the pattern of the pressure wavefrom, various defects could be noticed by changes in distinctive shape of portions of the pressure cuve. Faults that could be diagnosed included coked nozzles, high and low nozzle opening pressure, leaking nozzles, nozzles stuck open or closed, incorrect or worn delivery valve, defective pump plunger and spring and worn cam or camshaft bearings. Diagnosis was shown to be better with the pressure transducers at the nozzle as opposed to the pump end of the line.

A slightly different approach was taken in [3]. Instead of using one transducer for each injector, the measurement of peak pressure was made at the closest common point, the fuel delivery valve, which required only one transducer which could reveal catastrophic injector failure.

However, since peak injector pressure was affected by factors other than injector condition, a technique had to be devised to detect only deviations from the average. Consequently this method would only find major faults and could only be used on certain types of fuel injection equipment.

An extensive study was made into diesel injection fault diagnosis of both distributor and unit injection systems [44–47]. It was found that the peak fuel pressure in the unit injector pressure chamber and the peak fuel pressure in the line near the injector on the distributor injection system were very useful parameters for fault diagnosis. The peak fuel pressure on the unit system was measured with a strain gauge on the rocker arm and on the distributor system a pressure transducer was fitted just before the injector.

Using peak fuel pressure, a number of typical injector faults could be diagnosed taking measurements at two engine speeds under no-load conditions. Although a dynamometer was therefore not needed, the need for a pressure transducer or strain gauge was apparent, and further work included developing a removable clamp-on line pressure transducer.

The undersirability of modifying an injection system or breaking into the injection line was noted in [48]. It was felt that diesel operators would be opposed for a number of reasons:

- the effect of in-line transducer would have on the precise nature of fuel dynamics
- the high cost of pressure transducers
- the comparatively short life of such transducers

An alternative type of transducer was presented which could sense the strain in the fuel injection line. In order to eliminate the necessity of cementing a strain gauge directly to the injection line, a technique using the cold flow of cured resin held captive in a steel enclosure was developed. The transducer, Fig. 5, consisted of two identical steel halves, which provided rigidity and damping, whereas the cavity contained the resistance element, resin and perforated rubber. The resin could flow freely at the low strain levels compressing the perforated rubber and providing strain greater than that present in the surface of the line. However, installation did involve thoroughly cleaning the injection line, and applying resin to each half of the transducer before clamping the halves together on the line to cure.

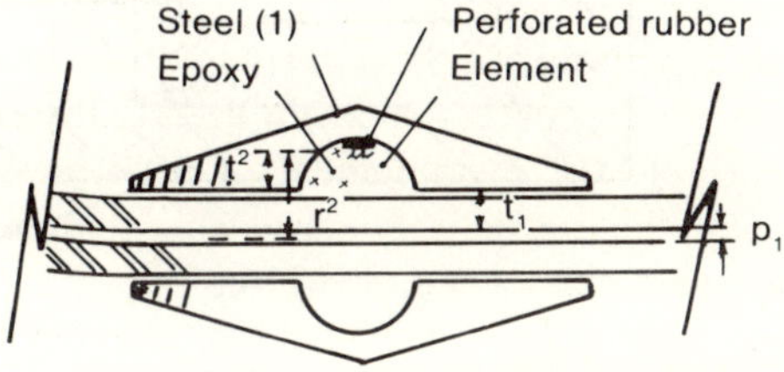

Fig. 5 – The transducer has the element remote from the line and working cavity is relatively large.

The signal from the transducer was used to diagnose fuel injection faults. A master signature of a healthy system was taken at a particular load and speed and retained for comparison. Subsequent observations were made at the same carefully chosen speed and load conditions. Nine common fuel injection faults were deliberately induced and the resulting signatures compared to the master.

Another type of clamp-on fuel line transducer is presented in [49]. The A.V.L. clamp-on transducer, Fig. 6, consisted of a pick-up casing of two halves, inside which was an elastic support which pressed the piezo-electric measuring element onto the outer wall of the injection line. A clamping device bringing the two halves together enabled a stable mounting of the sensor on the injection line. The quality of the signal obtained was claimed to be as good as that given by a pressure transducer. However, a limitation is that the pressure measurement obtained was only relative and the start of delivery was adopted as the point where a fraction of the peak pressure was reached. Special electronics ensured that spurious signals, caused by vibration, were ignored.

In addition to measuring injection pressure using conventional means, the system J.E.N.A. [50] made use of the ultrasonic pulses emitted from an in-line injection pump, which corresponded to the mechanical shocks developed by the plunger during injection. Using a 'condition monitoring' sleeve, simply clamped

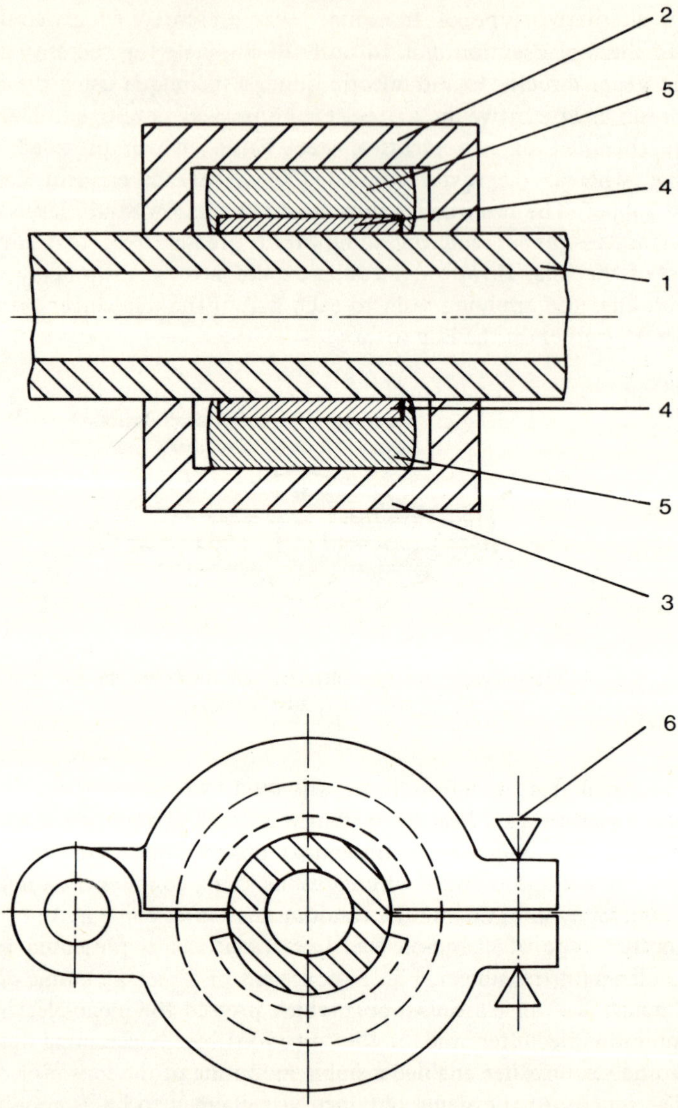

Fig. 6 – Construction of clamp-on transducer: 1. Injection line; 2. Pick-up arising; 3. Pick-up casing; 4. Piezo-electric measuring element; 5. Elastic support; 6. Clamping device.

to the discharge union of the pump, both ultrasonic signals and temperature measurements were made. The temperature of the discharge union was found to be a useful condition indicator of the high pressure part of the injection system.

The amount of fuel delivered per stroke was determined by measuring the angle between two pulses characteristic of the beginning and end of fuel delivery as measured using the ultrasonic transducer. In addition, the presence of the ultrasonic signals between the beginning and end of fuel delivery pulses indicated the mechanical condition of the injection pump.

6.6 Air Charging System

It is not unusual to find air flow monitoring equipment on marine diesel engines, since it had been realized for some time that insufficient air flow to the cylinders was often a cause of combustion anomalies [51].

The air flow was known to be affected by turbocharger efficiency (when fitted), fouling of the system and the temperature of the intake air [42].

Nearly all marine diesel engines are fitted with turbochargers and a typical monitoring system could comprise the following parameters:

- efficiency of turbocharger turbine
- efficiency of turbocharger compressor
- air cooler pressure drop
- value of air cooler effectiveness

In order to determine these parameters, a number of measurements could be recorded [51]. The turbocharger turbine service condition could be determined from the pressure ratio, turbocharger speed and inlet gas temperature, and the compressor condition could similarly be calculated.

The condition of the air cooler was usually determined from measurements of cooling water temperature, before and after the air cooler, and in addition the pressure drop across it could give an indication of contamination to the cooler. As pressure drop depended on the air flow through the system, it was considered necessary to measure this as well. A number of measuring methods are described in [42].

On smaller, high-speed diesel engines, it was considered useful [7, 14, 21] to measure the drop across the air cleaner which could indicate a clogged or missing filter and also monitor the boost pressure.

A technique for evaluating turbocharger performance was used in Dieselsense [22, 23]. This consisted of a snap acceleration from low idle to high idle. During this time the manifold pressure, the pressure rise as a function of time and engine speed were recorded. The values were compared to previously established limits for the particular engine–turbocharger combination.

On naturally aspirated engines, it was only necessary to make one pressure measurement, between the air filter and the engine [22, 23].

The above methods describe steady state analyses. Other techniques based on pulsation analysis are discussed later.

6.7 Exhaust System

There are a number of parameters that may be measured in the exhaust system. These are normally exhaust pressure, exhaust temperature (either individual exhaust ports or one measurement downstream), exhaust smoke and gaseous exhaust emissions (CO, CO_2, O_2, NO_x, HC).

Techniques using exhaust pressure pulsation are discussed in a separate section later.

Although it was felt that gaseous exhaust emissions were of less diagnostic significance than other exhaust measurements on diesel engines [52, 39], they were a very useful indicator of carburettor and ignition system function on petrol engines. In the Autosense system [18], CO and HC could be measured at no-load conditions and any speed.

A high concentration of CO could indicate a high fuel/air mixture, whereas a high HC concentration usually meant a fault with the ignition system or a mechanical problem.

Steady state exhaust pressure was used to detect any back pressure due to a faulty exhaust system in the P.R.D. system for diesel engines [21].

The most useful parameters for diesel engines were thought to be exhaust temperature and smoke. For marine use it was considered essential to measure individual cylinder exhaust temperatures to monitor variations in performance caused by the valves and injectors, for example see [39].

A system consisting of an instantaneous exhaust gas temperature and instantaneous smoke opacity sensor was developed [52] intended for use with U.S. Army's diagnostic equipment, STE/ICE, discussed earlier.

In order to measure instantaneous exhaust gas temperature a sensor was designed, based on the change of speed of sound with temperature. A smoke-meter was also developed to respond to instantaneous smoke. The two sensors were used together and fitted at the end of the exhaust pipe of a vehicle. Measuring the characteristics of discrete exhaust pulses from individual cylinders could be used to isolate a fault to a particular cylinder.

It was found that the instantaneous exhaust temperature was more useful than smoke data for identifying injector faults, particularly as the tests were carried out under no-load conditions.

6.8 Blow-by

Blow-by is generally recognized as a useful measure of engine wear. It was felt that crankcase breather flow was a better measure of blow-by than crankcase pressure [39] to indicate when seizure occurs.

However, the Dieselsense system [22, 23] checked blow-by by restricting crankcase ventilation, and monitoring the rise in pressure. This test could be carried out at high engine speed under no-load conditions.

6.9 Pressure Pulsation Analysis

This technique is discussed here since it has relevance to inlet, exhaust and crankcase pressures.

Spectrum analysis of the intake and exhaust manifold pressure signals in S.I. engines was used [17] to try to identify faults in the ignition system, leaks and blockages in the intake ane exhaust manifolds and malfunctions in the valve gear. It was found that good engines had a strong frequency component at the firing frequency with relatively little signal power below it, whereas a fault leading to a reduction in power from a cylinder would introduce strong low frequency signals into the exhaust signal at a sub-harmonic of the firing frequency. The intake manifold signal was a good indicator of proper intake valve action.

The sub-harmonic ratio was later introduced [37]. This was used to detect valve leaks and blocked fuel injectors taking measurements in the exhaust system.

In addition, an evaluation of the diesel engine's full power output could be made using the free acceleration test and using the cylinder firing frequency (C.F.F.). However, special equipment had to be developed, including a tracking filter, and details of this system are given in [37].

Further work on this method [53] showed that not only was the sub-harmonic ratio useful for identifying faults but also the harmonics of the cylinder firing frequency. Also, the greater the engine load, the larger the ratio, it was shown.

A number of spectral and amplitude discriminants were introduced [54] which were found useful for fault detection. These were the impulse factor, defined as the ratio of signal peak values to the average amplitude, and a discriminant which was the ratio of the D.C. pressure component to its A.C. average value.

These discriminants could be used on any of the four working gas processes, intake pressure, cylinder pressure, exhaust pressure and crankcase pressure. This work was developed in [55] to show that by measuring the A.C. and D.C. components of each pressure process and forming discriminants, much information could be gathered concerning the condition of the engine. Analogue equipment was developed in order to process these discriminants during a free acceleration test.

In particular it was felt that intake pressure pulsations were suited to inlet faults, crankcase pressure for engine performance and blow-by, and cylinder pressure was also useful for performance and quality of combustion.

6.10 Electrical System

The electrical system of a vehicle generally comprises starter motor, battery, solenoids, generator or alternator, regulator and various switches and cables. It is one of the easier systems on which to make measurements. The majority of diagnostic systems therefore incorporate voltage and current sensors to diagnose faults on electrical systems.

The VW system [16], for example, used voltage measurements to test battery, regulator and generator. The STE/ICE system [7, 14] used approximately 15 connections to check the condition of battery, starter, starter circuit, alternator, regulator and various cables and connections. The measurement of voltage could usually be accomplished with the aid of electrical clips whereas current could be measured with a Hall effect current probe (usually clamp-on).

As part of the harness in the Autosense system [18], there were six electrical clips and one clamp-on current probe which could be quickly fixed in position to identify all components of the starting system and the charging system, in addition to the compression balance check discussed earlier.

6.11 Lubrication System

Lubricating oil pressure is monitored on almost all engines as standard practice and has been used as a rough guide to bearing condition.

The depot M.A.I.D.S. system [56] used a number of pressure, temperature and flow measurements to check the condition of the oil pump, cooler, filter, relief valve and also to detect restrictions in the system.

The Dieselsense system [22, 23] used one pressure transducer inserted between the oil pump and filter, and measured the engine speed during a free acceleration. From this information the pressure relief valve setting could be checked and also the pumping performance and lubrication system flow capacity.

For naval marine engines [39] it was considered essential to measure oil filter differential pressure and lubricating oil pressure at the farthest end of the main oil gallery from the oil pump and preferable to take measurements at additional positions. In addition, oil temperature should be measured before and after the oil cooler.

It is useful to measure oil temperature in any case when performing diagnostic tests because it has a bearing on such measurements as power and friction, and could cause discrepancies.

6.12 Cooling System

The most useful parameters for the oil cooling system were considered to be coolant temperature and coolant level [7, 14]. This is fairly universal for all sizes of engine.

One diesel diagnostic system [3] used a modified radiator cap with a temperature sensor and a second sensor housed in a magnetic assembly attached to the inlet side of the thermostat housing to detect a faulty thermostat.

On the Dieselsense system [22, 23], pressure and temperature sensors were fitted just before the thermostat. By running the engine at low idle and high idle speeds, the thermostat operation and water pump performance could be studied.

6.13 Bearings

Although there are a large number of papers discussing a variety of techniques to identify bearing defects, there are no known existing systems up to now that are used for monitoring the bearing condition of internal combustion engines. This is not to say it is not possible, but owing to the relatively difficult access and small number of failures [39], this area is not considered cost effective for condition monitoring.

7. TREND MONITORING TECHNIQUES

The aim here is to observe and/or record the changes basically using a combination of the condition checking techniques discussed before. In this way it is possible to obtain accurate information about the mechanical health of the engine. The following techniques are in common use.

7.1 Vibration and Noise Monitoring

Signals are created by the movement of the components of a machine. Therefore, vibration monitoring is used for diagnosis, whereas electromagnetic signal detection and similar signals created by movement form the basis of other established diagnostic tests.

Using noise monitoring is also effective in certain cases and conditions, especially when using acoustic emission and acoustic intensity techniques.

The measurement of vibration is useful in this connection because it gives some characteristic information on the condition of the engine. It is well known that a high vibration level is both unwanted and probably harmful in the long run. Also, it can be expected that an engine which runs almost without vibrations will be able to work for a reasonably long time without trouble. Measurement of these vibrations can be made and analysed in various ways but the main difficulty has been to draw the right conclusions from the measured results in a simple way. Considerable activity has taken place at the international level concerning this question and we are now in the fortunate position of being able to see standard procedures developing, both from theoretical considerations and practical experience. Some of these procedures are now available in documents such as the VDI 2056 recommendation, the DIN Standard 45665 and BS 4675. These will give the user sufficient background to use vibration measurement, an excellent tool for judging a running condition, and above all has made fault diagnosis by vibration measurement and analysis even more reliable than conventional diagnostic tools such as temperature, pressure, etc.

The Mechanism of Vibration Generation in an I.C. Engine and Methods of Measurement

The origin of vibration of an I.C. engine is rather complex. The complexity arises from the fact that the whole engine structure is simultaneously excited by a number of forces of widely different characteristics. These forces are:

(i) *Primary exciting force*: that is the gas force in the cylinder resulting from combustion.

(ii) *Secondary exciting forces* of considerably different characteristics which are generated by the operation of the crank mechanism but of course related to gas force, resulting in piston impacts, impacts in bearings, impacts in timing gears, etc.
(iii) *Other exciting forces* in accessories not usually related to the primary gas force such as valve gear system, fuel injection system, etc.

These forces excite the engine structure and produce vibration response that is related both to the type of force and engine structure impedance at the point of force application.

Therefore there are two main aspects of I.C. engine vibration which the diagnostician must be able to identify, namely [76].

1. Vibrations which, although unwanted, are in no way detrimental to the operation or the life of the engine.
2. Vibrations due to faults or malfunctioning of the engine which must be recognized before serious damage occurs.

To achieve these two aims, specialized instrumentation has been developed to cope with various degrees of sophistication, depending on the required accuracy of fault detection. Normally, vibration measurements are made using special transducers (displacement, velocity, acceleration or strain, depending on type of engine structure and frequency of interest). For engine fault diagnosis it has been found that vibration acceleration measurements give better results in identifying most of the operational events. The next step is to choose the associated instrumentation to amplify, record and analyse the vibration measurement to enable comparisons to be made. Here again these instruments can range from the simplest (accelerometer, single reading vibration meter) to the highly sophisticated (accelerometer, amplifiers, frequency spectrometer, real time analyser, X–Y plotter, discriminator). Excellent literature by Brüel and Kjaer [77], and others is available to give guidance to engineers as to the best type of measurement, optimum measuring set-up, and the appropriate associated measuring instrumentation. In the present context, therefore, we are to assume that an optimum measuring set-up has been chosen and so proceed to describe increasing levels of sophistication by means of practical examples, where possible.

7.1.1 *Level 1 – Overall Vibration Measurements*

This is basically the simplest type of vibration measurement and so extra care must be taken in selecting the measurement position and in interpreting the results [74, 75]. Overall vibration is normally taken either to be linear or 'A'-weighted. In this sense the overall value gives the total value to include the effect of all operational events and so can be used simply as a general pointer to the onset of faults. However, when this overall value is displayed as an oscillograph (amplitude–time) then very useful results can be obtained in relation to the severity of engine operating parameters.

7.1.2 *Level 2 – Third Octave Frequency Analysis of Vibration Measurements*
At any engine speed most of the operational engine events can be identified at known frequencies and so can be observed individually. However, for most practical purposes it is both economical and neater to group these events in bands by using special analyser filters. Octave filters (71%) are too crude but the $\frac{1}{3}$-octave filters (23%) are considered adequate to detect abnormal engine events. Figure 7 shows a case of excessive piston slap leading to scuff, and Fig. 8 shows a case of a worn-out big end bearing leading to bearing damage. In these

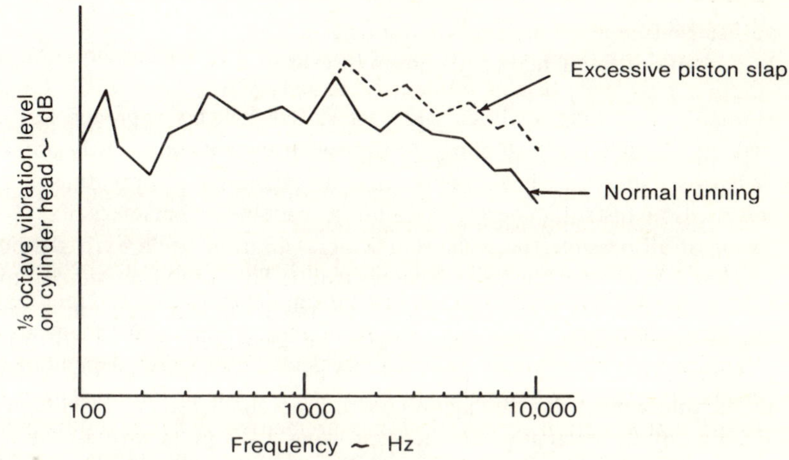

Fig. 7 – Vibration analysis showing the detection of excessive piston slap in an N.A. diesel engine.

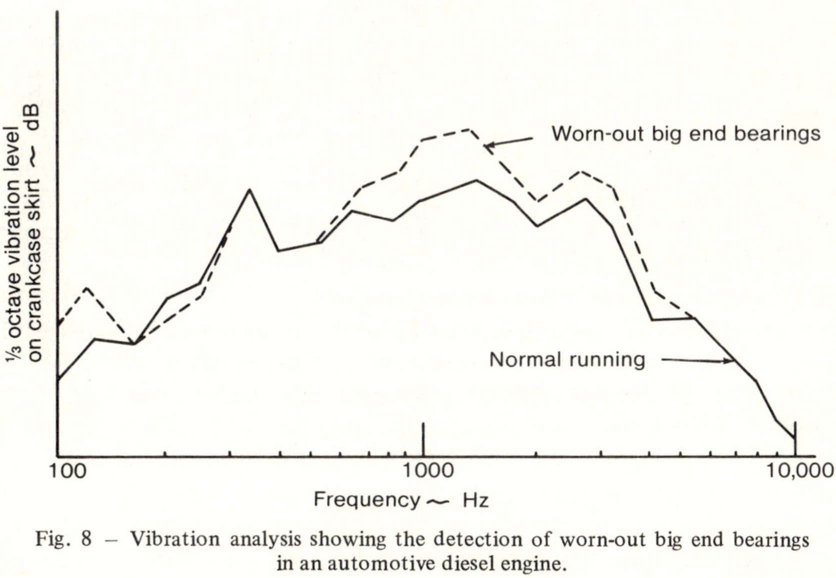

Fig. 8 – Vibration analysis showing the detection of worn-out big end bearings in an automotive diesel engine.

case studies it is useful to perform approximate calculations of the events or faults to show reasonable agreement with the measured results shown in Figs. 7 and 8.

7.1.3 *Level 3 – Narrow Band Analysis of Vibration Measurements*

This method provides a more accurate approach to identify events, especially those critical ones. In this case, constant narrow bandwidth filters are used, depending on requirements; that is whether:

a. single frequency monitoring of one event, or
b. a band of frequencies for a group of events

Normally, real time analysers are best suited for this approach in this case; it may not be necessary to compare the spectrum with previously obtained data. A narrow band real time analyser has the capability of providing the spectrum information instantaneously, revealing a machine's characteristics. A complete listing of all possible frequencies at which vibration could exist should be compiled and kept with the machine's portfolio. This would provide a rapid means by which the spectrum data could be resolved, and frequency components where high vibration levels exist could be identified. Soon a portfolio of case histories and vibration trend data would be compiled and used to make accurate predictions concerning the engine's operational characteristics and life expectancy. Records of these trends may be kept using $X–Y$ plots with all the rest of the information for reference purposes.

7.1.4 *Level 4 – Normalized Narrow Band Analysis of Vibration Measurements*

There are some practical difficulties which make the job of condition maintenance by interpretation of spectrum data very difficult. The major problem in interpreting data is that each of the frequency spectra must somehow be normalized to some typical operating condition so that it may be directly compared.

The technique of signature ratio analysis allows a comparison of multiples of running speed. This is done by dividing the frequency by the rotational frequency. But what about the amplituded information? Can the measured vibration levels be taken at their 'face values'?

To successfully reduce vibration and diagnose problems, the engineer should know the dynamic transfer functions of the components making up the structure. Without this information, the machine can be given 'aspirin' or 'band-aids' but the heart of the problem remains: How does the machine react to its self-generated vibration?

There is a very basic equation associated with this analysis:

$$\text{Inputs} \times \text{Transfer Functions} = \text{Response}$$

So far, only measurement of responses has been dealt with. The responses are only a 'surface' indication, however, in that they represent the internal forces (which do the actual damage) as modified by the path(s) between the forces and the measurement locations. In order to give the best insight into how to correct the input vibrations the transfer functions must be determined. This is done by providing known artificial inputs to the system and measuring the system's response.

This process is known as mechanical impedance analysis and is analogous to impedance in electronic circuits. It is not possible to determine how much vibration is present at critical points without the transfer function data. It tells what parts of the structure are particularly weak and therefore, respond to the structure's self-generated vibration. The proper location for vibration transducers is critical when measuring responses. There are case histories of machines tearing themselves apart while the vibration transducer was indicating that practically no vibration was present. This was due to the fact that there was very little vibration present at the point where the transducer was mounted. The severe vibration was somewhere next door. The opposite case exists where the vibration transducer happens to be on a particularly limber part of the structure and indicates high vibration at that point, but the vibration is not doing any damage because energy is not being transferred to critical components.

Because a vibrating force can produce deflections 30 or 40 times greater than the same force applied statically, lack of sufficient dynamic stiffness can be a serious problem. The technique of transfer function analysis can accurately identify the vibration frequencies at which the structure is weak. This technique involves plotting the ratio of the force applied to the structure and the motion produced in the structure by that force as the forcing frequency is slowly changed.

Based on the dynamic characteristics of the system, it is important that the transfer functions between the critical parts and the measurement locations are known.

To fully normalize data between different measurements, the following should be done:

- Determine the critical components in the piece of engine. These are the components which are prone to failure – those that have to be replaced on a periodic basis, etc.
- Measure the transfer functions between these critical components and the monitoring points in the system.
- Perform the condition monitoring on a periodic basis, but modify the real time analyzer spectra data for frequency using the signature ratio, and for amplitude, using the transfer function.

Since a very slight change in operating speed can account for large changes in the transfer function, the incorporation of the transfer function data should serve to remove the tremendous scatter from the data.

This vibration analysis method is the ultimate in fault detection in engines and for that matter in any machine where the need arises for critical and accurate diagnosis. Of course it requires more sophisticated hardware and software with associated peripherals. Spectral dynamics and others have successfully developed this technique to the extent of producing automatic on-line computerized monitors that monitor, analyse, record, compare, display and actuate alarms at the onset of faults.

These sort of systems should find applications for engines on board ships and such like when the cost of sophistications justifies the end result.

Other developments in vibration monitoring are specifically related to more sophisticated failure modes. Thus, shock pulse monitoring is particularly relevant to rolling element bearing monitoring; side-band analysis is particularly relevant to deteriorating gear teeth; kurtosis, cepstrum, autocorrelation, mobility/impedance/coherence and other methods have applications to particular problems.

7.2 Debris Analysis in Engine's Circulating Fluids

Wear and corrosion debris in a machine is transported by circulating fluid such as coolant or lubricating oil or working gas. These contaminants can be examined and the amount present in the fluid determined quantitatively and identified by type and shape. Debris analysis is an integral part of early failure detection, and the techniques principally in use are spectrometric oil analysis procedures (S.O.A.P.), magnetic plug inspection, microscopic examination with other techniques such as ferrography, X-ray fluorescence, thin-layer chromatography, blotter spot tests applied as conditions require. Spectrometric oil analysis procedure (S.O.A.P.) uses either an emission or atomic absorption spectrometer to measure the analytical 'spectrum' of a liquid sample. Magnetic plugs capture ferrous and some non-ferrous debris in circulation.

For characterizing wear, the most important properties of the debris are its:

- quantity
- size distribution
- morphology
- composition

The quantity and size distributions of the debris quantify the extent of the wear whereas the morphology and composition of the debris give information about the type of wear occurring.

Ferrography can give some information about all four of these properties. Since the technique depends upon magnetic precipitation, it is most sensitive to ferrous debris. However, non-ferrous debris can sometimes be detected because during the wear process the debris may pick up small inclusions of steel and therefore become weakly magnetic.

7.2.1 *Equipment (Ferrography)*
Two types of ferrograph exist: the direct reading, D.R., ferrograph and the analytical ferrograph [82].

Direct reading (D.R.) ferrograph: in the D.R. ferrograph the oil sample, diluted with a solvent, is passed through an upward sloping galss tube which is positioned above the poles of a magnet. The downward velocity of a ferrous particle in the magnetic field is approximately proportional to the square of its size. Thus the horizontal distance reached by a particle before it is deposited depends on its size and its height above the bottom of the tube when it enters the magnetic field. Hence there is a maximum distance that can be travelled by a particle of a certain size.

Analytical ferrograph: in the analytical ferrograph the debris is deposited on a glass slide (rather than a glass tube used for the D.R. ferrograph) and it can be examined by optical and electron microscopy. Quantitative measurements can be made with a simple optical densitometer. The shape of the particles depends on the mechanism of wear that produced them [84] and by studying the debris it is possible to distinguish between various wear mechanisms such as rolling contact fatigue, scuffing and corrosion.

The significant sources of the contaminants in an engine are as follows:

Iron	– Cylnider liners or piston rings, commonly associated with high wear rate, blow-by and scuffing.
Iron and Silicon	– Resulting from the ingress of airborne dust due to inefficent or choked air filters.
Copper and Lead	– Failure of copper/lead bearings.
Lead and Tin	– Failure of white-metal bearings.
Copper and Tin	– Failure of bronze and bushes.
Chromates	– Leakage of coolant containing anti-corrosion inhibitors.
Silver	– Deterioration of platings, silver-soldered pipe fittings.

Finally, it is important to bear in mind that the debris catch-efficiency of the various analysis techniques varies according to particle size. Figure 9 shows general catch-efficiencies for S.O.A.P., ferrography, magnetic plugs, filter retention, on-line particle counters and chromatography related to particle size – for broad guidance only.

7.3 Performance Trend Monitoring
In general this includes some of the following:

- Power units performance curves, e.g. power, torque B.M.E.P., fuel consumption, efficiencies, etc.
- Pressures of compression, combustion, blow-by, oil, air, exhaust, etc.
- Temperatures of bearings, liners, pistons, exhaust, etc.
- Pressure drop in filters, induction system, etc.
- Fluid levels, e.g. water in fuel, etc.

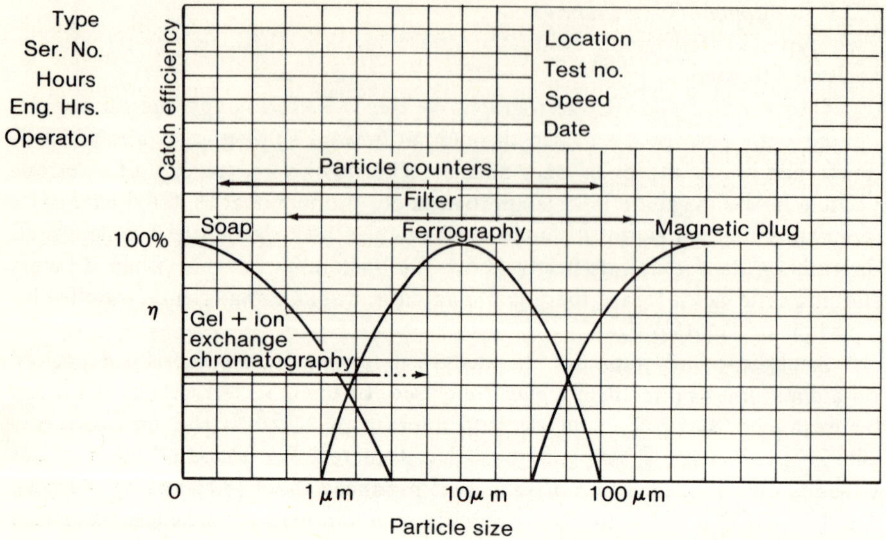

Fig. 9 – Comparison of available debris analysis techniques for diesel engines.

Of course the above have been discussed before under condition checking techniques. However, in this context the emphasis is on the trend of change in a certain parameter.

7.4 Some Non-destructive Testing Techniques

Most industrial machinery has scheduled idle time during each day or week so that the opportunity occurs for it to be examined and tested by methods of a non-destructive type. Visual examination is probably the most basic examination and can be aided by the use of image-intensifiers to 'light up' the dark interior of machines, flexible fibre optics to examine inaccessible corners and close-circuit television for multiple-viewing of cramped locations. A selection of the various techniques is as follows:

Corrosion-probe	– Material loss by corrosion measurement.
Dye penetrant	– Crack penetration using visible or fluorescent dyes.
Eddy current	– Surface defects detected by electro-magnetic induction.
Ultrasonics	– Which uses transmission characteristics of ultra-high frequency sonic waves to locate defects.
Microwave	– Infra-red inspection of non-conducting materials.
Optical aids	– Borescopes, fibre optics, C.C.T.V.
Radiography	– Penetration to show up defective and burnt-out parts.

Although N.D.T. methods are normally used when the machine is idle, some of these have been developed and adapted for effective use whilst the engine is actually running to give continuous condition and trend checking information.

7.5 Other Methods

These include some 'sophisticated' and 'under development' techniques:

- Wear and strain tests using wear gauges and strain gauge arrangements.
- Free acceleration testing.
- Corrosion monitoring and thin-layer activation using radioactive layers.
- Flywheel speed for cylinder power balance.
- On-line magnetic particle detectors, holographic and photoelastic techniques.
- Cylinder power by evaluating PV diagrams.
- Force measurements, e.g. on fuel injection systems.
- G.E.L. chromotography for lubricating oils.
- Telemetry (mechanical linkage, radio or slip rings).

It is important to realize that normally a combination of the methods given in sections 7.1 to 7.5 would prove most effective in producing a condition monitoring system for diesel engines.

Figure 10 shows a block diagram of a typical complete C.M. system.

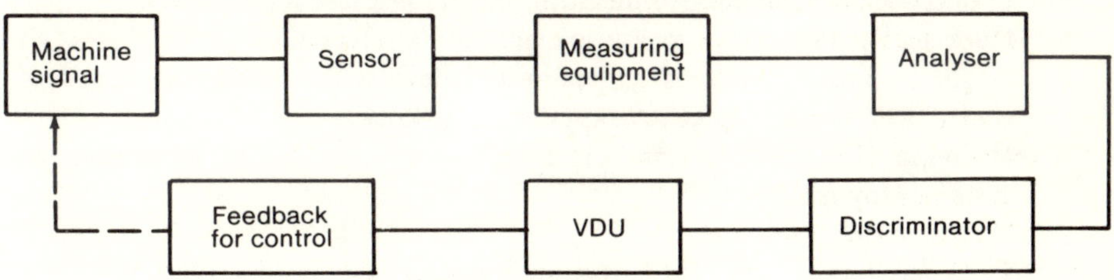

Fig. 10 – Typical C.M. System.

Such C.M. systems are discussed in detail in several papers by the author, and in other textbooks. The sensor, however, is a very important part of any C.M. system and so is briefly discussed here.

8. C.M. SENSORS

These are mainly of two types:

1. Invasive Sensors: that is when the sensor interferes with the machine set-up.
2. Non-invasive Sensors: that is when the sensor tends to present little or no interference with the machine set-up.

Naturally it is preferred to use non-invasive sensors whenever possible. The choice and selection depend on many factors.

In general, electrically-based sensors or transducers can be classified as follows:

(i) Non-generating (or passive) transducers
- variable resistance ex. metallic strain gauges
- variable inductance ex. position measurement
- variable capacitance ex. proximity gauges

(ii) Generating (or active) transducers
- thermocouples ex. temp. measurement
- piezo-electric crystals ex. pressure transducers
- inductive types ex. needle lift, tachos.

9. COMMON SENSORS USED FOR CONDITION MONITORING APPLICATIONS (IN DIESEL ENGINES)

1. Piezo-electric pressure transducers (diaphragm type, washer type, high temperature type) for measuring combustion pressure development (Kistler).
2. Reluctance gauges for measuring needle lift of fuel injectors (C.A.V.).
3. Normal temperature and pressure gauges (any make).
4. Strain type low pressure transducers for measuring around 1 bar pressure fluctuations in intake, exhaust and crankcase (any make).
5. Condensor microphones for measuring sound pressure level (B&K).
6. Piezo accelerometers for measuring vibration levels (B&K).
7. Magnetic pick-up systems and photocell devices for measuring 'top dead centre' and 'engine degree' marks (various makes).
8. Magnetic chip detectors for catching ferrous particles in lubricating oil systems (any make).
9. N.D.T. sensors (any make).
10. Load cells (strain type) for measuring forces in valve systems (C.A.V.).
11. Inductive gauges for proximity measurements of pistons (A.E.D.).
12. Strain gauges for measuring fuel line pressure (C.A.V. and A.V.L.).
13. Exhaust sensors for measuring O_2, NO_2, etc.
14. Performance sensors, power, torque, s.f.c. (conventional).

10. COST OF C.M.

It is perhaps useful to give some indication of the cost of some C.M. methods for some common machine applications, as shown in Tables 1 and 2, extracted from a report published in 1979.

To conclude it is appropriate to mention that the adaptability, complexity and cost effectiveness of the C.M. systems discussed rely a great deal on different engine size, class, usage and other parameters. Many of the references listed in this chapter give considerable guidance on this aspect of diesel engine diagnostics.

Table 1.
Monitoring equipment for rolling contact bearings (1979)

Rating	Machine	Access	Possible equipment	Approx. cost, £
	Simple bearing systems	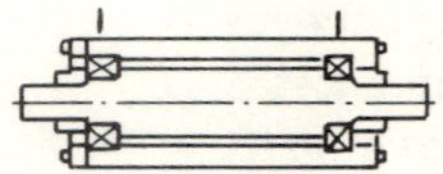	RMS level of vibration (10–20 KHz)	1000
	e.g. Line shafting		Shock pulse	800
	Spindles		Kurtosis measurement	–
Simple	Elect. motors		Crest factor analysis	1500
	Non-cavitating pumps		Damage detectors	2000
	Paper-making machinery		Oil film resistance methods	200
			Magnetic chip detector (with debris tester)	500
	Gearboxes	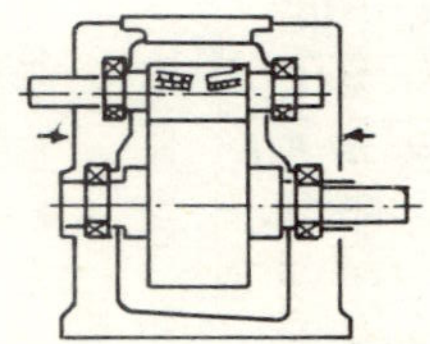	Shock pulse	800
	Cavitating pumps		Kurtosis measurement	–
Quite difficult	Screw compressors		Enveloping/P.S.D. methods	8000
	Machines with poor access to bearings		Damage detectors	2000
			In-line particle detector for oil scavenge line	1500
			Magnetic chip detectors	500
	Gas turbine mainshaft bearings	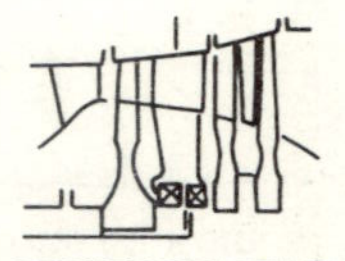	Advanced, computer-based methods of vibration analysis	20000
Very difficult	Epicyclic gearbox		Oil/debris analysis using instruments such as the Ferrograph, stereoscan, etc.	15000
	Planetary bearings		In-line particle detector (detection only)	1500
	Very low speed machines (less than 20 rpm)			
	Sinter plant mills etc.			

Table 2.
Monitoring equipment for gearboxes (1979)

Technique	Form of the implementation	Typical faults detected (Det) or diagnosed (Diag)	Approx. equipment cost, £
Vibration	Monitoring of the overall level (BS 4675 or DIN Std)	Component imbalance, alignment (Det) Damage at a fairly late stage (Det)	1500
	● Figure of merit analysis ●	Imbalance, misalignment, eccentricity (Diag)	25000
	Damage detector (only for faults underlined)	<u>*Tooth fracture (Diag)*</u>, <u>*Tooth damage of a fairly severe form (Diag)*</u>, Tooth damage of a more distributed kind (Det), Tooth separation (Diag) General problems (Det)	
	P.S.D. Analysis	General problems (Det	8000
Oil/debris analysis	Magnetic chip detectors	Tooth damage of all kinds (Det) Tooth damage of all kinds (Diag)	500 25000
	In-line particle detection	As yet unproven, but likely to detect many kinds of tooth damage (Det)	1600
Temperature	Thermocouples	Bearing problems, load distribution faults	–
	Infra-red	Not yet available, could detect problems associated with lubrication (e.g. scuffing)	–
Optical	Endoscopes	Problems affecting tooth bedding	–
	Eyeball (difficult with gear rotating)	Tooth surface distress, chipping, etc.	

REFERENCES

[1] Hambright, R. N. and Wood, C. D., 'Vehicle Diagnostic Systems – The State of the Art', Paper 73–144, presented at *1973 Meeting of American Society of Agricultural Engineers,* Kentucky, 17–20 June 1973. Publ. by ASEA St. Joseph, Michigan.

[2] 'Standardization of Automotive Diagnostic Systems', from *Proc. of the 1st Int. Conf. on Automotive Safety,* Dept. of Transportation, San Francisco, California, 17–19 July 1972.

[3] Janess, D. C. and Manning, H. K., 'Advanced Diesel Power Testing Techniques', *SAE 700666.*

[4] Staton, G. R. and Chalfin, A., 'Cost Effective Solution Using Technical Diagnosis of Reciprocating Engines', *2nd Int. Symp. on Automation of Engine Testing Performance Emissions and Diagnostics,* London, September 1973.

[5] 'Computerised Diagnostic Equipment Checks Tank Engine Malfunctions', *Automotive Engineering,* **80**:55, June 1972.

[6] Sama, D. S., Pradko, F. and Ancona, D. F., 'Diagnostic Equipment for Army Vehicles', *SAE 720236.*

[7] Teixeira, N. A., *et al.,* 'Diagnostic Instrumentation for Military Vehicles', *SAE 730658.*

[8] 'The Future for Automotive Diagnostic Systems', *Automotive Engineering,* **80**:9, September 1972.

[9] Fineman, H. E., 'Go/No Go Indicator System for Vehicle Readiness Testing', *IEEE Automatic Support Systems Conf.,* Philadelphia, November 1972.

[10] 'Automatic System Monitors Truck Performance Condition', *Automotive Engineering,* June 1973.

[11] Obits, W. and Ancona, D., 'Maintenance Indicator System for Military Trucks', *SAE 730657.*

[12] Teixeira, N. A. and Pradko, F., 'Test Equipment for Automotive Vehicles', *Automotive Vehicles, Automatic Testing 73 Conference,* Electronic Engineering Association, Brighton, England, November 1973.

[13] Fineman, H. E., *et al.,* 'Simplified Automotive Test Equipment through the Use of Advanced Electronics, *NEREM 74 Record,* October 1974.

[14] Sama, D. S. and Steyaert, J. W., 'Diagnostic Equipment for Military Vehicle Applications', *SAE 780029.*

[15] Faller, E. K., 'Automatic Diagnostic System for Military Trucks', *SAE 760835.*

[16] Spengler, E., 'Vehicle Diagnostics as an Integral Part of Modern Checkout Systems', *SAE 720235.*

[17] Kremidas, J. R. and Rothfurz, R. W., 'Automatic Vehicle Diagnostic Systems', *SAE 730205.*

[18] Parker, K. and Miller, R. D., 'The First Digital Automotive Diagnostic System – Autosense', *SAE 760144.*

[19] Germann, R., *et al.*, 'An Electronic Analyzer for Diesel Engines', *SAE 690474.*

[20] Forsman, C. R. and Schwindeman, W. R., 'Benefits of Modern Diesel Engine Diagnostic Tools in Fleet Maintenance and Engine Testing of Fuels and Lubricants, *SAE 730678.*

[21] Fuller, E. J., 'The Practical Realization of an Automated Diagnostic System for Diesel Powered Vehicles', *SAE 760838.*

[22] Garrett, K., 'Diesel Engine Diagnostics – The Problems, Principles, Practice and Economics', *S. African Mech. Engineer,* **30**, n.30, March 80.

[23] 'Vehicle Engine Diagnosis', *Diesel Engineering,* Spring 1978.

[24] Sarsten, A., *et al.*, 'Thermal Loading and Operating Conditions for Large Marine Diesel Engines', *IMAS,* London, 1969.

[25] Sandtorv, H., *et al.*, 'Instrumentation and Methodology for Computer Based Testing and Condition Monitoring of Diesel Engines', *3rd Int. Symp. on Automation of Engine Testing: Performance Emissions and Diagnostics,* London, September 1974.

[26] Sandtorv, H., *et al.*, 'Datatrend – A Computerised System for Engine Condition Monitoring and Preventive Maintenance of Large Bore Diesel Engines', *IFAC/IFIP Symposium on Ship Operation Automation,* Oslo, 1973.

[27] Langballe, M., *et al.*, 'Condition Monitoring of Diesel Engines', *Norwegian Maritime Research No. 3/1975.*

[28] Langballe, M., 'The Computer System for Engine Monitoring and Control Aboard "Høegh Multina" ', *Shipping World and Shipbuilders,* April 1972.

[29] Eberle, M. K., 'Diagnostic System for Diesel Engines', *Sulzer Technical Review 4/1976.*

[30] Hatschek, R., 'Measuring Technique for Monitoring Piston Ring Conditions', *Diesel and Gas Turbine Progress Worldwide,* March 1976.

[31] Hammerstrand, J. N., 'Condition Monitoring of Diesel Engines', *Diesel Engineers and Users Association,* Publication No. 382.

[32] Sletmo, K., 'Condition Monitoring of Marine Diesel Engines', *ASEA Journal,* n.6, 1978.

[33] Ibstedt, J., 'ALSY–8, A Modular Alarm and Measuring System for Marine and Industrial Applications', *ASEA Journal,* **4**, n.4, 1977.

[34] Sturesson, S., 'ASEA's Digital Electronic System, DS–8', *ASEA Journal,* 5, 1976.

[35] Tidekrans, H., 'Two New Sensors for the Condition Monitoring of Marine Diesel Engines', *ASEA Journal,* 6, 1978.

[36] Hanson, R. E. and Fischer, H. L., 'New Techniques for Automated Engine Diagnostics', *RCA Eng.,* **21**, pt. 6, Apr/May 1976.

[37] Hadden, S. C., *et al.*, 'Non-contact Diagnosis of Internal Combustion Engine Faults through Remote Sensing', *SAE 760146.*

[38] Jdanovsky, N. S., *Automotive Engine Diagnostics,* Kolos Publ., Leningrad, 1977.

[39] Harston, T. M., 'A Review of Diesel Engine Health Monitoring', *NGTE M79136,* December 1979.

[40] Allwood, A. I. S., 'A Simple Telemetry System for Acoustic Strain Gauges', Shell Research Ltd.

[41] Pelta, E. R., 'Discussion of Signals which can be Derived from Unmodified Standard Automotive Engine Ignition Systems for Test and Diagnosis Purposes', *3rd Int. Symp. on Automation of Engine Testing: Performance, Emissions and Diagnostics,* London, September 1973.

[42] Fagerland, H., *et al.*, 'Monitoring and Diagnosing Process Deviations in Marine Diesel Engines', *Norwegian Maritime Research No. 2,* 1979.

[43] Bergmann, S., Nye Sensoreelementer', *Teknisk Ukeblad,* n.8, 1974.

[44] Henein, N. A., *et al.*, 'Characterization and Simulation of a Distributor Injection System', *ASME Paper 74–DGP–12,* 1974.

[45] Henein, N. A. and Singh, T., 'Effect of Imposed Faults on a Distributor Injection System', *SAE 740531.*

[46] Henein, N. A., *et al.*, 'Characterization and Simulation of a Unit Injector', *SAE 750773.*

[47] Henein, N. A. and Rozanski, J. D., 'A Technique for the Diagnosis of Malfunctions in Diesel Injection Systems', *SAE 780033.*

[48] Heggie, W. S., 'The Use of Plastic Cold Flow in the Development of an Externally Connected Transducer for Recording Pressure–Time Histories of Diesel Fuel Injection Phenomena and its Application in Fault Diagnosis', *ASME Paper 76–DGP–3,* 1976.

[49] Zeiringer, H. and Denk, H., 'Experiences Using Clamp-on Transducers on Injection Lines', *ISATA,* 1979.

[50] Hanova, B., 'Semi-automatic Equipment for Compel Diagnostics of Diesel Engines', Paper D68, *13th CIMAC,* Vienna, 1979.

[51] Engja, H., 'Proposal for Turbocharger System Condition Monitoring', *European Shipbuilding,* **20**, Pt. 3, 1971.

[52] Hambright, R. N. and Benson, H. S., 'Diagnostic of Diesel Engines Using Exhaust Smoke and Temperature', *SAE 760833.*

[53] Cempel, C., *et al.*, 'Determination of Engine Condition by Means by Pressure Pulsation Frequency Analysis', *Proc. 2nd Conf. on Vehicle Diagnosis,* Tuczno, Nov. 1980.

[54] Cempel, C., *et al.*, 'Determination of Engine Condition by Using Pressure Pulsation Discriminants', *Proc. 2nd Conf. on Vehicle Diagnosis,* Tuczno, Nov. 1980.

[55] Cempel, C. and Haddad, S. D., 'Engine Fault Diagnosis by Monitoring the working Gas Pulsations', Dept. Transport technology, *Report TT81R01*, Loughborough University, 1981.

[56] Goodfriend, H. J., 'Automatic Engine Testing – Performance and Diagnostics', *2nd Int. Symp. on Automation of Engine Testing: Performance, Emissions and Diagnostics*, London, 1973.

[57] Iseki, Y., 'Diagnosis of Diesel Engine Troubles', *3rd Int. Symp. on Automation of Engine Testing: Performance, Emissions and Diagnostics*, London, 1974.

[58] Deiss, H. D., 'Synthetic Application of Diagnostic and Test Routines in Automobile Servicing', *SAE 730206*.

[59] Rath, H., 'A Diagnostic System for Vehicles Based on the Monitoring of Off-normal Conditions during Day-to-day Operations', *SAE 740534*.

[60] Bushnell, J., 'Diagnostic and Quality Control Equipment', Paper C111/72, I. Mech. E., *Auto. Div. Symp. on Automotive Electrical Equipment*, September 1972.

[61] Maringer, A. and Moeller, O., 'Automotive Diagnosis as a Feature of Modern Checkout Systems', *3rd Int. Symp. on Automation of Engine Testing: Performance, Emissions and Diagnostics*, London, 1974.

[62] McGlew, J. F., 'A Multimode Vehicle Performance Instrument', *SAE 780149*.

[63] Blackman, W. D. H., 'Maintenance and Repair of Military Vehicles', Paper No. 679, *14th FISITA Conf.*

[64] 'Diagnostic Testing', *Highway and Heavy Construction*, February 1981.

[65] Menard, R. W., 'Diesel Engines: Diagnostic Testing and Problem Analysis', *Mining Engineering*, **33**, April 1981.

[66] 'Automated Diagnostic Testing for Diesel Engines', *Pit and Quarry*, **66**:42, August 1973.

[67] Sturdivant, V. R., *et al.*, 'A Diesel Engine Analyzer', *I.S.A. Trans.*, **17**, No. 1, 1978.

[68] Cempel, C., 'Diagnostically Oriented Measures of Vibroacoustical Processes', *Journal of Sound and Vibration*, **73**(4), 1980.

[69] Latsch, R., *et al.*, 'Experiences with a New Method for Measuring the Engine Roughness', *ISATA Cont.*, 1979.

[70] Hatschek, R. A., 'Sonic Signature Evaluation for Diesel Engines', *Diesel and Gas Turbine Progress Worldwide*, July/August 1976.

[71] Mannix, T., 'Diagnostic Levels Required for Heavy Duty Vehicles', *SAE 730659*.

[72] Collacott, R. A., *Mechanical Fault Diagnosis*, Chapman and Hall, London, 1977.

[73] *Handbook of Loss Prevention*, Allianz Versicherungs AG, Springer Verlag, Berlin, 1978.

[74] Haddad, S. D. and Corless, M. J., 'Vibration Measurements to Monitor Faults in Rotating Machines', *Internoise 78,* 8–10 May 1978.

[75] Corless, M. J., 'A Condition Monitoring Approach to the Detection by Vibration Methods of Faults Occurring in a Rotating Machine', *M.Sc. Thesis,* Loughborough University of Technology, September 1977.

[76] I. Mech. E. Conference: *Acoustics as a Diagnostic Tool,* 20 October 1970.

[77] Brüel and Kjaer Publications on Preventive Maintenance, Condition Monitoring and Relevant Instrumentation Manuals.

[78] Haddad, S. D. and Pullen, H. L., 'Piston Slap as a Source of Noise and Vibration in Diesel Engines', *Journal of Sound and Vibration,* **34**(2), 249–260, (1974).

[79] Haddad, S. D., 'Mechanically-induced Noise and Vibration in the Automotive Diesel Engine', *ASME Paper 77–DET–37,* 1978.

[80] Kerfoot, R. E., Hauck, L. T. and Palm, J. E., 'On-line Vibration Spectrum Monitoring', *Mechanical Engineering,* February 1974.

[81] Pocock, G., 'Wear Debris Analysis for Machinery Health Monitoring', *British J. of NDT,* **21**, 125–130, (1979).

[82] Bowen, R., Scott, D., Seifert, W. W. and Westcott, V. C., 'Ferrography', *Tribol. Int.,* **9**, 109–115, (1976).

[83] Pocock, G. and Courtney, S. J., 'Some Quantitative Aspects of Ferrography', *Wear,* **67**, 287–301, (1981).

[84] Bowen, E. R. and Westcott, V. C., 'Wear Particle Atlas', *Naval Air Engineering Centre Report No. N00156–74–C–1682,* July 1976.

[85] Pocock, G. and Courtney, S. J., 'Ferrography as a Health Monitor and a Design Aid for the Development of Helicopter Gearboxes', *ASLE Transactions,* **24**, 555–560, (1981).

[86] Scott, D. McCullagh, P. J. and Campbell, G. W., 'Condition Monitoring of Gas Turbines – An Exploratory Investigation of Ferrographic Trend Analysis', *Wear,* **49**, 373–389, (1978).

[87] Hoffman, M. V. and Johnson, J. H., 'The Development of Ferrography as a Laboratory Wear Measurement Method for the Study of Engine Operating Conditions on Diesel Engine Wear', *Wear,* **44**, 183–199, (1977).

Index